wildfires

wildfires
PREVENTION AND CONTROL

Harry P. Gaylor
State and Extension Forestry
Kansas State University

ROBERT J. BRADY COMPANY
A Prentice-Hall Company
Bowie, Maryland 20715

ISBN 0-87618-131-0
Library of Congress Catalog Card No. 74-14269

A Prentice-Hall Company
Bowie, Maryland 20715

Printed in the United States of America

84 85 86 87 88 89 90 10 9 8 7 6 5 4

To fire fighters everywhere, past, present, and future,
as they give of themselves for the benefit of everyone

CONTENTS

PREFACE

Ultimately, wildfire prevention and control seek to manage the interaction of man and the four elements of his environment: air, water, earth, and fire. The objective of this text is to provide fire fighters with a better understanding of the relationships among these elements.

The term "wildfire" refers to all uncontrolled fires that burn surface vegetation (grass, weeds, grainfields, brush, chaparall, tundra, forest and woodland fuels); often these fires also involve structures. Regardless of the type of vegetation, basic methods of wildfire control remain constant. Techniques, which are applications of these basic methods, vary according to terrain, fuel, fire climate region, equipment, and local custom.

This text has been developed for use by federal, state, municipal, surburban, rural, and wildland protection units, whether their members are paid or serve voluntarily. It is addressed primarily to fire crews and their immediate supervisors. It may be used either as a training manual or as a supplementary reference on the organization, functions, and desired capabilities of a wildfire protection unit. The information in this text has been sifted and selected from material published by the U.S. Forest Service, state forestry services, and other forest protection groups. It has been tempered with my experience as a teacher and as a wildfire fighter.

In the past, wildfires have caused disastrous losses of life, property, and natural resources. Although some strides have been made in the area of prevention, the threat of wildfire and resulting destruction persists. There is a vital and continuing need for regular and supplemental fire-fighting forces who have technical knowledge as well as training in effective wildfire-control techniques.

ACKNOWLEDGMENTS

It is a pleasure to thank those who have contributed to making this text possible.

I wish to acknowledge my debt to those authors cited in the bibliography whose books have served as valuable source material for my own work. For their interest and dedication, I commend the thousands of fire fighters who serve from coast to coast, many of whom I have met while conducting training courses in wildfire control. I am also grateful to a great many federal and state forestry personnel with whom I have studied and fought fire. I wish to thank my colleagues in the International Society of Fire Service Instructors who have contributed to a better understanding of the total fire service. To my teachers, mentors, and associates both in and out of the "Service," I am deeply appreciative. In particular I would like to thank James A. McCain, A.A. Brown, R.E. McArdle, Merle S. Lowden, C.A. Yates, John M. Pierovitch, George D. Post, Louis J. Amabili, John W. Hoglund, C.R. Biswell, and H.G. Gallaher. This manuscript has benefitted from the skillful review of Gerald E. Wadlow and Harvey P. Gibson to whom I am grateful. I would like to thank Joe Barani who produced many of the preliminary sketches for the text artwork. Finally, to my wife Freda Harryette Gaylor, I express deep gratitude for many reasons.

While preparing this text, I have felt the impact of all those named above, and have felt fortunate for their association.

wildfires

1 PREVENTION

CONTROLLED AND UNCONTROLLED FIRE

Fire can be good or bad. Fire heats our homes, gives us light, drives our cars, and even sends us to the moon. This type of fire is beneficial because it is under control. When fire gets out of control, it is damaging and dangerous and can be disastrous.

Wildfire is the term used to identify uncontrolled fires that burn over the different types of surface vegetation that cover the land—forests, brushlands, grasslands, rangelands, and grainfields.

Controlled fire is used in several ways to manage the composition of the vegetative cover of the land. This use is termed "controlled burning" or "prescribed burning" and is done under safe burning conditions with strict controls to accomplish a specific purpose. Some uses of controlled burning are: (1) controlled burning of ground vegetation under pine forest stands in the South to eliminate competition and thus provide faster growth of the trees, (2) controlled burning of rangelands to eliminate the heavy thatch of old growth so that the new growth of grass is more readily available to grazing by livestock, and (3) controlled burning of brush fields or small areas of forest to make more grass available or to change the cover to species of vegetation more desirable for wildlife in the area. These are only a few of the ways that fire is used as a tool by managers of the land.

Figure 1.1. Prevention of wildfires benefits everyone in the community.

Therefore, when we speak of the prevention and control of fire or wildfire, we are referring to unwanted, uncontrolled, or escaped fires caused either by man or by lightning. Wildfires may occur in any vegetative cover type when conditions are favorable for burning, and they may also occur in dry peat bogs or dried-up lake beds.

It is important to understand that controlled fire in surface vegetation can be beneficial when the burning is correctly done by a professional for a specific purpose. Even Smokey Bear, the national symbol for forest fire prevention, does not say that all fires are bad. He says, "Use fire with caution" and "When you burn, follow my rules."

THE IMPORTANCE OF FIRE PREVENTION

Prevention of unwanted or escaped fires is a never ending job. Wildfire is an ever present danger, and, to be effective, fire prevention must be constantly practiced. Fire prevention is often said to be the most important function of a fire control service.

Rule of Thumb: There is no honor in fighting a fire that could have been prevented.

THE FIREMAN'S RESPONSIBILITY

Fire prevention is the most important part of every fireman's job. Your fire prevention work requires time and effort; you must be knowledgeable to do a creditable job. As a wildfire fireman you are a key man in bringing about fire prevention consciousness in your community. Preventing uncontrolled fires is the principal way to reduce the millions of dollars that are annually lost as a result of fire.

Prevention is not as glamorous as fighting fire, and it may seem at times to be ineffective; but continued efforts in prevention will reduce the loss of human life and property from fire. Prevention work is a basic part of every fireman's job.

WHAT IS WILDFIRE PREVENTION?

Basically, wildfire prevention means keeping all unwanted man-caused wildfires from starting. Actually, prevention covers a wide range of activities and must be patterned to the causes of fires in a particular area. The best tools we have for prevention are education, elimination of hazards, and fire law enforcement.

Risk is defined as (1) the chance of the start of a fire as determined by the presence and activity of a causative agent and (2) a causative agent—people, people's activities (such as smoking and starting fires), equipment (such as stoves, mufflers, and electric appliances), and lightning. A risk is a fire starter.

A *hazard* is a fuel complex defined by kind, arrangement, volume, condition, and location that forms a special threat of ignition or of suppression difficulty. Areas covered with grass, brush, and forest fuels are examples of hazards.

Figure 1.2. *Risk* (human use) plus *hazard* (accumulation of fuel) may cause *fire*.

In organizing wildfire prevention in a particular area, one must first know the usual causes of fires and the risks and hazards involved. The prevention effort of education should be effectively tailored to the cause and to the risks and hazards to be eliminated or reduced. The objective is for everyone to know how to prevent wildfires. The causes, risks, and hazards will vary in different sections of the country.

IDENTIFYING WILDFIRE PREVENTION PROBLEMS

There are usually several sources of information that will help in breaking the problem down into its component parts. Before a fire prevention plan can be made and action can be taken, it is necessary to find out what local fire problems your prevention effort must deal with.

Sources of information are:

- Local fire statistics and records.
- Prefire surveys, fire hazard surveys, or similar inspections that may have been made by a county or state fire service, or some other organization. (It may be advisable to make your own survey of hazards and risks in your area by inspection.)

- Local knowledge of fire occurrence and loss.
- State or federal fire protection agencies, especially if your area is adjacent to them.
- State rating bureaus and state fire marshals.
- Adaptation of national statistics available from your state forestry agency.
- State fire-training organizations.

WILDFIRE PREVENTION PLANS

Fire prevention plans organize the work of fire prevention. The plan need not be cumbersome or overdone; in fact, the simpler it is, the more effective it will be. The written part of the prevention plan should include maps, tables, and graphs, as desired by the individual fire service. The material should be updated at least yearly. To develop a fire prevention plan, the following steps should be taken.

- Develop the background information that is indicated.
- Analyze the needs as dictated by risks, hazards, and resources involved.
- Determine the most common fire causes and concentrate prevention efforts on them.
- Assign responsibility for the actions to be taken; that is, decide who will give talks to groups, who will contact the media, who will work with schools, who will do inspection and prefire planning, who will work on hazard reduction, etc. Decide when each job will be done.
- Obtain feedback information from all who are contacted.

A sample fire prevention plan follows, with explanations and suggestions for each item. It is by no means complete, and all items will not be adaptable for each fire district or department.

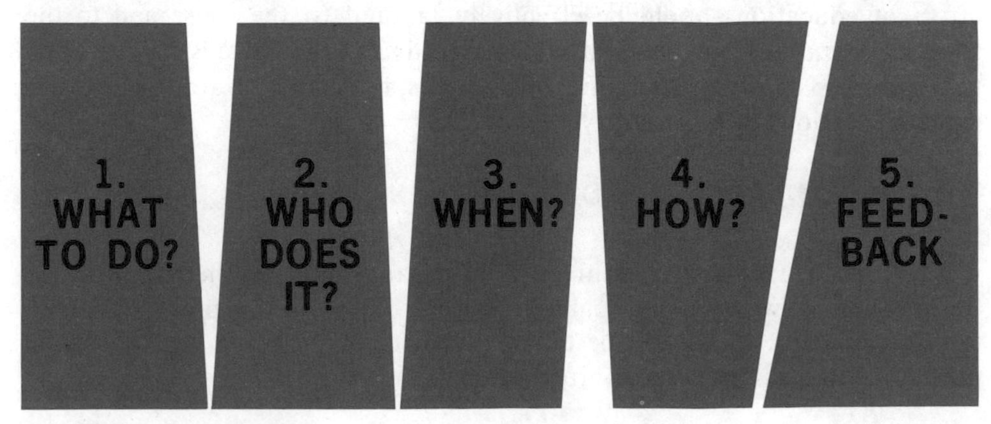

Figure 1.3. Plan of action.

SAMPLE WILDFIRE PREVENTION PLAN

I. Basis of the plan

A. Fire occurrence map

B. Fire statistics

C. Fire risk map

D. Industrial operations map

E. Hazard map

F. Sign map

II. Prevention objectives

III. Summary of problems and action to be taken

IV. Prevention contact plan

V. Public education

VI. Closures and restrictions

VII. Industrial operations

VIII. Reduction of physical hazards

IX. Sign-posting plan

X. Prevention training

BASIS OF THE PLAN

Every fire prevention plan needs information and statistics on a particular area on which to base its goals and methods of prevention. From the local sources of information, prepare a fire occurrence map, fire statistics, a fire risk map, an industrial operations map, a hazard map, and a sign map.

Fire occurrence map. Show causes of past fires and the year of occurrence on a map of your area. The year can be shown by color, and the cause can be shown by a letter inside a small circle, for instance R for railroad, M for miscellaneous, etc. If the fire covered a large area, show an area of damage around the circle. Fires may be shown by colored pins if the map is posted on the wall with suitable backing. Depending on the frequency of occurrence, the map can be made for each year or for a five-year or longer period. The more years represented on this map, the greater the possibility that patterns will begin to become evident.

Fire statistics. Tables, bar graphs, or line graphs may show fire statistics. List the causes of wildfires that are appropriate for your area. Table 1.1 is an example of a fire statistics table.

TABLE 1.1
Number of Wildfires by Causes

Year or month	*Lightning*	*Smoking*	*Railroad*	*Campfire*	*Debris*	*Incendiary*	*Total*	*Total man-caused*

Fire risk map. Outline problem areas by causes and indicate the priority of action that should be exerted on these areas. The protection area should be divided into zones by types and degrees of risk as determined by the results of the analysis. Boundaries of risk zones should be marked with heavy black lines and should be located to separate types of risk by wildfire causes. Different colors can be used to outline the various zones by the type of cause. If necessary, cross-hatching can also be used. A priority number should be assigned to each area and placed in that area on the map.

Industrial operations map. On a map, show locations of all industrial operations with an appropriate symbol and a number from a list prepared for industrial operations. Be sure to take into consideration new resource use, new industrial development, and construction projects such as highways, power lines, pipelines, and dams.

Hazard map. The hazard map shows the different areas that are covered with wildfire fuels in the protection district. The fuels will be shown in different colors for the areas they cover. Particularly hazardous areas should be cross-hatched over the color. High hazard areas include the following: slash; highly flammable flash fuels such as grass, brush, and grainfields; extensive heavy fuels mixed with fine fuels; disease- or insect-killed timber; areas of blow down with high fuel concentrations or any hazardous fuels indigenous to your area. Zones along highways, railroads, or areas of people concentration where fire occurrence has been high in the past should also be shown if they have not been delineated on the risk map.

Sign map. On a sign map, identify the location of each posted fire prevention sign with a number. This number may be placed on any one of the other maps if space permits.

PREVENTION OBJECTIVES

In a fire prevention plan, logical goals must be determined. How much reduction of man-caused wildfires can be logically expected? How much reduction of incendiary fires can be expected if these are a problem? What will be the goal of the unit?

SUMMARY OF PROBLEMS AND ACTION TO BE TAKEN

When the statistics on your area have been organized and the prevention goals have been set, the answers to the following questions will help to formulate the fire prevention plan for your area.

- What are the leading causes of wildfires (smokers, debris burning, etc.)?
- Where do wildfires occur?
- Can wildfire zones be identified on the risk map?
- What methods of fire prevention are planned (individual contact, group contact, hazard removal, etc.)?
- What are the other causes of wildfires (campfires, machinery, railroads, etc.)?

- How will prevention efforts be directed to these causes (mass media, improvement of laws, ordinances, zoning, contact with local and division railroad officials, etc.)?

PREVENTION CONTACT PLAN

Table 1.2 is an example of a prevention contact plan.

TABLE 1.2
Prevention Contact Plan

Who	*Location/ occupation*	*Contact by*	*When*	*Objective*	*Approach*

Who: Give names of individuals or groups to be contacted.

Location/occupation: Give the affiliation of individuals or groups to be contacted, for example, Grange, store operator, farmer, school, railroad, law enforcement officials.

Contact by: Name a specific member of the fire service.

When: Give the month of the year when contact would be most effective.

Objective: Indicate the purpose for contacting this individual or group—to prevent specific causes of fire, enlist cooperation in obtaining prevention, change ideas on burning, obtain satisfactory operation of dumps, acquaint the public with damage caused by wildfires, etc.

Approach: Determine the individual's or group's interest, such as hunting or fishing, and approach them through that interest. Discuss the importance of fire prevention to the community, conduct information and demonstration trips, use slides and movies at appropriate times, and enlist the support of the local governing body identified with the individual or group you are endeavoring to reach.

PUBLIC EDUCATION

A number of methods can be used to contact the public on fire prevention: information and demonstration trips, fire prevention letters, exhibits, and mass media (send and follow up on Smokey Bear and Sparky material to local press, radio, and television). In your fire prevention plan, outline the specific methods that are to be used to contact the public, assign specific individuals to accomplish the contact, and indicate the groups and organizations to be contacted.

CLOSURES AND RESTRICTIONS

Decide where and when closures and restrictions will be necessary and why they should be put into effect. Decide how they will be put into operation and how they will be enforced. Discuss alternate actions that may be taken.

INDUSTRIAL OPERATIONS

Table 1.3 is an example of the information that should be gathered on industrial operations in your area.

TABLE 1.3
Industrial Operations

				Inspection		
Number	*Type of operation*	*Operator responsible*	*Extent*	*By whom*	*When*	*What to check*

Number: Assign each operation a number for identification on the map.

Type of operation: Indicate power lines, manufacturing plants, sawmills, railroads, road construction, housing construction, mining, and those operations specific to your area.

Operator responsible: Specify the firm and the manager or individual to be contacted.

Extent of operation: Give the location of the operation and its size in terms of acres or miles.

Inspection: Decide who will inspect the operation and when. Draw up a checklist. Check for adherence to contract requirements and fire prevention requirements. Make sure that fire equipment is usable. Follow up to obtain compliance.

REDUCTION OF PHYSICAL HAZARDS

Table 1.4 illustrates the organization of information needed for the reduction of physical hazards in an area.

Kind of hazard: Specify flammable fuel buildup around campgrounds; dumps; fence row accumulations; fuel accumulations in shelterbelts; slash areas; fuel pileup along road construction projects; accumulations along rights-of-way such as power lines, pipelines, railroads, and highways; land clearing; construction projects; insect and disease infestations; and any hazards peculiar to your community.

Who is responsible: Indicate whether natural causes, a company, a person, or a community is responsible for the hazard. Discuss whether the cause can be eliminated in the future. Consider any recent influx of people and expansion of towns and communities.

TABLE 1.4
Reduction of Physical Hazards

Kind of hazard	*Who is responsible*	*Ownership of land*	*Extent (acres, miles)*	*Proposed disposal date*	*By whom*	*Date disposed*

Ownership: Give the name and address of the individual with authority and responsibility for the hazard.

Extent of hazard: Indicate the location of the hazard, and extent in acres or miles.

Method of disposal: Determine the best method for disposal of the hazard such as reduction of volume by removal, composting, controlled burning, chemical treatment, firebreaks, fuel breaks, chipping, shredding, railing, plowing, or replacement with more fire-resistant cover (with or without irrigation to maintain the longest possible time in the green stage).

Date of disposal: Program a definite time by which the hazard must be removed or eliminated.

By whom: The fire service, jurisdictional authority, owner, or public assistance may be responsible for the disposal of the hazard. Decide who will supervise the operation and determine its completion.

Date of disposal: Give the final date of disposal, initialed by the supervisor.

SIGN-POSTING PLAN

Table 1.5 is an example of a sign-posting plan.

TABLE 1.5
Sign-Posting Plan

Where posted	*Number*	*Type of sign*	*Source*	*Date posted*	*By whom*

Where posted: Indicate where the sign was posted, such as a highway, store or other business location, chamber of commerce, fire station, recreation area, etc.

Number: The number assigned to each sign is used to identify its location on the map.

Type of sign: Indicate the type of sign, e.g., smoker caution, campfire caution, fire danger, specific hazard or group, etc.

Source: The sign may be from the Smokey or Sparky program or from a state fire prevention program. It may be a commercial sign or may have been designed and produced locally.

Date posted and by whom: Assign someone the responsibility for posting the sign by a particular date. Provide for checking of its condition and for its removal, if necessary.

PREVENTION TRAINING

In planning your training program, consider the following questions:

- Who will be trained in prevention work?
- What subjects will be given in training?
- Where will the training be done?
- Is experience necessary, and, if so, how much and in what disciplines?

Determine the necessary qualifications for the various fire prevention activities. Decide how much help will be needed from outside the fire service—building inspectors, electricians, foresters, fire-training agencies, arson investigators, flammable liquid specialists, and any others who can assist with their speciality. Program the training so that individuals assigned to specific actions will be proficient.

FEEDBACK

After the prevention plan has been in operation each fire season, try to analyze the fire occurrence during that time to determine what effect your plan has had. Discuss with your group, and with others who may have helped, the reaction of people with whom they have talked. Get the pros and cons from all possible sources, and then digest the information obtained to improve your wildfire prevention action.

WILDFIRE CAUSES AND RISKS

Most wildfires are caused by risks not associated with structures, flammable liquids, or transportation (other than railroads). However, some wildfires spread *from* these sources rather than *to* them. Therefore, all types of risks and hazards in the protection area should be considered in a wildfire prevention analysis. The following wildfire causes are not necessarily presented in order of frequency of occurrence, because this varies in different parts of the country.

SMOKERS

Careless smokers are known to be a leading cause of wildfires. Picnickers, campers, hikers, fishermen, hunters, travelers, or local residents who smoke can cause a disastrous blaze by carelessly discarding a cigarette, cigar, pipe ash, or match. To help reduce this cause of wildfires, each smoker should be made aware of his use of smoking materials. Smokers should take the following precautions:

Figure 1.4. Careless smoking is a leading cause of wildfires.

- Extinguish smokes and matches in a safe place.
- Never throw away smokes or matches without first making sure they are out.
- Use the ashtray in vehicles.
- Smoke only in safe places when you are around hazardous fuels, e.g., in a cleared area or on a gravel bar, next to a stream or lake, or on a bare roadway.
- Avoid smoking in hazardous areas during windy conditions.
- Do not smoke while you are riding or walking; if you must smoke, stop in a safe place.
- Never smoke around flammables like gasoline.

The smoker who causes burns in clothing, upholstery, or rugs will probably be just as careless out in wooded areas or in open country; such a person is a great risk.

LAND OCCUPANCY

Land clearing and debris and rubbish burning are the major causes of wildfires in many areas. Such wildfires are the result of failure to select the proper time, place, and method of burning or to properly supervise and control burning.

Passing local ordinances that require burning permits or allow burning only in designated areas is one approach to prevention of escaped fire. Many states now have definite regulations that are enforced by environmental agencies. Many towns allow no trash burning. Where outside burning is still allowed, the public must be trained to do the following:

- Burn only during safe conditions of no wind or after a rain.
- Remain with the burner until the fire is completely out.
- Locate the burner where blowing sparks cannot ignite other fuels.
- Burn debris from land clearing only in safe locations and under safe weather conditions.
- Have suppression equipment available to prevent the fire from spreading.

Figure 1.5. An unsafe incinerator (*left*) presents a fire potential, but a safely maintained incinerator (*right*) can help to prevent wildfires.

CROP FIRES

Grainfield and other agricultural crop fires are particularly hazardous when the fuel is in the cured condition. Successful prevention measures have included:

- No smoking in or near the field
- Mufflers on vehicles protected by an extra sheet of metal to prevent them from touching the fuel
- Satisfactory spark arrestors.
- Fueling equipment outside the field

DUMPS

Dumps are a prime cause of fires in many areas. They are a risk that can definitely be controlled by local authorities. The local fire service should assist by recommending and obtaining suitable installations and operations. The incinerator method is expensive, but it provides excellent control, provided that sparks from the stack are arrested. Most communities use the open dump, the trench method, or the landfill dump.

Open dump. The open dump has some objectionable features, but it is the dump that communities can most easily afford. The dump area is located below an embankment, and the rubbish is unloaded at the top of the embankment. Additional dumping space can be provided by raising the top of the embankment, but care must be taken so that the top will not cave in; suitable barriers must be in place so that cars and trucks do not drive off the top.

Tree limbs and lumber, garbage, and flammable rubbish should be separated before dumping. Garbage should be covered with gravel or soil as soon as it is dumped. Special care must be given to eliminate rodents, flies, and fires at dumps. The prevailing wind should blow across the top of the bank and into the dump area; if the wind blows from the dump area against the embankment, mass transport of embers could easily occur.

Trench dump. The trench dump is at least 5 feet deep; it is dug with a bulldozer, and all waste material–including garbage, tree limbs, and rubbish–is dumped into the trench and burned. When the trench is filled to within a foot of the top, soil is bulldozed onto the waste material to fill the trench back to the level of the original surface. It is compacted with the bulldozer in the process. The hazard of mass transport of burning material is reduced by burning in the trench because the wind cannot blow directly on the fire. The trench method is often more expensive to operate than the open dump, and over a period of time the trench method requires more space.

Landfill dump. The landfill method is mostly used by large operations. All waste materials are mixed together and dumped in an area created by a gully, wash, or ravine; in pits formed by gravel removal; or in a pit dug specifically for dumping. Each trench is dug immediately adjacent to the previous one, and bulldozers compact the waste as it is dumped. When the desired level is reached, about 2 feet of soil is compacted over the trench to effectively seal the waste area. There is less fire hazard with landfill dumps.

The dump site should be owned by the city or town rather than rented, year-round accessibility should be provided, and the dump should be out of

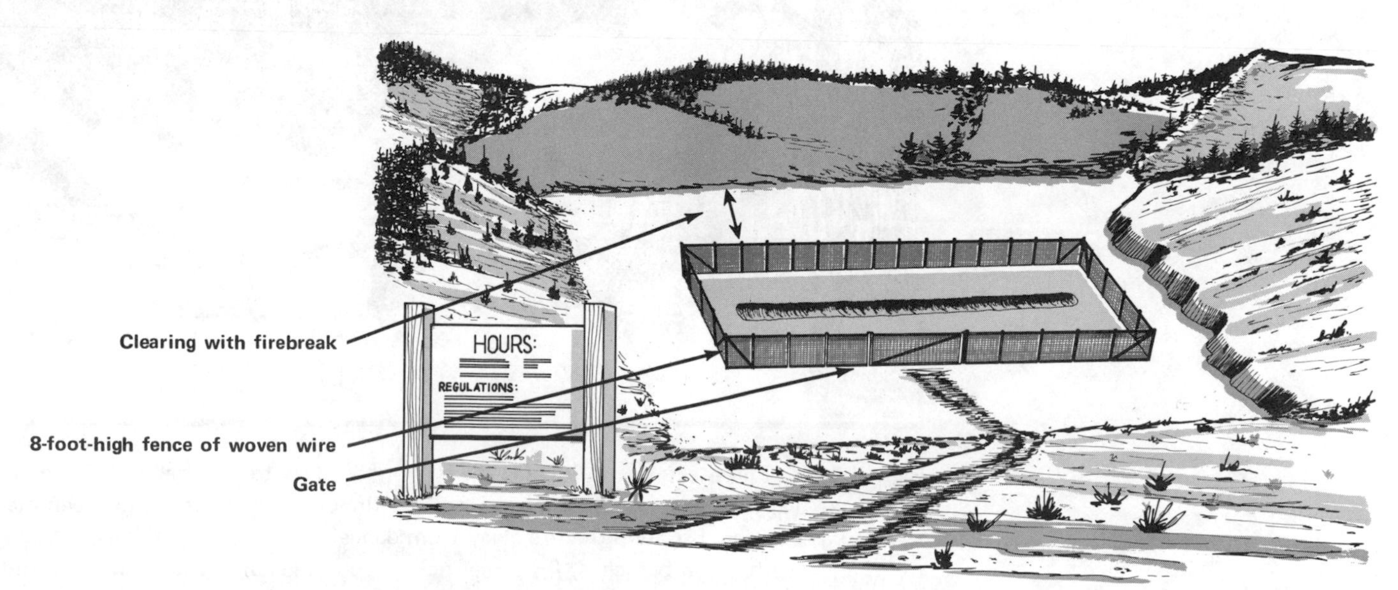

Figure 1.6. Dumps should be properly planned and maintained.

sight from main routes of travel or suitably screened. The site should be located so that it will not pollute any water; if it is adjacent to a wooded area, there should be a 50-foot-wide clearing around the dump, and adjacent trees should be pruned to at least 10 feet above the ground. If possible, a water supply should be nearby, but water can be provided in other ways. The entire area should be surrounded by a firebreak that is maintained to mineral soil. The intervening vegetation between the firebreak and the edge of the dump should be burned off annually or more often if necessary. The prevailing wind direction should be considered. The dump area should be fenced with a sturdy wire mesh, preferably 8 feet high, both to provide controlled use of the area and to catch blowing paper and other light items. A locked gate will provide control when the area is not attended. If possible, the use of the area should be supervised by an attendant. This supervision would require a schedule governing dump use, and appropriate signs should be posted to inform the public. Water should be provided if only in the form of a few 55-gallon drums and a backpack pump or several pails. If a shelter is not available, a weatherproof box should be provided, containing a pitchfork, rake, shovel, and ax. Wise operation will eliminate fires and other objectionable features of dumps.

CAMPFIRES

Campfires are a frequent cause of wildfires in those areas where camping, hunting, fishing, and picnicking are popular. The campfire should be built in a constructed fireplace, if one is available, and should be kept small. The fireplace should be away from overhead vegetation. If there is no constructed fireplace, the fire should be built in the center of an area cleared of all vegetation so that the mineral soil is exposed for a diameter of at least 10 feet. It should be well away from concentrations of flammable fuels—the wood pile, steep slopes, rotten logs, trees, and overhanging branches. A camp or warming fire should never be left unattended because the wind can blow sparks into adjacent fuels. Before the camper leaves, the fire should be extinguished with water, and the burning material should be thoroughly mixed with water or soil to make sure it is dead. The camper should feel the

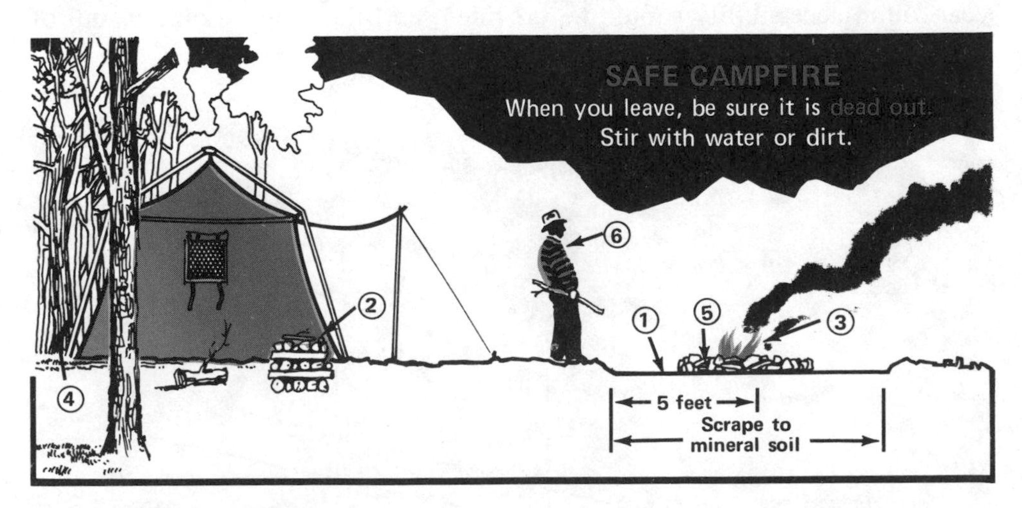

Figure 1.7. For a safe campfire, (1) clear ground cover to mineral soil 5 feet in all directions, (2) keep the woodpile away from the fire on windward side, (3) keep the fire as small as possible, (4) build the fire away from dangers, such as steep slopes, rotten logs, trees, and overhanging branches, (5) place rocks around the edge of the campfire, and (6) never leave the campfire unattended.

fire with his hand to be certain it is out. Many wildfires have started from abandoned campfires that were thought to be extinguished but were still smoldering.

RAILROADS

Railroads can cause fires in many ways. Some lines are very effective in their prevention efforts; other are very lax. With the passing of the steam locomotive and the use of diesels, railroad-caused wildfires have generally decreased. The diesel locomotive, however, can throw sparks and burning embers and does cause many fires in some areas. The incidence is worst on grades where additional pressure is forced through the stack. Arrestors are provided on most equipment, but, if they are not properly maintained, they are not effective. Newer and more effective types of arrestors are being installed, but they are not in universal operation. When diesels idle, carbon sludge builds up on the screens, and later, when full power is applied, the carbon particles are blown out, often in full glowing condition. If the right-of-way has not been cleared of burnable vegetation, many fires are started. The use of higher grade fuels helps to reduce carbon accumulations. Definite action is being initiated in many areas to reduce this risk, and local fire services should check with state fire marshals and other state regulatory agencies for information and possible assistance. Fires caused by hot boxes and pieces of brake shoes can be eliminated by proper maintenance. Usually, the local railroad officials are amenable to fire prevention efforts.

Figure 1.8. Trains are a fire risk. Exhaust sparks, sparks from brake shoes, and flares may cause fire to start.

Burning of old railroad ties and other waste materials and carelessness of passengers and railroad employees account for many fire starts. Educating individuals, obtaining cooperation of officials, and enforcing burning regulations and local laws are among the efforts that can be used to reduce this cause of wildfires.

SAWMILLS AND LOGGING OPERATIONS

Sawmills and logging operations are often a major cause of wildfires. Sawdust and slab burners, when they are placed in hazardous areas, have often caused wildfires. However, the operation and design of these burners are coming under more rigid controls, and, with proper enforcement, their potential for starting fires should be eliminated. Carelessness by employees, especially around the smaller mills, is a threat that requires constant prevention effort. The use of approved spark arrestors on tractors and other power equipment in the woods will reduce this cause of fires. The forest service maintains a testing laboratory for spark arrestors of all kinds,

Figure 1.9. The use of approved spark arrestors on tractors and other power equipment will reduce the chance of fire.

including those on motorcycles. A list of approved spark arrestors can be obtained from the Forest Service Equipment Development Center, 444 East Bonita Avenue, San Dimas, California 91773.

Welding operations on machinery used in the woods are another cause of wildfires. Clearing an area to barren soil where the welding is to be done prevents sparks from falling into duff, punky logs, and other flammable fuels. A shovel and backpack pump filled with water or a suitable fire extinguisher will eliminate any fires that do start. Power saws should not be refueled until they cool, and then they should be refueled only on bare ground. They should be started at least 10 feet from where they were filled because of the possibility of igniting spilled gasoline. Power saws should be fitted with approved mufflers and should be adequately maintained.

ARSONISTS

Figure 1.10.
Arsonists are a major cause of wildfires.

Tragically, in many parts of the country, arsonists are a major cause of wildfires. Arsonists burn for kicks, revenge, personal gain, spite, sexual stimulation, or the blind urge to destroy. Psychologists believe arsonists are sick. They torch resources worth millions of dollars every year and account for considerable loss of life. In rural communities, they are usually soon suspected and can be watched and apprehended. Law enforcement is the usual deterrent.

CHILDREN

Children playing with matches or with other forms of fire cause an increasing number of wildfires each year. Training, realistic education, and proper parental supervision are necessary to control this cause of wildfires.

Figure 1.11. Children playing with matches cause an increasing number of wildfires each year.

LIGHTNING

Lightning is one cause of wildfires that is not yet preventable, although there is research in progress to reduce or eliminate this cause in areas of severe occurrence. Usually, lightning is accompanied by rain, but occasionally a "dry" lightning storm will produce many starts. Lightning strikes may smolder for days before conditions become favorable for the spread of the fire. Constant detection is required to locate these "sleeper" fires. Lightning storms usually follow definite paths across the terrain. A map with lightning fires for a ten-year period plotted on it will usually show

Figure 1.12. Lightning is one cause of wildfires that is presently unpreventable.

the lightning fire pattern. Fuel types that contain a considerable amount of tall dead snags are susceptible to lightning starts. Prompt detection is the best defense against lightning-caused wildfires.

SECONDARY WILDFIRE CAUSES

Secondary wildfire causes include almost all the other causes of fire usually associated with inhabited areas, such as burning structures that spread fire to grass, grainfields, brush, and forests. The basic cause may be faulty electrical wiring or appliances, defective chimneys and flues for stoves or furnaces, improper storage of flammable liquids and gases, poor housekeeping, improper crop storage, spontaneous combustion, or careless use of drying equipment. Electric fences, vehicle wrecks, train wrecks, airplane crashes, broken power lines, and a number of other causes usually listed as miscellaneous also trigger wildfires if the original fire occurs near susceptible wildfire fuels and escapes. The wildfire prevention effort of a fire service must therefore consider all possible types of fire causes in its protection area.

FIRE HAZARDS

Fire hazards are both natural and man-made. There are many types of fuels that create hazards; these include:

- Fuels from brush, field, and right-of-way clearings; slash accumulations in timber cuttings; grass and debris accumulations in fields, along fence rows, around buildings, and along road and railroad rights-of-way; and similar fuel pileup in other areas

Figure 1.13. Flammable materials around buildings are fire hazards.

- Large accumulation in forests of flammable leaves, dead trees, dry brush, etc.
- Exceptionally dry fuels due to prolonged dry spells
- Large areas of high hazard fuels with no access by road or trail
- Trash and rubbish accumulations in or around residences, warehouses, shops, equipment sheds, storage areas, and other buildings

- Improper storage of flammable gas and liquids in or near buildings, warehouses, and storage areas
- Fuel and trash buildup in poorly administered community dumps
- Bird nests in electrical power installations
- Intentional burning or prescribed burning operations that escape to adjacent areas

Wildfire fuel hazards may be dealt with by breaking up into smaller parcels, scattering to break up concentrations, removal, treatment with compacting or mowing machinery, burning at an appropriate time with adequate controls, or treatment with chemicals. Fire hazards can be isolated by construction of firebreaks or fuel breaks. Often the only sure way to handle an extreme fuel hazard is to patrol the area during periods of high fire danger. Some vegetative fuel hazards can be replaced with less hazardous or more fire-resistant vegetation. Closing hazardous areas to use during periods of extreme fire weather conditions may be the only solution.

Figure 1.14. The public must be contacted on the importance of fire prevention.

METHODS TO USE IN PREVENTION

The objective is to *fireproof* the public. Every fireman should "talk up" fire prevention wherever he goes; he need not talk it to death, but he should be sure to offer a few words of caution or advice to appropriate people at appropriate times.

PERSONAL CONTACT

Personal contact is probably the most effective method of fire prevention, if it is done correctly. The best place for demonstrating prevention technique is at the site of a potential fire. Here one can demonstrate how to build a campfire correctly and care for it, how to use smoking materials carefully, and how to prevent different causes of fires. Be open, direct, and friendly in your approach. Try to be helpful and put people at ease. Don't be overbearing or appear overly conscious of your own importance. Above all, don't misuse your authority or misrepresent your organization. Use tact and understanding. The most permanent accomplishments will be won through public understanding and cooperation, depending on the awareness, interest, attitude, opinion, and belief of the individual citizen.

Awareness: First gain the individual's attention. Show him that wildfires do affect him and that he can cause them.

Interest: If the individual is aware of the problem, he will want to know more about it. Increase his interest with local information.

Attitude: A person weighs and evaluates the merits of what you said on the basis of his own experience and the way others around him think. He should develop an attitude toward fire prevention—if he doesn't, try again, but don't give up.

Opinion: The individual's attitude may develop into an opinion about the problem. If it is positive, you are fortunate, because an opinion is often more difficult to change than an attitude.

Belief: Belief is the final step in adopting an idea. The individual should now believe that he should do something to prevent wildfires. Every contact is different. Success in making prevention contacts depends on your ability to vary the approach or conversation to meet many situations. Planned, systematic contacts are the best approach.

PERSONAL APPEARANCES

Giving prevention talks and demonstrations before organized groups is an important method of disseminating fire prevention information. Always prepare and practice your presentation in advance. Contact groups such as Boy Scouts, Girl Scouts, 4-H clubs, Future Farmers of America clubs, women's organizations, service clubs, chambers of commerce, farmers' and stockmen's organizations, fraternal organizations, granges, environmental groups, wildlife clubs, sportsmen's associations, campers' organizations, summer homeowners, watershed groups, soil and water conservancy districts, church groups, and many others. Furnish material and suggestions to speaker bureaus and toastmaster groups.

All groups can be effective in the detection of fires as well as in prevention. The more people you have on your side, the more effective you will be in eliminating human carelessness as the cause of wildfires. Visual aids make a talk more effective. Use slides, movies, overhead projectors, graphs, exhibits, and handouts to outline how to prevent particular types of wildfires. Emphasize the need to tell the public how to prevent fires. An analysis of statements by convicted fire starters showed that they tried to put their fires out but didn't know how.

MASS MEDIA

Mass media includes radio, television, newspapers, and various other publications designed to reach the general public or specific groups. Use of mass media is one of the best means of public education for prevention of wildfires.

Each of these media devotes a percentage of its time and space to public service free of charge. Also, it is often possible to encourage a paying advertiser to sponsor a fire prevention message in his time or space. Many interests and activities compete for the public service time, so your fire prevention message must be timely, well prepared, and interesting in order to be used. News articles in local newspapers are very effective. Encourage radio, television, and newspapers to use the Smokey and Sparky material that is sent to them periodically. Spot messages at local cinemas are useful reminders.

Provision for advertising high and extreme fire weather danger periods and for advising against open burning over radio and television and in newspapers will help to keep down escaped fires. These alerts can be arranged through your local weather service.

SCHOOLS

Fire prevention training in schools is an important part of any prevention effort. Such training fits in well with a number of subjects taught in all elementary, junior high, and senior high schools. How to introduce the material will depend on the particular school system. The best approach is to first contact the superintendent and find out what type of presentation the school system uses. You might work entirely through the teachers by informing them and furnishing material to them that they can incorporate into their curriculum, or you may present a program for a general assembly. If it is not possible to contact all grades, endeavor to contact the fifth and sixth grades each year. Yearly contact with these grades will produce good results. The benefit of school contacts is that you reach not only the students but also the parents through the message that the students carry home. Use slides, movies, overhead projectors, and other visual aids as much as possible. Many kinds of handouts are available from national or state wildfire prevention programs.

PREVENTION LETTERS

A well-prepared letter aimed at a specific cause and sent with a personal message to a group is very effective. Such a letter should be timed according to the fire season. For example, send a letter to local hunters just before the hunting season. Give information on hunting prospects and conditions in the fields and woods with emphasis on precautionary measures for smoking and camping. Use the mailing list from the local sportsmen's club. Send a letter to farmers in early spring emphasizing precautions to take when they are burning debris. Use the county agent's mailing list.

PARADES, FAIRS, ETC.

Participation in parades and fairs by your fire department or district is usually appreciated. For parades, use your fire truck with a catchy slogan, accompanied by handouts or a loudspeaker message. Smokey and Sparky costumes are available (be careful that the images of these symbols are not abused). A float may be designed to show how to prevent a specific cause of wildfires. For fairs, prepare an exhibit and have a fireman in attendance to answer questions, give information, and distribute handouts. This is also a good opportunity to get information, ideas, and feedback.

NATIONAL FIRE PREVENTION WEEK

National fire prevention week (the first full week in October) presents an excellent opportunity to tie in with national publicity and to correlate with adjacent fire departments. A sincere effort should be made to identify with this program.

INSPECTION

Inspection is an effective method of preventing wildfires, but its use will be dependent on local laws, ordinances, and the authority vested in your fire organization. Its effectiveness will be directly proportional to how you use it. When an inspection is made in a helpful manner, most operators appreciate the assistance. If it is used, inspection should be objectively and impartially applied.

Inspection can be used for almost all kinds of risks and hazards, but it is most effective when it is used for agribusiness, power line rights-of-way, petroleum or gas pipeline rights-of-way, road construction, industrial construction, sawmills, logging operations, reservoir construction, mining, short-term housing, permanent housing, dumps, railroad rights-of-way, airports, and the maintenance of each of these operations.

The inspection should include housekeeping, area cleanup, slash disposal, refuse disposal, adherence to electrical codes, flammable liquid storage, approved spark arrestors for equipment, prescribed burning operations, up-to-date extinguishers and water supply where applicable, correct heating installations, firebreaks (where necessary) and their maintenance, care in smoking, use of warming and brush fires with provision to check that they are completely extinguished, control of welding operations, maintenance of rights-of-way, elimination of trees that might fall on the line, prevention of fuel buildup, compliance with burning permits, compliance with state laws and local ordinances, cleanup around bridges and trestles, safe blasting procedures (including use of electric blasting caps), adequate patrol where it is advisable after daily operations, and adequate fire prevention and fire control plans where they are applicable.

FIREBREAKS AND FUEL BREAKS

Firebreaks may be either natural barriers (such as a road or a stream) or barriers especially constructed to check the spread of fires and to provide an established control line in case a fire occurs. The usual firebreak is a strip of land that has been cleared of brush and trees. A trail is dug to mineral soil along this strip. The width of the cleared strip and the width of trail on it will depend on the kinds of fuels, the location in respect to the lay of the land, and the fire weather conditions that can be expected at the time of fire occurrence. Usually, the width of clearing will be not less than one-half the height of the tallest fuel, and the trail dug to mineral soil will not be less than 18 inches wide. These firebreaks are sometimes called fire lines. They are constructed before a fire occurs. To be effective, firebreaks must be well maintained.

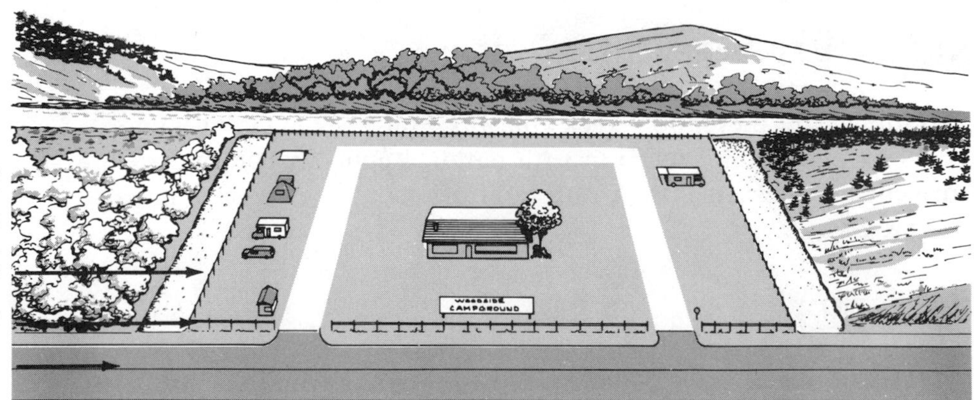

Figure 1.15. Firebreaks may be constructed to prevent a wildfire from spreading into or out of an area.

A fuel break is a wide strip or block of land on which the native vegetation has been permanently modified so that fires burning into it can be more readily extinguished with relative safety to the fire fighters. It may or may not have fire lines constructed in it prior to fire occurrence. Fuel breaks are generally located strategically along ridges and valleys and include a road for access.

Firebreaks and fuel breaks may be constructed to prevent a wildfire from burning into an area or to prevent its spreading from an area. They have been effective around campgrounds and areas of heavy public use, between dumps and surrounding vegetation, along roads where wildfires have occurred, in areas of high wildfire incidence, on railroad routes, and around storage areas, particularly around flammable liquid storage areas and other property that might be the site or origin of a serious wildfire. A greenbelt is an adaptation of a fuel break in which the vegetation is kept green and living through irrigation.

SIGNS

Fire prevention signs can be used to inform the public of fire regulations, restrictions, and procedures on how to be careful. Many posters are available that may be placed on a wood background frame, or special signs may be painted for specific uses. Signs should be posted where they will be most effective—in carefully selected places, along roads, at camp and picnic grounds, stores, filling stations, and places of people concentration. Timing is very important; a sign warning of extreme fire danger should be removed as soon as the danger has passed. Don't "Cry, 'Wolf!' "

Figure 1.16. Roadside signs can be effective in fire prevention.

A few well-placed signs will be most effective. Avoid plastering the countryside. Two few signs are more effective than too many. Place each sign so that it will be clearly seen, neat, distinctive, and not involved with other items. Avoid bulletin boards where the poster can be partially or completely covered up. Avoid placing a sign in a store window where it can slip down into a curled position and be covered with fly specks by August. Damaged signs should be repaired or replaced.

A sign at the fire station or other appropriate places giving the daily class of fire danger is often effective. But do not use such a sign unless it will be changed each day and taken down or covered in the off-season. Nothing is worse than a high fire danger sign with snow on it.

Stress the HOW in prevention signs. A sign that says "Drown your campfire" is more effective than one that says "Be careful with fire."

SMOKEY BEAR PROGRAM

The Cooperative Forest Fire Prevention (CFFP) program began in 1942 as a wartime emergency program, and Smokey Bear became its symbol in 1950. His heir was recently installed at the National Zoo in Washington. The program is supported by the National Association of State Foresters, the U.S. Department of Agriculture, Forest Service, and the Advertising Council, Incorporated. The Advertising Council, Incorporated, is a private nonprofit organization supported by American business and advertising agencies to conduct advertising campaigns for the public good. Since 1942 it has secured the contribution of $7 billion worth of services and facilities to Smokey Bear programs. In 1973 the estimated mass media donations made to the Smokey Bear campaign were $37,457,825. This forest fire prevention campaign is credited with saving more than $17 billion worth of natural resources since 1942 and has reduced man-caused wildfires in our forests from 5,000 fires per million visitors in 1942 to 176 fires per million visitors in 1972. The program is also used in Canada by the Canadian Forestry Association, and Smokey is known as "Simon El Oso" in Mexico. The Southern CFFP program is a part of the national program, but is designed particularly for southern wildfire problems.

Figure 1.17. Smokey Bear represents the campaign for wildfire prevention.

An act of Congress protects the use of the name Smokey Bear for commercial items or other uses. However, licenses are granted for the use of the name on products approved by the Smokey Bear headquarters in Washington, D.C., where he has his own zip code—20252. The proceeds from royalties help to support the program financially. The policy of the program is to actively cooperate with any group interested in preventing wildfires. It has also been used to aid in the prevention of range and trash fires.

Each fire department or district should make use of the Smokey Bear program to the fullest extent possible for that particular area. Contact your state forester for assistance and free materials that you can use in your district. Materials include posters, car cards, radio scripts and prepared spots, television films, bookmarkers, rulers, pocket planners, several comic books, several pamphlets, coloring sheets, pins, shoulder patches, stickers, bumper stickers, and many others. In addition, there are a number of commercial items. An illustrated guide showing the availability of Smokey Bear items can be obtained from your state forester or from your regional USDA Forest Service office.

The program seeks to obtain national coverage so that locally you can identify with it to assist in your prevention efforts. It is the best source of wildfire prevention material.

SPARKY PROGRAM

Sparky the Fire Dog originated with and is sponsored by the National Fire Protection Association (NFPA). Sparky stands for personal and home fire safety, while Smokey Bear represents wildfire prevention. The Sparky program produces posters, films, streamers, coloring sheets, telephone stickers, timely folders, booklets, comic books, book covers, games, and many other materials. Many of these items will be helpful in your wildfire prevention efforts. Contact your state fire marshal or state fire authority for catalogs and assistance. Many fire marshals maintain a library of films to loan to fire districts. The address for NFPA is 470 Atlantic Avenue, Boston, Massachusetts 02110.

STATE "KEEP GREEN" PROGRAMS

A number of states conduct fire prevention programs tailored for their particular state, such as the "Keep Oregon Green" program, "Keep Pennsylvania Green" program, and others. Very useful material and assistance for wildfire prevention are produced in these programs. Contact your state forester for information and materials.

FIRE WEATHER FORECASTS

The fire weather forecasting service is available in all states during the usual fire season. This information is transmitted daily on the National Weather Service TWX circuits. Arrangements can be made with radio and television stations to include this information on their regular weather forecasts, particularly when the fire danger reaches very high or extreme; the media can be used to prevent outdoor burning and to make people more aware of the dangers of wildfire. This is a very useful service both for prevention and for suppression. Contact either your local weather service or state forester for information and assistance.

Figure 1.18. Fire weather forecasts are a useful service during the fire season.

LOCAL LAWS

State and local laws vary by locality. A compilation of the laws and ordinances currently in effect in your community is basic to developing a program of fire prevention action. It may be that you will want to begin a campaign to obtain legislation to improve the prevention process in your community.

Building codes and regulations of any kind relating to fire and zoning ordinances should be known and understood by the local fire service. Some fire departments have the authority to enforce these codes and regulations, and others do not, because the enforcement is the responsibility of other agencies. Where the latter is true, it is important for the fire service to obtain close cooperation. Where the fire service is responsible, inspection is required and should be provided for both in manpower and in the budget.

Obtaining compliance with building codes and zoning ordinances during the construction period can eliminate many later problems for the fire service. The local fire service should insist on being included in any planning and development in its area, and it is just as important for the fire service representatives to be knowledgeable and accurate in their recommendations. Assistance can usually be obtained from your state agencies that have fire or insurance responsibilities. It is important for communities to take fire prevention and fire suppression capabilities into account during planning and zoning.

Rural water systems should be planned to allow for maximum use for fire suppression. Road construction projects can often be located to be of maximum value in fire suppression; so working with the project formulation can pay good dividends. Real estate developments of all kinds should include the necessary fire prevention requirements in the planning and construction stages.

FIRE LAW ENFORCEMENT

Federal, state, and county laws and regulations for smoking, campfires, and debris burning are important prevention measures. Such laws and regulations should be impartial and should be aggressively enforced. Collection of suppression costs is good prevention treatment.

Figure 1.19. Law enforcement is basic to a successful fire prevention program.

Successful law enforcement has more of a widespread prevention effect when it is applied to incendiary situations than to any other cause. It is most effective when it is supported by public opinion. Cooperation from such agencies as the Federal Bureau of Investigation, state investigative and enforcement agencies, the county sheriff, the local police, local magistrates, state fire marshals, and others is often required. This cooperation is possible and effective only if it is arranged in advance.

Fire law enforcement requires preparatory training and correlation with law enforcement agencies. If a fire service can develop this law enforcement capability, it will be very beneficial to the fire prevention effort.

FIRE INVESTIGATION

Regardless of how the enforcement is to be conducted, immediate investigation and protection of evidence at the fire site are necessary. The first fireman to arrive on the scene should be responsible for protecting the evidence. Usually, an investigator arrives somewhat later, and he should be given full cooperation. It is important that the scene be preserved in its original condition. Nothing that might be evidence should be moved unless it is absolutely necessary to prevent it from burning.

On the way to the fire, the fire fighter should:

1. Make note of anyone or anything that could relate to starting the fire.
2. Observe vehicles near the fire area and those moving away from it.
3. Record license numbers, description of vehicles, number of people, personal descriptions, and location or direction of travel.
4. Preserve tracks on the way to the fire and in its vicinity (horse, foot, and vehicle) of any travelers suspected of being involved with the fire. Protect any clues made on the land surface by blocking off the area or boxing in with logs, brush, or limbs, being careful not to disturb the clues. It may be necessary to assign one of the crew members to this task while the rest of the crew proceeds to the fire.

Figure 1.20. Evidence at the fire site must be protected for proper investigation.

At the fire, the fire fighter should:

1. Record and preserve evidence as described in points 3 and 4 above.
2. Determine point of origin from wind direction and the direction in which the fire has spread.
3. Keep fire fighters away from protected clues except where necessary to effect control.
4. Keep on the lookout for evidence on how the fire started and clues to who started it. Evidence might include cigarette butts, matches, lunch remains, evidence of a campfire, scraps of paper, newspapers and magazines, bottles or other containers, store receipts, etc.
5. Handle evidence so as not to destroy fingerprints (they may be found on most anything).
6. Record the time and place where each piece of evidence is found, put the objects in suitable containers, number or identify each container, and place your initials on the container.
7. Prepare a map of the fire area and the area around it. Show the point of origin and show where evidence was found.
8. Record all information obtained, including names of anyone contacted and a resumé of any conversations with them.
9. Obtain statements from witnesses or from anyone who was connected in any way with the fire. Statements should be complete as to detail, and they should be dated, signed, and witnessed as to signature.

Only those with the proper training and authority should attempt to make an arrest. Each fire fighter should obtain the most accurate information possible and turn this over to the proper law enforcement authority.

2 PRESUPPRESSION

INTRODUCTION

Presuppression includes all the actions required (with the exception of prevention) before the actual control of a wildfire. It includes all the preparation, organization, development, and maintenance of equipment in addition to planning, cooperation, mutual aid arrangements, and training. Suppression will only be as effective as the quality and continuity of presuppression action. Most of the work in a wildfire service is required in the area of presuppression.

JURISDICTION AND STRUCTURE OF A FIRE SERVICE

Organization is basic to any kind of effort, but, because it is necessary to have quick and sure responses in the control of wildfires, it is of paramount importance that everyone concerned act with accuracy and precision in their part of the team effort. Effective preparation and training are the only way in which this goal can be accomplished.

There are wide variations in the way wildfire services are organized. The laws, regulations, and authority creating and operating the service will vary by state, federal, county, local, and private jurisdictions. Generally, state and federal organizations will have clear-cut lines of authority and will have published the duties and functions of each position. Planning will be complete, and training will be well along. County and local jurisdictions are organized under state and local legislation and will generally have good understanding among their members, but some units may be in need of more clearly defined rules of operation. A study of existing laws and regulations will indicate whether additional or clarifying legislation is needed.

If the wildfire suppression contingent of a fire service is part of an overall fire service organization, either municipal or rural, consideration must be given to the requirements of equipment and organization peculiar to wildfire control in that area.

PLANNING

To develop objectives, policies, organization, equipment, and total operations, planning is required. Planning or research, or a combination of both, by whatever name it is called, is the logical basis for appraising what is involved, what will be done, why it is necessary, and how, by whom, and when it will be accomplished. Good fire district planning provides for continuity of programs, establishes coordinated effort, eliminates disruptions due to changes in personnel, and generally establishes the proficiency of the service. The plan should be written, and accomplishments should be recorded and updated periodically. It may be written for a five-, ten-, or twenty-year period or with short-range and long-range objectives and how to accomplish them.

The following outline is by no means complete—it is intended to suggest to the individual district or department those aspects of fire fighting that need analysis in the course of planning for adequate protection. Hazards, risks, and values will provide the controls in planning and will dictate justifiable expenditures.

BACKGROUND DATA

Map of protection area. The best map available should be chosen to represent the area under your protection. It may be a county map, a map from a government agency (such as the Soil Conservation Service), or the U.S. Geological Survey quadrangle map. It should not be less than ¼ inch to the mile and preferably should be drawn on a larger scale. It should be used as a base map for overlays; a copy of it can be used for other maps. On it, show the protection boundaries; adjacent fire service locations; areas where initial attack is assumed by more than one service, especially along boundaries; and any other jurisdictional information. These areas can be shown by cross-hatching or by different colors.

Identify by class of construction all roads; trails; truck trails; railroads; bridges and their capacity; interstate exits and crossings; water travel routes and docks, canals, and streams too deep to ford with your equipment; natural barriers to cross-country travel (such as swamps, box canyons, and tail water ponds); topographic barriers; and locked gates and private jurisdictions blocked to travel. From this information the best route and method of travel to the various hazards and risks can be determined. Safe speeds for the different roads should be established for the kinds of equipment in use. If space permits, show all cultural features, such as airports, heliports, marinas, recreation areas, residences, farms, ranches, industrial plants, schools, hospitals, nursing homes, pumping plants, power lines, flammable liquid storage, dumps, dams, water supply sources, and any man-made installations. These features can be shown on a transparent overlay if space does not permit on the base map.

Map of hazards. The map of hazards may be a second map, an overlay of the protection area, or, if space allows, actually part of the protection area map. It should show the areas of hazard as related to fuels; e.g., types of vegetation should be shown in different colors. To delineate forest types, show broadleaf and evergreen separately as well as timber-cutting areas with hazardous slash; areas of heavy fuel concentration due to insects, disease, or age; barren or rocky areas with little or no fuel; and areas of grass, grainfields, and brush. This map will assist with the prevention effort—it will help to plan routes of travel and to predict the expected fire occurrence. (Also see the fire prevention plans in Chapter 1.)

Fire occurrence by causes and years. The information on fire occurrence by causes and years should at least be tabulated and may be made into graphs. If the locations of individual fires are placed on a map or an overlay, the information may be more distinct and might reveal definite patterns (Also see the fire prevention plans in Chapter 1.)

OBJECTIVES OF THE PLAN

From the information obtained and outlined, develop an action plan outlining what will be done in given situations and clearly outline the objectives:

- What category of fires will be suppressed?
- Will rescue service be provided—if so, how and with what equipment?
- How much reduction of man-caused fires is expected?

Include a statement of cost related to benefit in the plan.

ORGANIZATION

There should be a statement of the laws and regulations that provide for the service. Most fire services have an organizational structure like that shown in Figure 2.1. Include a written statement on organization in your plan and a chart to show the flow of authority.

EQUIPMENT

Prepare an inventory of the equipment on hand, including location, housing, kind and amount of apparatus, hand equipment, and vehicles. Also list equipment that is planned and the approximate dates for its acquisition. This inventory can be in written form or can be shown on a chart.

The amount and kind of cooperative equipment available from other jurisdictions and private or governmental sources should be listed and shown for the several areas of the protection district. This information may be shown on the working map developed for use at fire stations. Provisions for housing and maintenance of all the equipment should be developed.

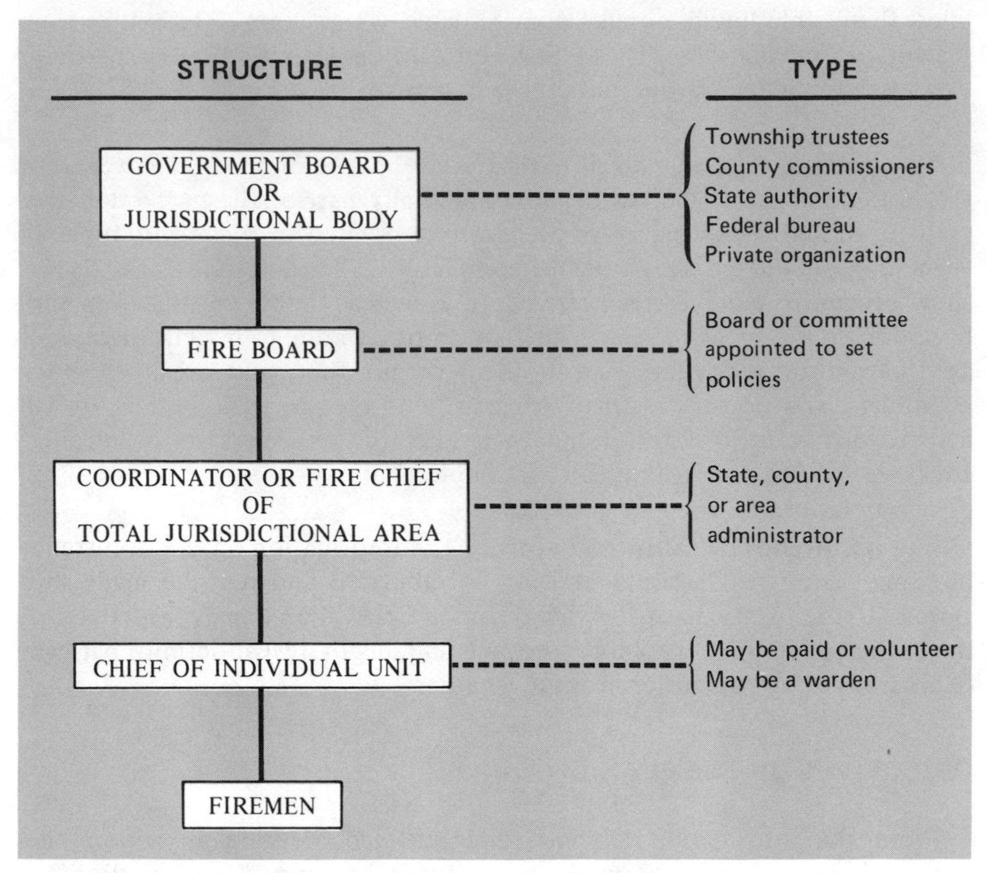

Figure 2.1. Organizational structure of a typical fire service.

TRAINING

Specific training programs must be provided for in the fire service plan. The section on training in this chapter outlines the objectives, methods, and sources of training.

PREVENTION

All plans must detail the functions to be performed in the effort to prevent the occurrence of fires. *Inspection* is one of these functions. Plan for the necessary inspection work with provision for the required training of local inspectors. Include prefire planning along with inspection. The participation of the fire service in *hazard reduction* should be determined. The hazard and risk map can be used to plan where and how hazards will be reduced or eliminated. The chapter on prevention contains suggestions regarding hazards, risks, fire laws, and zoning. *Public relations* will be included in the prevention plan but should also be considered in overall operations.

MUTUAL AID

Very few fire services can afford or can justify owning all the equipment and maintaining all the man power necessary for unusually large fires. Therefore, most fire districts are dependent on aid from fire units adjacent to their jurisdictions to cope with those fires beyond their capability. This mutual aid takes many different forms but is a necessary part of the planning of any fire protection unit.

Important considerations in programing mutual aid include the following:

- Necessary arrangements for mutual aid must be made in advance of the need, preferably through the use of cooperative agreements.
- The fire boss in any situation must be defined. Normally, all other fire chiefs report to the fire chief of the jurisdiction in which the fire occurs. However, there may be situations where this rule would be changed. These situations must be understood by all units and personnel.
- The type of response, financial arrangements, indemnity, and responsibility must be clearly stated and understood.
- Definite arrangements must be made on the method for requesting mutual aid and on the identification of such requests as legitimate.
- Arrangements must be made for backup protection when units are away from their jurisdictions.
- Response by adjoining units to an alarm along a mutual boundary should be agreed to before such an alarm is received.

Other items of mutual interest may also be important to arrange for, depending on the local situation. State and federal agencies can help in developing mutual aid agreements in accordance with local laws.

FIRE DISTRICT OR UNIT ORGANIZATION

The unit organization for wildfire is managed by the *fire chief*, or *fire boss*. Chief and boss are synonymous, but boss has been used traditionally by wildfire organizations, probably because of its connotation of firm control.

In any fire control situation, there can be only one fire boss, regardless of the size of the fire. As the organization structure builds owing to the size of the fire, the fire boss will delegate more and more authority and responsibility to assistants, but the fire boss is always the final authority and bears the total responsibility.

BASIC ORGANIZATION

The fire boss should be appointed by the governing body on the basis of merit and ability for an indefinite term, and he should be removable only for just cause.

Assignment of personnel might be made according to Figure 2.2, depending on the geographical location and the types of fuels encountered. The organization may include any or all of the crews shown in whatever multiples are necessary.

Rule of thumb: The capability and effectiveness of a fire control unit are largely dependent on the interest, ability, and management of the fire boss.

DUTIES

Fire boss (chief). The fire boss is in charge of the entire operation, including assignment and supervision of outside forces.

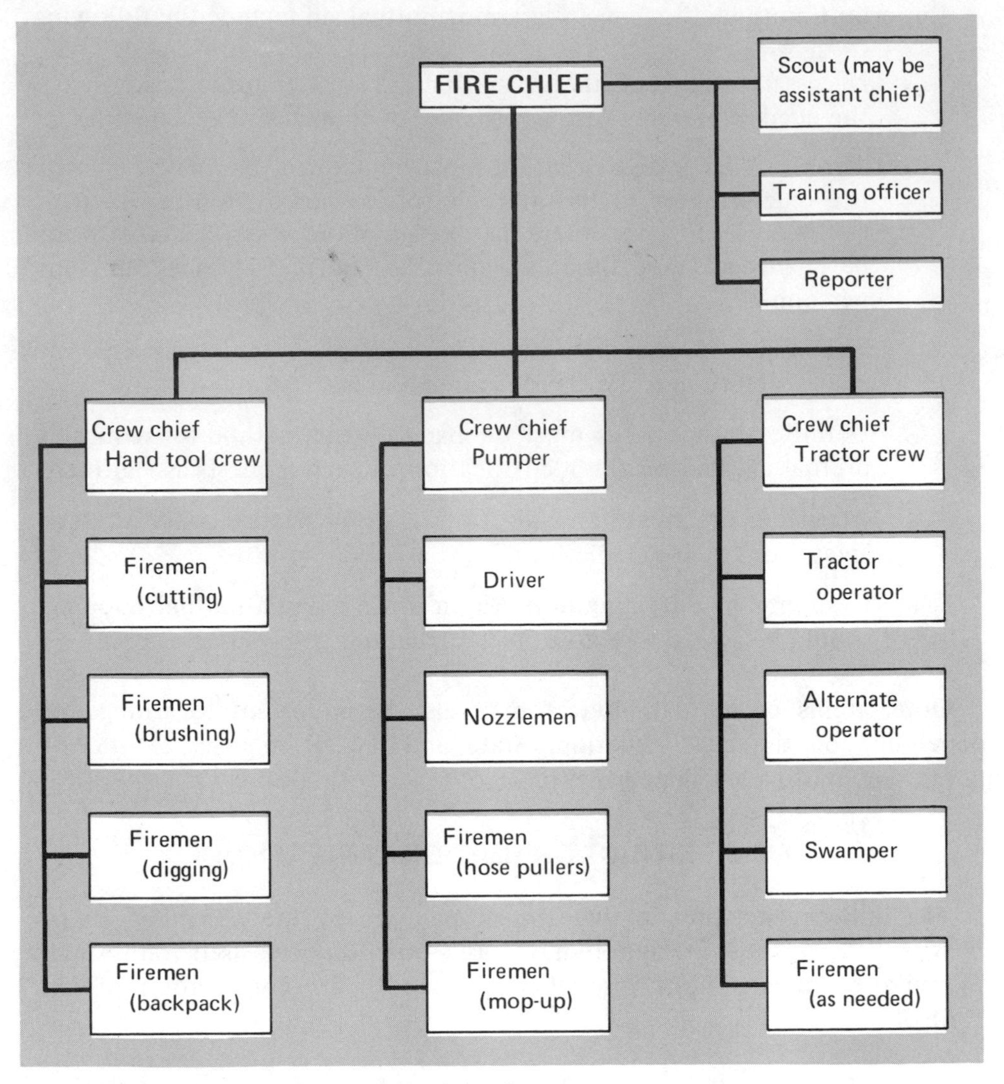

Figure 2.2. Organization of a fire service unit.

Scout. On a going fire, the scout obtains full detailed information on the location, condition, progress, behavior, and safety requirements. He reports at times and locations arranged by the fire boss.

Training officer. The training officer may have any assignment on a going fire. He is responsible for developing and conducting the training program. He should have completed an instructor training course or otherwise should be knowledgeable about training processes. He is responsible for developing a training library and scheduling of self-help courses, training programs by state and other agencies, first aid courses, and practice sessions.

Reporter. The reporter develops information necessary for fire reports and prepares the reports. He prepares news releases and maintains records. He may also be secretary-treasurer for the group, or his job may be a separate assignment.

Crew chief (crew boss). The duties of a crew chief and the duties of a squad boss are essentially the same. In a large fire organization the squad boss supervises the work of six to ten firemen, and three squad bosses are, in turn, supervised by a crew boss. A crew is the basic unit of organization; in ordinary situations the crew boss works directly under the fire boss. If the size of the fire requires more than three crews, the perimeter is divided into sectors, and a sector boss directs the work of crew bosses.

The crew chief has the following responsibilities:

- Explains job to be done, issues instructions, and organizes the crew to accomplish specific tasks
- Provides for on-the-job training and constantly checks on safe practices, correcting as necessary
- Provides for first aid and for the welfare of his crew
- Inspects the assigned area frequently and obtains required standards of performance
- Reports directly to the fire boss or on larger fire organizations to the sector boss, pumper boss, or tractor boss

Driver. The driver is responsible for the truck, pumper, or tanker assigned to him. He must be able to drive safely under all conditions and to keep his unit functioning. He is responsible for the safety of anyone riding on his unit. He may be assigned as crew chief for the unit.

Nozzleman. The nozzleman should be experienced and proficient in fire fighting, especially in the proper use of water. He must be adequately clothed for his assignment. He supervises the hose pullers assigned to him. He is the key man in application of the water. He functions under the direction of the crew chief, or he may be the crew chief.

Tractor operator. The tractor operator must be experienced and knowledgeable in the operation of his equipment. He directs the work of swampers and helpers assigned to the equipment. He is responsible for the operation, maintenance, and safety of his piece of equipment. He may be the crew chief.

Torchman. The torchman is preferably an experienced fireman who works under the direction of the crew chief of the crew following the tractor or the initial line crew or pumper. He must be able to follow instructions to the letter.

RECRUITING AND SCREENING

Candidates for membership in an organized fire crew should be subjected to a thorough physical and medical examination. Those recruited at the fire as "pickup labor" should be as thoroughly screened as possible by the hiring officer. If there is any doubt of their physical ability to withstand the rigors of wildfire suppression work, they should not be hired. Those who are overweight, overaged, suffering any physical ailment, or insufficiently dressed should not be allowed on the line.

Age limits should be established—both minimum and maximum. Generally, those 50 years of age or older should not be employed on the line. Those below the minimum age limit for your state should have work permits and should be assigned to nonhazardous tasks such as headquarters and camps and patrol of quiet sectors.

Aptitude tests to establish physical and personality characteristics are quite useful in selecting members of a fire department. Requirements should be set for education and character and should be made known to candidates. Candidates should be checked for these requirements from the information sources available. A standard of probationary work and training should be established and adhered to for all new recruits.

OPERATIONS

COMMUNICATIONS

Adequate and reliable communications are necessary for detection, reporting, and dispatching of aid before, as well as during, suppression action.

Telephone services can usually be relied upon for detection, reporting, and dispatching, but radio is also being used for these purposes. Some difficulties accompany the use of dial systems, but these can usually be overcome by cooperation from the local telephone service.

Where more than one vehicle or more than one crew is assigned to a fire, quick communication becomes mandatory. Two-way radio is the best medium for such communication; but it is somewhat costly and requires periodic maintenance, and assignment of frequencies is becoming difficult. When they can be afforded, radios are the best investment a fire service can make after procurement of equipment and safety items. Prompt dependable communications are a necessary part of good fire control work.

Telephone. Some good telephone practices when you are speaking for the fire service are:

- Be courteous and answer promptly.
- Speak up and speak directly into the mouthpiece.
- If you are receiving a report of a fire, try to keep the caller on the line to get all the information possible. Often the caller is excited and will hang up before giving the location. Try to get the location of the fire, its size, what is burning, who is calling, the telephone number the caller is using, and any other information he can give.
- Talk deliberately, and thank the caller.

Radio. Fire service radios are to be used only for official business. Radio use is systematically monitored by the Federal Communications Commission. Do not transmit personal messages. Confine your transmission only to messages relevant to the fire control operation. Good practices in using a radio are:

- Answer promptly and be courteous.
- Use proper language. Swearing and obscene language are prohibited.
- Think about what you want to say before you transmit. Be concise but explicit. Do not do your thinking on the air.
- Be as brief as possible—use the ten code (see next section). Do not say please, thank you, over, or clear.

- If others are using the air, wait until they are finished unless your message is an emergency.
- If you have an emergency message, break into the conversation and ask for the air.
- Pronounce your words distinctly.
- Hold the mike at an angle to your mouth—about 1 or 2 inches from your mouth.
- Push the "talk" button and hold the button while you talk. As soon as you have finished talking, release the button. The receiving party will then hear the "chuh" sound and will know that you are ready for him to answer. Remember to release the "talk" button to listen.
- Speak clearly, at a constant speed, and with the rhythm of ordinary conversation, about sixty words per minute.
- Speak with normal volume for conversation—your message will be distorted if you speak too loudly.
- The words affirmative and negative should be used instead of yes and no.
- If one transmission lasts over thirty seconds, the transmitter should say break, wait for acknowledgement of reception, and then pause for ten seconds before continuing. This allows for interruption for emergency messages. Remember, as long as you are pressing the "talk" button, you cannot receive messages.
- When you are receiving a message that you are to repeat to a third person, you should write down the message and then read it to the third party.
- When the message you wish to send is complex, you should carefully write it out before you send it; then read it on the radio.
- For greater clarity, most numbers should be transmitted as separate digits, such as three-six for 36 and one-four-three for 143. The following numbers should be transmitted as in normal conversation: whole units of 1,000—say one thousand, two thousand, etc.; day of the month—say January twenty one, December fifteen, etc.; code number—say ten twenty-two, etc.; time measurement—say thirty minutes, twenty-four hours, etc.; other numbers are read as separate digits.
- Disasters such as death or injury, either occurring or about to occur, are indicated by the international distress signal Mayday, which is derived from the French term *maidez*, a request for help. The person using Mayday should, if possible, give his name and location and should describe the situation. Upon receiving the Mayday signal, leave the air free of other traffic. Names of individuals who are victims of accident or fatality should not be used in radio traffic.
- The words fire flash are used only to report a fire.
- Priorities of messages are: (1) death, injury, medical aid, (2) reporting a fire for the first time, (3) fire control traffic, and (4) administration.

- When you are finished transmitting, give your call number. This indicates that you are clear of the air.

Radio messages should be clear, concise, and complete.

Ten code. The ten code is designed for reliability and speed. It has replaced the four code formerly used on fire control work. Some fire control agencies use a nine code. Selected parts of the police ten code are used, and they are shown here. Individual services often add their own designations.

10-1	Receiving poorly
10-2	Receiving well
10-3	Stop transmitting
10-4	Acknowledgement or ok
10-5	Verbal repeat
10-6	Standby will call
10-7	Out of service (radio turned off ________ minutes)
10-8	In service (radio turned on)
10-9	Repeat beginning with ________
10-11	Slow up (talking too fast)
10-13	Transmit weather information
10-19	Return to your station or am returning
10-20	What is your location
10-22	Disregard last message
10-23	Arrived at scene
10-25	Do you have contact with ________
10-33	Emergency traffic at this station
10-36	Correct time
10-49	Proceed to ________
10-71	Advise nature of fire
10-72	Report progress of fire
10-77	ETA (estimated time of arrival)

Time of day. Time of day designations are the same as those used by the military in order to eliminate confusion about A.M. and P.M. To convert to military time, add twelve to the P.M. time to get the first two numbers of the hour. The hour of 8 P.M. is twenty hundred (8 + 12 = 20). The first two digits denote the hour on a twenty-four hour basis, and the second two digits denote the minutes of the hour. Midnight plus one minute (12:01 A.M.) is 0001 (say oh-oh-oh-one). Midnight plus thirty minutes (12:30 A.M.) is 0030 (say oh-oh-thirty); 1:00 A.M. is 0100 (say oh-one hundred); 2:00 A.M. is 0200 (say oh-two hundred); and 12 noon is 1200 (say twelve hundred). Noon plus one minute (12:01 P.M.) is 1201 (say twelve-oh-one); 1:00 P.M. is 1300 (say thirteen hundred); and 1:30 P.M. is 1330 (say thirteen thirty).

Midnight (12:00 P.M.) is 2400 (say twenty-four hundred)—the last number in the series. For minutes past the hour, leave off the hundred and give the number of minutes. Remember, from noon to midnight is P.M., and from midnight to noon is A.M.

Numbers. Numbers are important for correct messages. The correct pronunciation as used by the U.S. Air Force (USAF), International Civil Aviation Organization (ICAO), and Associated Public Safety Communications & Officers, Inc. (APSCO) is shown in Table 2.1.

TABLE 2.1
Number Pronunciation Guide

Number	*USAF-ICAO*	*APSCO*
1	wun	wun with strong w and n
2	too	too with strong and long o
3	tree	th-r-ee with a slightly rolling r and long e
4	fow-er	fo-wer with a long o and strong w and final r
5	fife	fie-yiv with a long i changing to a short and strong y and v
6	six	siks with a strong s and ks
7	sev-en	sev-ven with a strong s and v and a well-sounded ven
8	ait	ate with a long a and strong t
9	niner	ni-yen with a strong n at the beginning, a long i, and a well-sounded yen
0	zero	zero with a strong z and a short ro

Phonetic alphabet. For spelling out unusual names of locations, say A-Alpha and B-Bravo, not A as in Alpha and B as in Bravo. The USAF, ICAO, and APSCO alphabets are shown in Table 2.2.

Safety precautions in the use of radio. High voltages build up in some circuits. Do not open or otherwise contact the inside of AC-powered equipment. Such equipment should be kept locked, and only radio technicians should have access to it.

Never use the radio during lightning storms, except from within a rubber-tired vehicle, if the storm is within 1 mile. Never extend the aerial on a pack set if the storm is within 1 mile. Do not use a radio transmitter within 300 feet of any electric blasting or in any areas where electric detonators are handled or stored.

Other methods of communication. Although they are not now commonly used because of the availability of radios, the portable field telephone may be helpful in some situations where radios are not available. This method requires laying out the line and keeping it usable, but, in cases where no other communication is available between the headquarters and telephone central, portable field telephones may be valuable. Often the headquarters must be located where telephone communication is available, and the portable can be used to tie to existing telephone systems.

TABLE 2.2
Phonetic Alphabet

Letter	*USAF-ICAO*	*APSCO*
A	Alpha	Adam
B	Bravo (Brah-voe)	Boy
C	Coco	Charles
D	Delta	David
E	Echo	Edward
F	Foxtrot	Frank
G	Golf	George
H	Hotel	Henry
I	India	Ida
J	Juliet (Joolee-yet)	John
K	Kilo (Kee-loe)	King
L	Lima (Lee-mah)	Lincoln
M	Metro (Met-roe)	Mary
N	Nector	Nora
O	Oscar	Ocean
P	Papa	Paul
Q	Quebec (Kay-beck)	Queen
R	Romeo	Robert
S	Sierra	Sam
T	Tango	Tom
U	Union	Union
V	Victor	Victor
W	Whiskey	William
X	Extra	X-ray
Y	Yankee	Young
Z	Zulu	Zebra

Written message. Since the radio has come into such wide use, the system of communicating by written messages has become passé; however, there may be instances in which it could still be used if radios are unavailable or out of order. When you are sending messages in this manner, be sure that they are written clearly and legibly and that a copy is kept. Beware of sending messages by word of mouth because the receiver often gets a garbled or incorrect transmission.

Hand signals. Hand signals are useful in pumper operations, tractor use, and helicopter and air operations. The signals used with each are shown in the chapter on tactics.

DETECTION

The capability of discovering and locating fire starts in the protection area of a fire service is basic to effective fire suppression. The occurrence of a wildfire must be observed and reported as soon as possible in order to begin suppression action while the fire is small.

Rule of thumb: The smaller the fire, the easier it is to control.

Discovery and reporting of uncontrolled fires is accomplished in a number of ways. Sources of detection are people living or working in the area, travelers through the area, patrol for the specific purpose of detection, aerial observation, and fixed lookouts.

For detection to be effective, everyone must know where and how to report a discovered fire. Develop a fire telephone number for reporting fires, and establish a dispatch system. Fifty to eighty percent of wildfires are reported by residents or users of the areas where fires occur.

State, federal, and many private agencies have definite detection systems in operation. These may have some capability of detection for adjacent areas. Regular travel operations should be enlisted in a detection program. These would include airlines, contract air operations, railroads, bus lines, mail carriers, county units, state and federal employees traveling through the area, highway patrolmen, and routemen of various kinds.

Of equal importance is the cooperation of all citizens in reporting "smokes," as soon as they are seen. This detection requires an educational program designed to obtain everyone's cooperation and to provide the necessary information on where to report fires. It is imperative that the reporting person give an estimate of the size and kind of fire and its location as accurately as possible.

Where prescribed burning is practiced, it is very helpful to report these to the dispatcher before they are started. Such a procedure can prevent many false alarms.

DISPATCHING

Closely allied with the detection capability are the requirements of dispatching, including receiving and recording calls made to report fires, activating the fire service to respond, and supporting the fire fighters while they are on the fire with information and reinforcements of either manpower or equipment when it is requested. Radio contact with the fire control units is almost mandatory in this day and age, but other means of communication may still be used.

The dispatcher may be a police dispatcher, a sheriff's dispatcher, a municipal employee, a telephone operator for initial dispatch, one of the firemen, or a fireman's wife who has been designated as the dispatcher. However the arrangement is made, the dispatcher's first job is to get the best possible information from the person reporting the fire. He immediately transmits this information to the person in charge of the initial attack crew so they can make an immediate response. An alarm system to alert the firemen is necessary. Subsequently, the dispatcher can perform tasks requested by the fire boss and can get reinforcements on the way by activating previous arrangements for support.

EQUIPMENT

HAND TOOLS

In the early days of wildfire control, hand tools were the only equipment available. Today, there is a great variety of motorized equipment and some of it is very sophisticated. However, hand tools are still needed to complete the fire control job. Using hand tools is still the only way to attack wildfires in many areas.

Many kinds of hand tools have been used on fire suppression, but only a few have passed the test of time. Some are used universally; others have found favor only in certain areas. The kind and amount of tools needed will depend on the particular fuel types and the soil conditions (amount of rock) encountered as well as local preference.

Only the most common hand tools are described here. They are available through most hardware dealers, fire equipment companies, the General Services Administration (GSA), or state and federal agencies.

Safety is important in using all hand tools. Chapter 4 reviews the safety rules that must be observed when hand tools are used.

Round point shovel. The round point shovel is probably the most universally used fire tool. Originally, the number 2 long-handled round point shovel was used, and in some places it is still used. Later the number 0 long-handled round point or "lady" shovel was quite popular. The U.S. Forest Service developed a standard for fire shovels that is now most commonly used. The blade is somewhat larger than the number 0, and the neck has a bend that makes it easier to use for scooping up dirt and cleaning the fire line. It has definite specifications and is usually listed as a fire shovel. Short-handled shovels are back breakers.

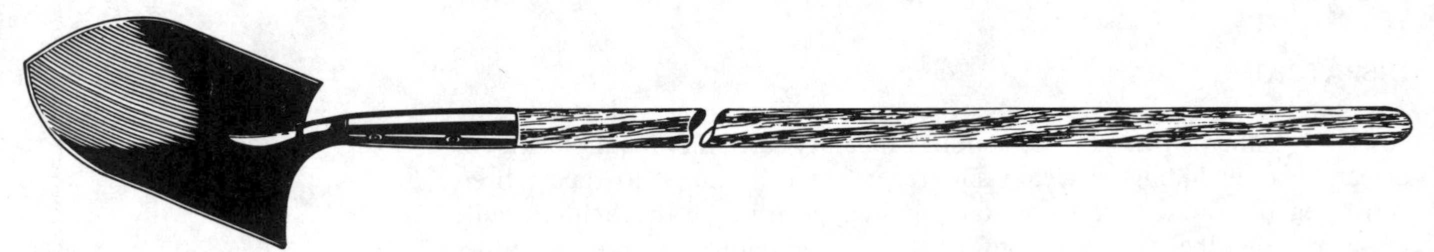

Figure 2.3. Round point shovel.

In the hands of a skilled worker, the shovel is a very effective fire tool. It can be used to dig, to swat, to throw dirt, to scrape the fire line clean to mineral soil, and, to some extent, to cut. The shovel also makes an ideal face shield for deflecting heat on a hot fire line. It can be used as a hook to pull material, and it is particularly effective in throwing dirt to check a running fire or to dig out burning material.

Two basic methods, fanning and casting, are used for throwing dirt, and crews should be drilled in the methods used in a particular location. Accuracy is a primary consideration. Dirt that fails to reach the target does not help put the fire out and results in wasted time and effort. Dirt to be thrown should be free of clods and burnable material.

The side arm throw (fanning). This method is most effective when the fire is burning ground fuels. The method is designed to spread each shovel of dirt rapidly and evenly in a thin layer over the greatest possible distance, thus extinguishing the maximum amount of fire but without burying the burning material.

The shovel is grasped firmly with one hand near the end of the handle and the other hand well down on the handle near the balance. Begin the swing with the shovel to the rear of the body on the side toward the fire and 12 to 18 inches above the ground. Bring the shovel rapidly around from rear to front in the same direction in which you are moving. The lower hand controls the direction of the throw as the arms and shoulders pivot the

Figure 2.4. The fanning action of the side arm throw spreads a layer of dirt rapidly and evenly over the greatest possible distance.

shovel around the body; at the same time the shovel is rotated until the throw ends with the shovel in an upside-down position. It is this rotation of the shovel that results in the fanning action of the dirt that is essential for covering the greatest possible area.

Because the shovel must be held on the side towards the fire and the dirt is thrown in the direction of movement, crew members must be drilled in making this throw from either the right or the left side. The side nearest the fire is referred to as the near side, and the opposite side is called the off side. Additional distance can be covered if the throw is begun with the off-side foot in a forward position. The body is then pivoted around this foot, bringing the near-side foot forward in a follow-through motion that results in obtaining the maximum possible reach while at the same time allowing the near-side hand to guide the shovel accurately along the fire line during the entire length of the movement.

The overhand throw (casting). This throw casts dirt on an overhead target or one that cannot be approached closely. To be most effective, the dirt should be kept in a compact unit until it reaches the target. Grasp the shovel with one hand near the end of the handle and the other near the middle of the handle. Keeping the loaded shovel handle parallel to the ground, raise it to shoulder height, with the shovel to the rear of the body. The end of the handle should be lined up exactly with the target. The hand balancing the shovel should be positioned so that the forearm is straight up and down, thus holding the handle directly above or immediately adjacent to the shoulder. This hand is then extended sharply upward and forward, using the end of the shovel handle as a pivot point. The dirt should be made to leave the shovel sharply at the end of the throw in order to hold its compactness until it reaches the target.

Figure 2.5. The overhead throw is used to cast dirt on an overhead target or one that cannot be approached closely.

When the overhand throw is used, the dirt is always thrown forward; therefore it makes little difference whether the shovel is held on the left or the right side of the body. Each man should be encouraged to practice this throw from the side that is most comfortable and effective for him.

It may be necessary for other shovel men and/or mattock men to dig up and stockpile dirt either for fanning or casting if the soil is tight and cannot be scooped up readily.

Scraping. The position and motion of the body are most important in using a shovel for scraping. To assume the proper right-hand position for scraping, first grasp the shovel firmly with the left hand near the end of the handle and with the right hand slightly below the center of the handle. Extend the right foot forward 12 to 18 inches, and place the right forearm firmly against the inner side of the right leg at the knee. With the knees slightly flexed, bend the body forward at the waist until the blade of the shovel is almost parallel with the ground and tilted very slightly either to the left or to the direction toward which you are going to scrape. Using the right forearm and knee as a fulcrum, scrape from right to left, keeping the right forearm firmly fixed in position as described above and using the motion of

Figure 2.6. Proper position for scraping.

the upper part of the body to pivot the shovel at the right knee. The shovel blade should not be tilted enough to cause it to dig in but should slide smoothly over the surface, picking up just enough material so that the blade will be carrying a full load at the end of the stroke. At the end of each stroke, the blade is returned to its original position and is advanced for the next sweep by taking a short step forward. Where a heavy layer of loose material is being removed, a shorter step forward will be required to prevent the blade from overloading and becoming too heavy to handle efficiently. When you are scraping from left to right, the above position is reversed.

Maintenance. The blade of the shovel should be sharpened on both edges to within 1½ inches of the top. The bevel should be on the inner face. Thus the shovel can be used as a cutting tool to some extent. Keep the handle smooth and its attachment to the blade tight.

Ax. The single-bit, or pole, ax is most popular in the South and East, while the double-bit ax is the one most used in the West. However, both types are used to some extent in all areas. The ax can be used to a limited extent as a scraping tool. The single-bit ax can be used to knock stumps, logs, and heavy fuels apart. It should never be used on steel wedges because its eye becomes distorted and will not hold the handle. A 36-inch handle is usually preferred. If double-bit axes are to be used in cutting roots or in rocky soils, one bit is left fairly dull and one blade is kept sharp. The sharp blade should be used only when it will not be damaged by rocks.

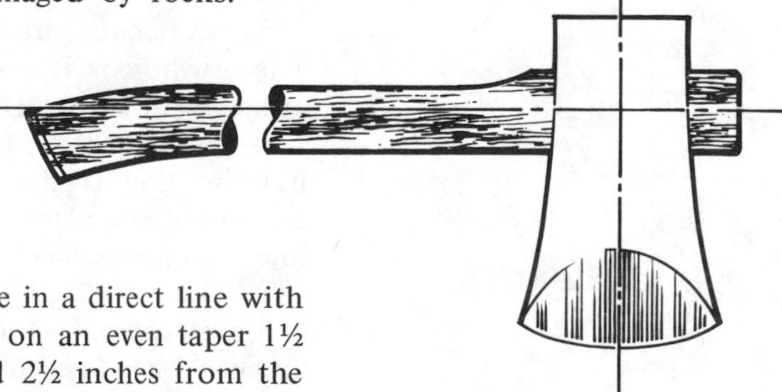

Figure 2.7. Single-bit ax.

Maintenance. The cutting edge of the ax must be in a direct line with the axis of the handle. The blade should be ground on an even taper 1½ inches from the cutting edge on a single-bit ax and 2½ inches from the cutting edge on a double-bit ax. Keep the handles smooth and tight. Use epoxy to fasten the head to the handle. When you are sharpening axes and similar cutting tools, always be careful not to overheat the cutting surface. Grind to within ¼ inch of the cutting edge and finish with a file or whetstone. Maintain the proper taper or bevel. All axes should have protective guards for transportation.

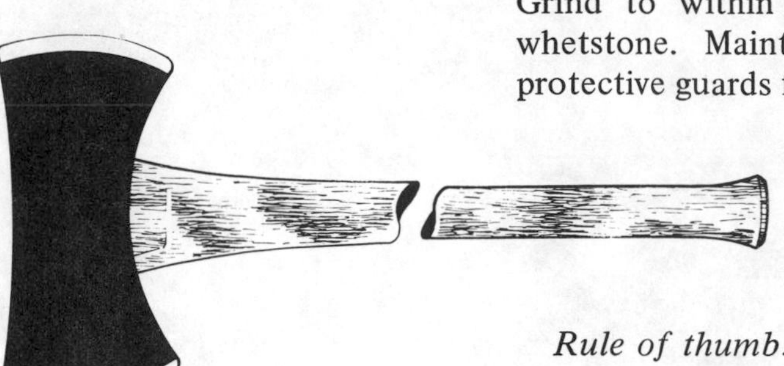

Figure 2.8. Double-bit ax.

Rule of thumb: A sharp tool is a safe tool. A dull ax is a dangerous tool.

Swatter. The fire swatter, or flail or flap, is used mainly in grass, needles, or light ground fuels. It is laid on the fire edge and moved progressively along. Hard vertical swatting should be avoided because it tends to spread the fire and is very tiring to the user. It should be used from outside the fire toward the fire area. Care must be taken that fire is not swept into the unburned area.

The tool should be ruggedly made because it is frequently in contact with burning fuel. Ordinarily, the head is fashioned from a piece of four-ply heavy rubber volcanized belting stock that is 12 by 15 inches and ¼ inch thick. This head is attached with a T-iron to a wooden handle at least 60 inches long; some are 16 feet long, but these are too unwieldy.

A shovel can also be used as a swatter, especially a scoop shovel. A wet gunny sack or bough can be used for the same purpose, but these are swung at about a 45-degree angle toward the fire, striking the fire edge.

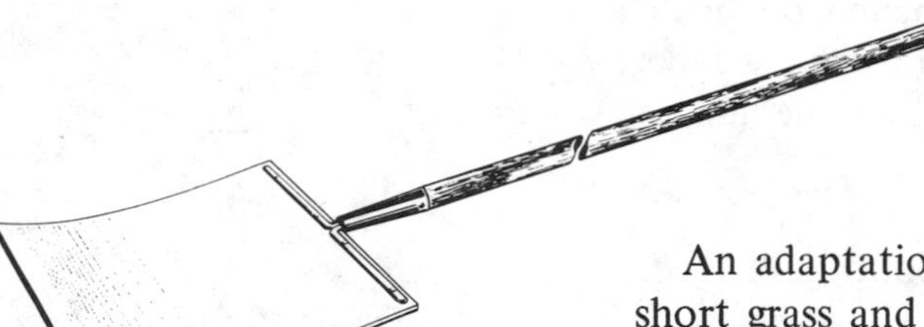

Figure 2.9. Fire swatter.

An adaptation of the swatter is the rotary flail, designed for range fires in short grass and stubble. Four beaters revolve around the drum and strike the ground with considerable centrifugal force ahead of the roller. The beaters also provide a track for the roller similar to a caterpillar tractor. It may be operated backwards or forwards with equal efficiency.

Maintenance: Keep the handles smooth and tight. Maintain attachment of the flap.

Brush hook. Brush hooks are produced as both single bit and double bit. The brush hook is designed for clearing dense brush in locations difficult to attack with an ax. It can be used for cutting heavy material, but an ax is preferred. When the brush hook is used with a straight overhead "swing and pull" motion, it shears off limbs and stems. Unskilled men can operate this tool more effectively than an ax with less chance of damage to the tool or injury to themselves.

Maintenance. Grind both sides of the long portion of the cutting edge in an even bevel back from the cutting edge at least 1 inch. Take care to keep the circular pattern at the throat. Carefully grind the throat back 1 inch on an even bevel. Grind the hook or point on a bevel approximately ¾ inch

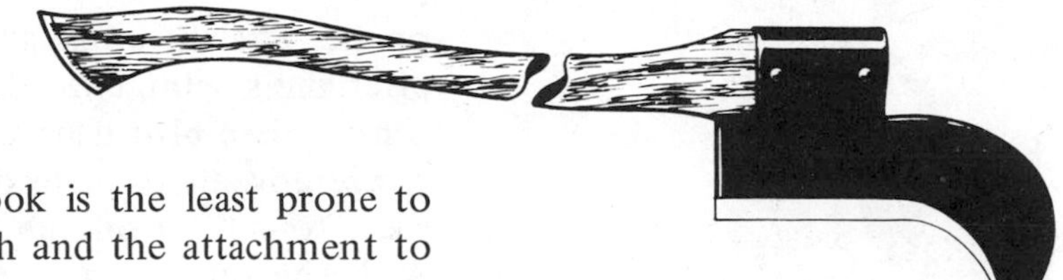

Figure 2.10. Single-bit brush hook.

deep. The handle of the ax eye type of brush hook is the least prone to breakage. On all brush hooks, keep handles smooth and the attachment to the blade tight.

Pulaski tool. The Pulaski tool was originally designed by Ranger E. C. Pulaski for use in Montana and the upper Idaho forests. It is a popular tool in most sections of the country because it serves the dual purpose of an ax and a grub hoe. It can also be used with a scraping motion to move dirt and clean fire lines. It is a very versatile tool. Practice is required to use the Pulaski tool as an ax because the balance is different from that of an ax.

Maintenance. The ax bit of the Pulaski tool should be ground with an even taper back from the cutting edge at least 2½ inches. The hoe side should be ground to a good cutting edge, with the bevel approximately ⅜ inch deep on the inside of the blade on the side facing the handle. Maintain the handle and make sure it fits tightly into the head.

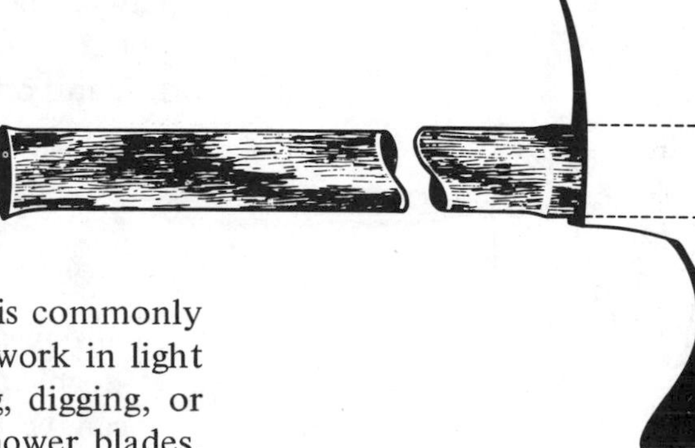
Figure 2.11. Pulaski tool.

Council rake. Also called the rich tool, the council rake is commonly referred to as a fire rake. It is highly efficient for trenching work in light brush, duff, and small roots, and it can be used for cutting, digging, or raking. It is also available with a scraper opposite the four mower blades. When you are carrying it, grasp the handle about 8 or 10 inches behind the rake head with the cutter sections pointed down toward the ground and along the side with the arm down. In this position, the rake head will be in front of the carrier. The rake head should be attached to the handle so that the flat side of the blade is on the inside, the side facing the worker.

The council rake is very versatile; it can also be used in mop-up. The council rake can be used to

- Rake embers away from the fire edge
- Shovel soil and sand on stumps and logs
- Scrape fire off standing snags and stumps
- Pull hot logs and tops from the fire edge
- Cut small trees or brush

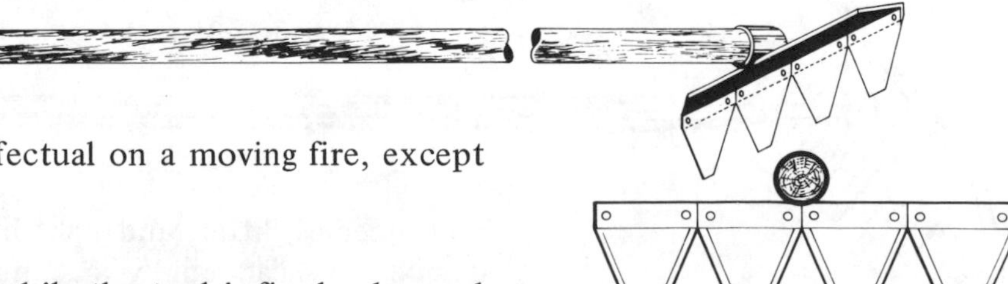
Figure 2.12. Council rake.

Strangely enough, this tool is fairly ineffectual on a moving fire, except for the hand labor of constructing a fire line.

Maintenance. Grinding should be done while the tool is firmly clamped into a special council tool grinder. A file may also be used but not the ordinary emery wheel. The square point on the blade is retained by grinding on the straight stone that comes with the grinder. This action takes place after the sloping edges have been ground on the bevelled rock. The straight rock is then placed on the grinder, and the four points are touched up. Do not remove the handle during grinding. Hold the handle at a point about 2 feet from the blade and apply pressure while the grinder is being turned. With the other hand, operate the slide to center the blades on the grinder. The slide and clamp that come with the average sickle grinder will not hold the council tool, although the slide and clamp from a council tool grinder will fit a sickle grinder. Do not discard clamps or slides, since they seldom wear out and sickle grinding stones may be purchased locally to go with them. Maintain handles and attachment to the blade.

McLeod tool. The McLeod tool is similar to the council tool. It is preferred in some areas, especially open pine areas of the West and Southwest and in the eastern and northern hardwood forests. Its larger blade and rake side work well in needles, duff, and leaf mold. In the fuels where it is used it is said to build line faster and with less effort than any other tool available. The hoe blade is very useful in cutting through deep litter, matted brush, squaw carpet, bear clover, and similar fuels. It can be used for trenching, grubbing, or slashing low limbs. The strong teeth are useful in pulling out brush and other cut material.

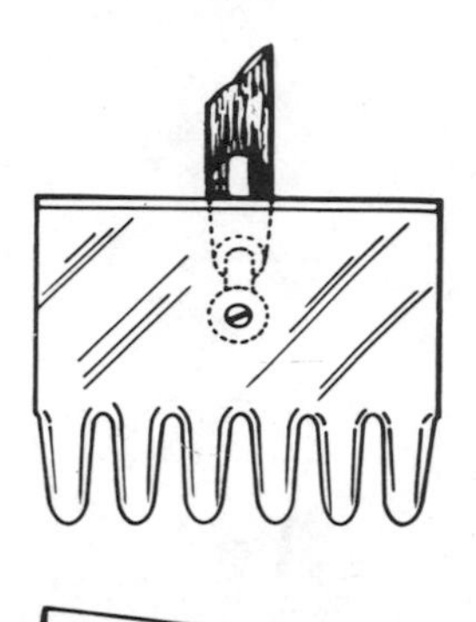

Figure 2.13. McLeod tool.

Maintenance: The hoe blade of the McLeod tool should be sharpened to a 45-degree angle. The head is attached to the handle by a tightening screw and can be removed. (The Kortick fire tool is similar to the McLeod tool, but it is heavier and the handle is attached to the blade with a threaded wing nut, which makes this tool much easier to transport. Maintain a sharp cutting blade, and keep the handle smooth and its attachment to the blade tight.

Figure 2.14. "Asphalt" pattern fire rake.

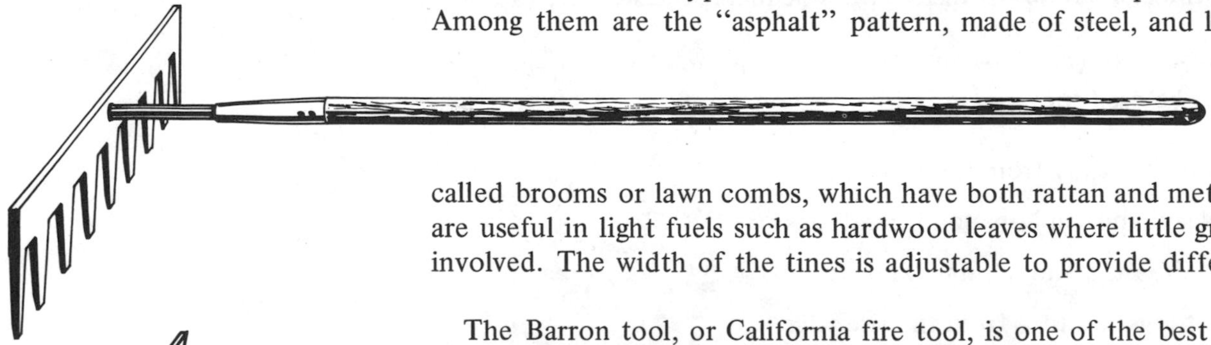

Rakes. Several types of rakes are used in different parts of the country. Among them are the "asphalt" pattern, made of steel, and leaf rakes, also called brooms or lawn combs, which have both rattan and metal tines. These are useful in light fuels such as hardwood leaves where little grass or brush is involved. The width of the tines is adjustable to provide different tensions.

The Barron tool, or California fire tool, is one of the best rakes. It is an ideal tool for fast spreading fires in light flashy fuels, such as grass, grain, pine needles, litter, and light brush. The hoe blade is set at an angle of 65 degrees to shear rapidly, cutting through light roots, matted grass, and small brush without cutting into the mineral soil. The tines are practically nonclogging, and the tool can be used to hold and drag burning material when you are setting fires for burnout or fighting fire with fire.

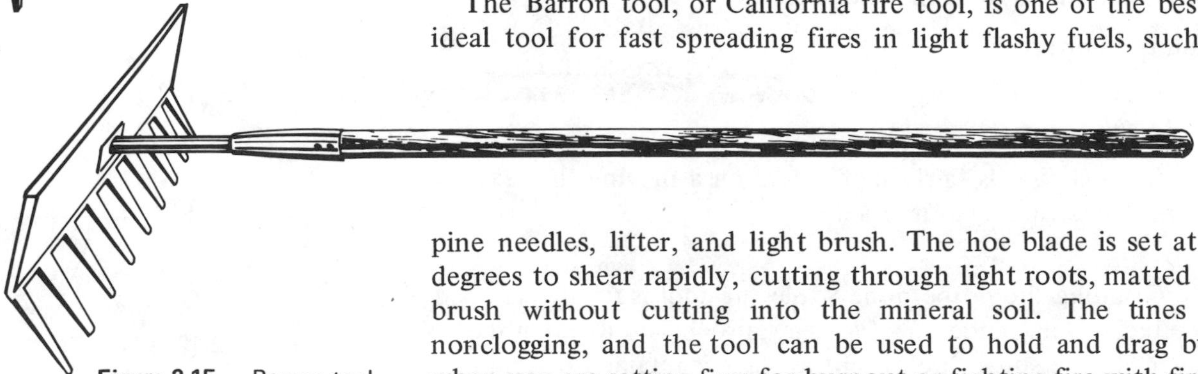

Figure 2.15. Barron tool.

Maintenance: Keep the blades on the Barron tool sharp, and keep the handle smooth and its attachment to the blade tight.

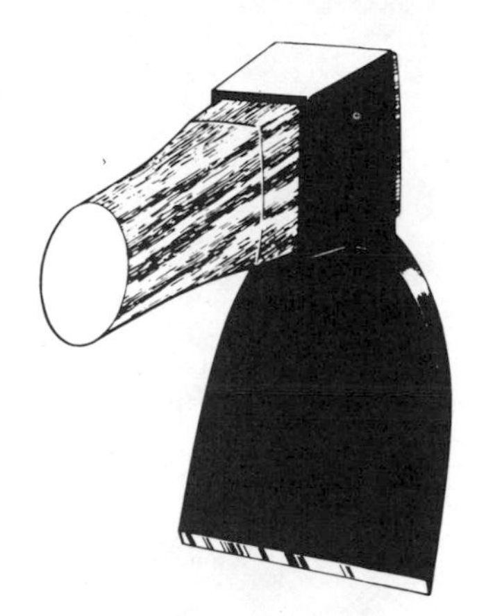

Figure 2.16.
Adze hoe, Osbourne pattern.

Hoes. The most popular hoe is the adze hoe (Osbourne pattern), used extensively in the northwest Douglas fir region and well adapted to the Great Lakes region. It is designed for heavy grubbing or trenching, and it is ideal where vine grape, swamp maple, and similar entangling fuels plus deep ground litter or deep duff are found. The blade is set at an adze angle so that the operator may straddle the line and skin the ground to mineral soil.

Maintenance: The blade of the adze should be sharpened to a 3/8-inch bevel. The handle should protrude 3/8 inch above the head.

Saws. Many types of saws are usable for wildfire control. Where any amount of sawing is required, most units now use power saws, which are covered in a following section on power equipment. However, there are a few places where hand saws are still appropriate and are in use. *Ribbon saws* can be either the felling or the bucking design. The felling saw is somewhat narrower and more flexible and is more desirable for all-around work. The length varies from 5½ to 7 feet. The bucking design has a wider blade. *Swedish, or bow, saws* are used where the timber is small. They can be easily carried. There are a number of sizes and types of teeth for these saws. In larger-size timber, suitable wedges and a hammer should accompany the saws.

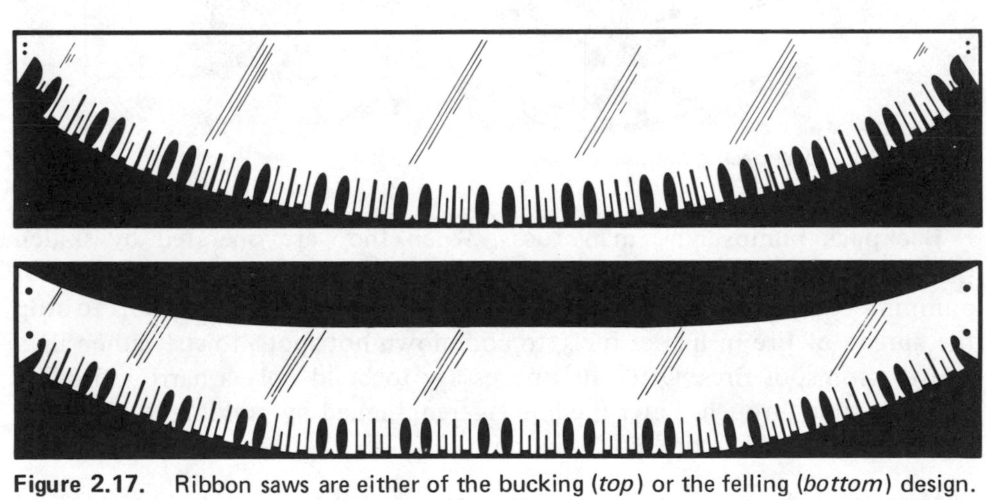

Figure 2.17. Ribbon saws are either of the bucking (*top*) or the felling (*bottom*) design.

Maintenance: Only the finest material and workmanship will provide satisfactory saws because they must withstand hard use with a minimum of filing. They should be equipped with hardwood handles that may be attached in either a vertical or a horizontal position. Saws should be equipped with guards for the teeth and should be treated with a rust preventative during storage. The teeth should always be kept sharp.

Backpack pumps. The backpack pump is one of the best wildfire tools, and it is universally used. The 4- to 5-gallon tank may be made of galvanized steel, stainless steel, brass, chrome, plated brass, or fiber glass. It is also available as a neoprene bag holding 5 to 6 gallons; this has the advantages of light weight, smaller storage space, and better fit to the back. The neoprene bag collapses as the water is used. It does not require an air breather opening and is less likely to leak water on the user's back. All models are equipped with a slide- or trombone-type pump of varying lengths. The pump is attached to the tank with a hose approximately 27 inches long. Some units have a pump attached to the tank and discharge through a hose and nozzle. The trombone pump comes equipped to deliver a straight stream, fog, or spray. A guide to performance indicates that the pump should project a stream of water vertically at least 17 feet above the nozzle at about ¼ pound of water per discharge stroke when it is operated at 17 strokes per minute. It is equipped with carrying straps and weighs 55 to 60 pounds when it is full, and so proper lifting procedures are necessary to prevent back injury. Carriers for holding the metal units on a vehicle are available.

Figure 2.18. Backpack pump.

Backpack pumps have many uses. When they are operated by skilled personnel, they are the most efficient, flexible, and economical type of pumping equipment. They are used as an adjunct to other hand tools to stop the spread of fire in lighter fuels, to cool down hot spots, to strengthen line, and to stop spot fires. Often it is necessary to build only a narrow fire line the first time through. Later the line is strengthened by building it wider and knocking down more fire inside the line.

Make each drop of water count—it's very precious when it has been lugged to the fire in a backpack pump. Hold the nozzle end steady and in place with one hand, and pump with the other hand. In this way, the nozzle can be most accurately pointed. The nozzle should be held close to the fire so that as much water as possible reaches the burning fuel. Apply the water as a spray in a sweeping motion at the base of the flames.

Maintenance: Maintenance of the pump is important. If the sliding parts of the pump become difficult to operate, apply a thin film of graphite to the outer portion. If the pump becomes clogged, check the nozzle first and then the hose. Use clean water and always fill through the screen provided. Protect the bottom side of the tank from punctures. Replace graphite packing or plastic rings as they become worn. If the barrel or plunger becomes bent or creased, replace the pump. Thoroughly wash all pump parts daily if retardants are used. Retardants corrode brass and aluminum. If the plunger barrel gets gummed up, clean it with solvent.

Firing devices. Firing devices include fusees, drip torches, pneumatic flame throwers, and power flame throwers. For the ordinary fire service, either the fusee or the drip torch is adequate and requires the least investment.

Fusees. Fusees are produced in three- to ten-minute and twenty-minute burning periods. They are self-contained and are started by striking the primer against the safety cap. They should be stored in a metal container in humid climates, or otherwise they will become soft and mushy. Dispose of the remains so that livestock cannot eat them; they are poisonous to livestock.

Figure 2.19. Fusees are adequate firing devices and require a small investment.

Drip torches. Sometimes referred to as a backfire pot, the drip torch is efficient, safe, and simple to operate. It is designed for firing semidry fuels that slowly ignite; when the burning oil is dripped onto the fuel, the operation can proceed without waiting for the fuel to ignite. Drip torches can fire a large area in a short time. These torches are equipped with a fuel

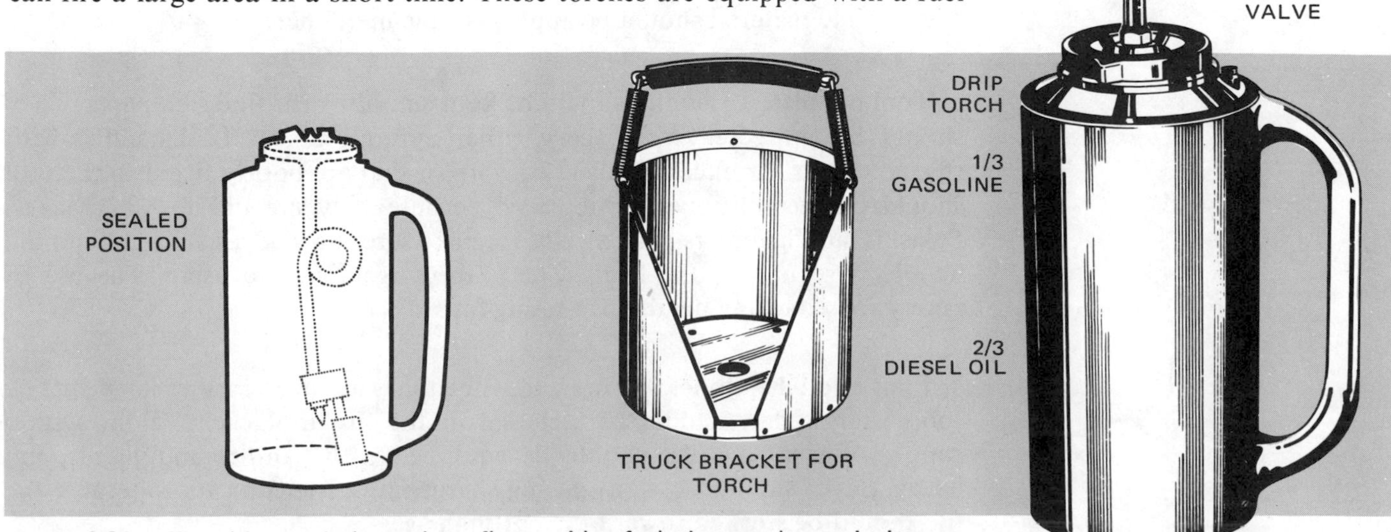

Figure 2.20. The drip torch is used to fire semidry fuels that are slow to ignite.

trap on the spout to prevent flashback into the joint and a check valve in the cover to provide double protection against flashback. A breather valve, oilproof gasket, and sealed outlets prevent slopping of fuel. They will operate best with a mixture of one-third gasoline and two-thirds diesel oil.

Do not fill near an open flame or hot embers or while anyone is smoking, as the empty tank contains vapors that may explode. To operate, tilt the torch forward until fuel flows over the burning igniter, and spread the flaming oil parallel to the direction of travel. To gain range, swing the torch forward and terminate with a snapping motion of the wrist. Make sure the igniter has completely cooled before putting it out of service. Vehicle brackets are available for transport.

Prolonged exposure to the sun in very hot climates may build up vapor pressure inside the sealed tank. To avoid this, a $^{1}/_{64}$-inch hole may be drilled through the bottom of the blind thread socket in the tank cover (top), but care must be taken to be certain that the sealing plug is screwed securely into this socket in order to prevent leakage when the torch is operated. Maintain the torch in good order, inspect it periodically, and replace any damaged parts.

Pneumatic and powered flame throwers. These firing devices are not recommended for the ordinary fire service because of their cost and their potential danger if they are not carefully and periodically maintained.

Marking fire tools and equipment. It's a good idea to mark the tools and equipment owned by your district or department. They can then be easily identified, especially if they may be used on fires where other units are also employed.

Hand tools are best marked with a band of paint 2 to 3 inches wide on the handle next to the head of the tool, or they may be stamped with an appropriate initial. Use some color other than red if the tools might be used near U.S. Forest Service tools.

Rust prevention. Axes, shovels, Pulaskis, and tools with sharp edges should be treated periodically to prevent rust. Oronite priming solution or a comparable material should be applied to the metal parts.

Tool handles. Handles should be kept smooth with fine sandpaper. They should be protected from the weather during storage. Tool handles with slivers, cracks, or excessive twist or warp or that are poorly fitted to the tool should be reconditioned or replaced because they are unsafe. Check axes, Pulaskis, and hammers periodically to make sure the wedges are tight and the handle is secure on the head. Check tools again before using. The use of epoxy for attaching handles is gaining favor.

Files and whetstones. Files and whetstones are necessary to keep cutting tools sharp. They should be included in the fire tool cache, along with a supply of wedges. Files should be equipped with handles and guards, and heavy gloves should be worn during sharpening of cutting tools because the file should be stroked toward the cutting edge.

Guards. Axes, Pulaskis, saws, and all sharp tools should have guards attached to them for transport. Do not transport tools with men in a truck unless the tools are boxed, and, if at all possible, the box should be secured to the truck bed.

TABLE 2.3
Suggested Minimum Tools for Fire Tool Caches

For fuels that are predominantly timber and heavy brush with no grass	*10-man crew*	*20-man crew*
	Number of tools	
Axes, double bit or single bit	2	4
Brush hooks, double or single edge	2	3
Shovels, round point FS (Forest Service) type	5	10
Rakes—or McLeod tools	3	6
Pulaski tools	3	6
Pumps, backpack	2	4
Head lamps	10	20
Power saw or crosscut saw	1	2
Torch	1	1
Files and whetstones	X	X
Canteens	10	20

For fuels that are predominantly light brush and small timber with grass interspersed	*10-man crew*	*20-man crew*
	Number of tools	
Axes, double or single bit	1	1
Brush hooks, double or single edge	2	4
Power saw or crosscut saw	1	1
Shovels, round point FS type	4	8
Adze eye hoes	4	5
Rakes—McLeod, council, or Barron	2	5
Pulaski tools	2	3
Pumps, backpack	2	4
Head lamps	10	20
Torch	1	1
Files and whetstones	X	X
Canteens	10	20

Fuels that are predominantly light duff, needles, grass	*10-man crew*	*20-man crew*
	Number of tools	
Brush hooks	1	2
Shovels, round point FS type	5	11
Rakes, council or Barron	5	10
Rakes, broom type	4	6
McLeod tools	3	5
Pumps, backpack	2	5
Swatters	2	4
Head lamps	10	20
Torch	1	1
Canteens	10	20

TABLE 2.3 (continued)
Suggested Minimum Tools for Fire Tool Caches

Fuels that are predominantly heavy duff, small roots, and matted brush	*10-man crew*	*20-man crew*
	Number of tools	
Brush hooks	2	4
Shovels, round point FS type	4	7
Adze eye hoes	4	7
McLeod tools	6	10
Pulaski tools	2	5
Pumps, backpack	2	5
Head lamps	10	20
Torch	1	1
Files and whetstones	X	X
Canteens	10	20

SUPPORT EQUIPMENT

Support equipment includes all the items that are desirable (and often required) to use hand tools, apparatus, and mechanical equipment effectively.

Headlamps. There are a number of different types of lamps that attach over the hard hat with a connecting wire to a battery or a box for batteries. This container can be placed in a hip pocket or attached to the belt. When the lamp is adjusted, light is thrown wherever the head is turned; thus both hands are free for working. Spare batteries should be carried.

Hard hats. The ordinary construction hard hat is a must in timbered areas and should be worn at any fire. Many types are available. They should be adjusted to fit comfortably and should have a chin strap. It may be necessary to attach clips for holding the head lamp. Regular fire service helmets are also used. Plastic face shields should be required when water is used in mop-up of hot stump holes and dozer piles.

First aid kits. A suitable first aid kit should be available for all crews, and firemen should be encouraged to use it for minor burns and cuts. In snake country, snakebite kits are a must. In chigger and mosquito country, suitable quantities of insect repellent will be much appreciated.

Drinking water containers. Good potable drinking water for wildfire firemen is almost as important as hand tools. Firemen perspire quite freely and soon become more or less dehydrated, and so drinking water is a necessity. It should be drunk in moderation, and salt tablets should be provided over a period of several hours.

Hand tool crewmen should carry quart containers of water on their belts. Resupply can be made by a water boy carrying a 5-gallon water bag. In some instances, 1-gallon canteens for each man are advisable.

On equipment such as bulldozers, plows, and pumpers the 5-gallon dispenser with paper cups that is used by utility crews is suitable. This can be carried on the pumper or in a support vehicle.

Miscellaneous equipment. Ladders, self-contained breathing apparatus, hot sticks, pike poles, stretchers, rope, forcible entry tools, and other regular equipment carried on fire department pumpers may become necessary if buildings and other improvements are involved in the wildfire.

Camp and headquarters equipment. If the fire control job runs past three hours, a meal is in order. This can either be provided as a hot meal or as sandwiches and coffee. In many of the heavier fuels, the campaign for control will last over several days and will require a camp, headquarters, and all that goes with them. These facilities are available from the federal and state forestry agencies, park services, the Bureau of Land Management, and other organizations for fires on or near their jurisdiction.

Most states can provide assistance through the Civil Defense organization, or assistance might be obtained from nearby Army, Air Force, or Navy installations. Some fire services, other than state and federal agencies, may have to stock camp and headquarters items, such as disposable bedrolls made of paper, disposable food service items, kitchen equipment, tents, dry rations, frozen meals, and perhaps other items. These are warehoused by GSA, and information on their procurement can be obtained through your state forester.

Storage and housing. Fire tools and equipment should be kept in a dry location—and for hoses the storage should also be kept as cool as possible. Usually, fire tools are stored in compartments on the pumper. Otherwise, a toolbox of sufficient size should be made for their storage. It should be equipped with sufficient handles and should be a reasonable size for ease of loading and unloading. It should be anchored to the truck.

When a fire station is being planned, provide for additional equipment needs of the future as well as sufficient room for training and maintenance sessions. It's much better to have very adequate room than to try to operate in cramped quarters.

Water additives. There are a number of chemicals available that can be added to water used on wildfires. These include retardants used in slurry drops from aircraft and wet water additives. Retardants are now available for use in ground tankers and back pack pumps. Firetrol is an example. They are useful in laying down a line as a fire barrier in the manner of a firebreak. They are effective until they are washed off by rainfall.

However, "wet water" additives are often worth their cost for use in ground equipment, especially in mop-up work. The additive should be stored on the pumper and only added to the tank when it is to be used. It should not be left in the tank or lines or pump during storage, because it may cause etching of metals. These additives break down the surface tension of the water so that it spreads farther and penetrates better than ordinary water.

Clothing. Wildfire firemen need to wear a heavy, sturdy type of work shoe, preferably a leather field boot. Low cut shoes, tennis shoes, and similar footwear are inappropriate and dangerous. Rubber boots should not be worn on wildfire control.

Wool clothing is the best from a flammability standpoint, but it is uncomfortable. Denim or tightly woven cotton is next best, but cotton can ignite quickly. Beware of synthetics because, when they burn, the material melts and causes deep burns. Fire-resistant shirts and pants are available.

Hard hats or helmets should be worn at any wildfire and are required in timber. A warm jacket is most welcome if the job lasts overnight. If the assignment is to last for several days, several changes of socks will probably eliminate blisters and sore feet.

Protective clothing should be worn by nozzlemen working with pumpers, especially when the direct attack is used. Goggles or face shields should be provided. Comfortable and sturdy work gloves are very desirable for any kind of work on the line.

POWER-OPERATED EQUIPMENT

Power saws or chain saws. Self-contained power saws are universally used, and, if they are maintained properly, they are the best means of cutting large material and felling timber and snags. Operators must have sufficient training and experience when they are felling trees. They should be required to observe the necessary safety precautions. Two men should work together.

Saws should be equipped with an approved spark arrestor. Each saw should be accompanied by a small fire extinguisher and a shovel. The motor should be stopped when you are moving from place to place. Saws should not be carried on the shoulder unless the chain is well padded and guarded. Protective chaps should be worn when you are operating a chain saw.

During refueling, allow the motor to cool at least five minutes, fill the tank on bare ground, clean off spilled fuel, and move at least 10 feet away before starting. Use approved safety fuel containers. Stop the motor for all cleaning and adjustments or if the saw becomes pinched or wedged.

Power trenchers. A number of power-operated trenchers have been developed, and a few are in use. Generally, they have not been accepted because of operational problems. In some fuel types with a moderate amount of rocks and roots, they perform adequately and may become more widely used as they are perfected. A good hand tool crew is preferred on almost every location. Where they are practical and the cost is justifiable, power trenchers should be considered.

Power mowers and brush hogs. These machines have limited use and are not recommended for ordinary fire service. Where they are available and in general use, arrangements for their use in fire control certainly should be considered.

HEAVY EQUIPMENT

Bulldozers. Bulldozers are excellent line-building machines (see the chapter on tactics) in areas where they can be operated. Many fire services cannot afford to own them only for fire control work, considering the amount of use made of them. However, bulldozers used on local construction or logging jobs are often utilized in wildfire control.

Where these machines can be made available for wildfire control, arrangements should be made to use them. Suitable transport equipment is necessary to get them to the job and servicing must be provided. Written contracts in advance of use are desirable.

PUMPING EQUIPMENT

There are a wide variety of pumps used on wildfire control. Each type and size have been designed for specific kinds of service. Pumps are generally classified in several categories such as:

1. Portable pump—this lightweight portable pump generally weighs 60 pounds or less and is rigged for backpacking. Other portable pumps are carried by one to four men to the water source to supply hose lays or resupply pumpers or tankers.

Figure 2.21. The portable pump generally weighs 60 pounds or less and is rigged for backpacking.

Figure 2.22. Floatable pump in use.

2. Floatable pump—this pump is mounted in a fiber glass container so that it floats on the water. A screened intake opening is located in the bottom of the unit, thus entirely eliminating the use of a suction hose. These are usually two-cycle air-cooled engines and a small gasoline tank attached to a centrifugal pump.

3. Mounted or fixed pump—this combination of a pump and related engine is mounted on a pumper truck and usually is not easily removed for portable use.

4. Power takeoff pump—this pump is operated by the engine of the truck and is mounted as an integral part of the vehicle. It includes midship pumps, front-mounted pumps, and power takeoff pumps. Unless this type of pump is used with a truck engine of sufficient horsepower and has a qualified performance, it should not be the only pump on a wildfire pumper. It is strongly recommended that a wildfire pumper use a pump with a separate engine, first, to provide a capability to pump water if the truck engine should stall or if vapor lock should occur and, second, to allow full power for operation of the truck.
5. Transfer pumper—this is used to pump water from storage tanks and reservoirs and to load pumpers and tankers. It is a high-volume unit that handles dirty water and other solutions. It develops low pressures but high volumes and is not suitable for use with hose lines.

Pumps are generally of two types, either centrifugal or positive displacement. The latter includes both gear and piston types. Since the typical fire line assignments for a wildfire pump are demanding, the pump must be a piece of rugged equipment. It must satisfy the following demands that will be placed upon it.

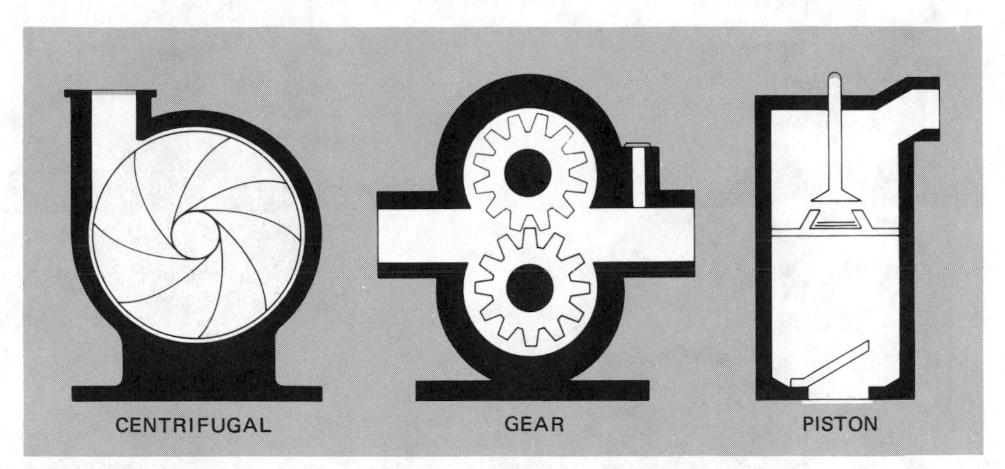

Figure 2.23. Pumps are generally of two types, either centrifugal or positive displacement. Positive displacement pumps include gear and piston types.

- High flows are often required, but the need is highly variable. Water conservation is also always important.
- Service is through lightweight hose lines with small diameters, where the friction loss is comparatively high.
- Hose lays are often long.
- Hoses are often laid up steep slopes, and high static head losses from pump to nozzle are the result.
- Water is customarily replaced under high static suction lifts from source to pump.
- Engine power is reduced as altitude increases.
- Temperatures are often high.
- Hours of work are often long.

- A long service life is required.
- Pumpers must be light, particularly portable ones.
- The available water is often abrasive.

Centrifugal pump. The centrifugal pump employs outward force from a center of rotation to move or impel water. It is usually larger than positive displacement pumps, and it provides higher volume and lower pressure. Centrifugal pumps may be single stage or multiple stage, depending on the number of impellers. They are not satisfactory for raising water to a high elevation—a positive displacement pump is much better for this type of operation.

One big advantage of centrifugal pumps is that they can pump dirty silt-laden water without wearing out quickly. However, any pump will wear when it is pumping abrasives such as sand and gravel, so this type of material should be kept out of the water in pumping operations.

Gallon for gallon, a centrifugal pump usually requires more horsepower than a positive displacement pump. Generally, they do not require a pressure relief valve, except those that develop extremely high pressures. Single-stage centrifugals control the volume and pressure by the engine throttle setting. Thus the centrifugal pump is simple to operate.

Most centrifugal pumps require priming. This is accomplished by using exhaust from the engine. The venturi action draws air from the pump, allowing water to fill the suction line and pump. The primer shown in Figure 2.24 has been modified so that the exhaust enters the primer body tangentially. This spinning action provides an efficient arrestor when it is carefully designed and fabricated.

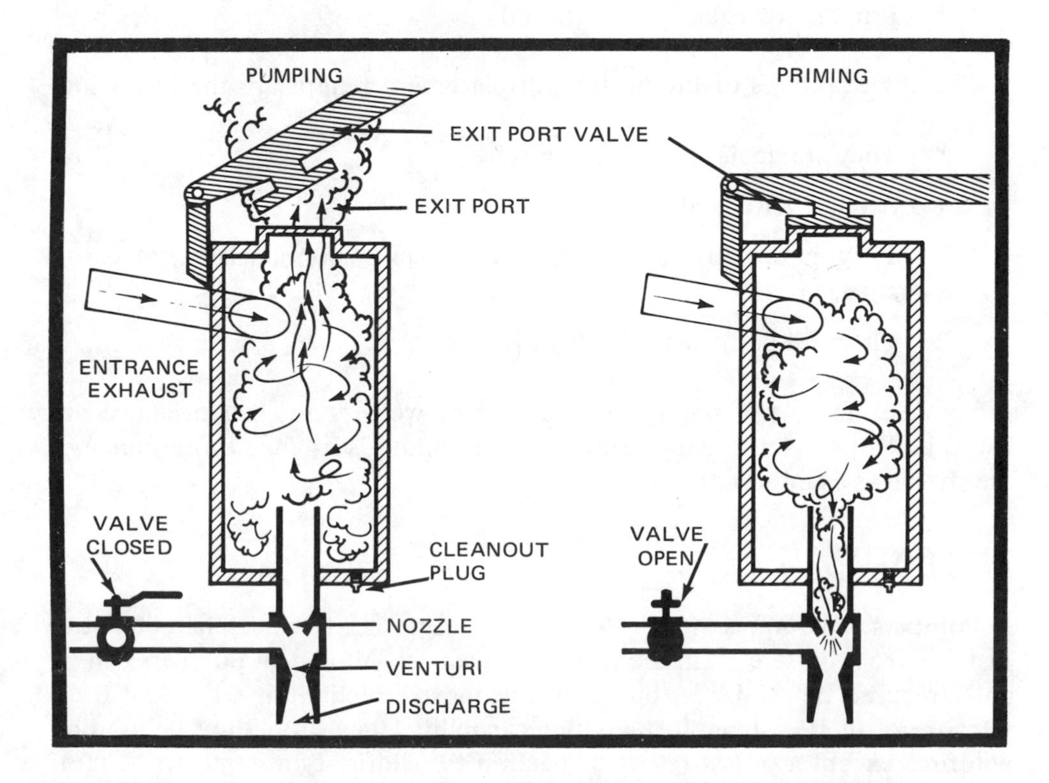

Figure 2.24. The venturi action of the exhaust primer draws air from the pump.

The advantages of the centrifugal pump are the following:

- The pressure can be changed by adjusting the revolutions per minute (rpm).
- The volume can be changed by adjusting the pounds per square inch (psi) and revolutions per minute (rpm).
- Relief valves are not required.
- Dirty water and small particles can be passed without damage.
- Refill performance is good.

The disadvantages of the centrifugal pump are the following:

- More power is required for higher pressures.
- Primer is usually required.
- To avoid heating, bypass is required when no water is moved.

Positive displacement pumps. Positive displacement pumps pass or move a definite quantity of water with each stroke or revolution of the impeller. Examples are rotary, gear, cam, and piston self-priming pumps. They are subject to abrasive damage by grit, soil, sand, and gravel and must use clear water. They also require relief valves to handle line surges, overloads, and flows not needed at the nozzle. They are generally preferred for constant flow, high pressure applications. Their volume depends only on speed (rpm).

The advantages of the positive displacement pumps are the following:

- Higher pressures can usually be produced with less power.
- Primers are usually not required.

The disadvantages of the positive displacement pumps are the following:

- They are damaged by dirty water.
- A relief valve is required.
- They have a fixed output, and the psi performance is not easily changed.
- The refill performance is low.

Some of the many pumps available, their weights and performance are listed in Table 2.4. No endorsement of any brand is intended. Brand names are shown as examples only.

APPARATUS

Pumpers. Pumpers for wildfire use have needs and characteristics distinct from those used in the regular fire service. Municipal pumpers can be and are used on wildfires, but they are most effective on structural fires. Because of their cost and high-volume capability, they should not be used on wildfires except as a last resort or backup to wildfire-type pumpers. Use of high-cost pumpers in off-the-road service is seldom justified.

Wildfire pumpers need not discharge large volumes of water under most conditions. Volumes of 6 gallons per minute (gpm) to 30 gpm are adequate for wildfires; however, the ability to produce larger volumes up to 250 gpm is often needed on structures exposed by wildfire. So the latter capability is desirable if it can be afforded and would require a high-volume pump in addition to the wildfire pump.

TABLE 2.4
Examples of Available Pumps

	Weight (lbs.)	*psi at pump*				
		50	*100*	*150*	*200*	*250*
				gpm		
Portable centrifugal (60 lbs. or less)						
Pacific Pumper Mark 3						
1 ½ in. suction hose	55	70	67	55	44	25
Gorman Rupp Model No. 61-½DF						
1 ½ in. suction hose	29	50	42	27	—	—
Waterous Floatable	50	100	20	—	—	—
Portable centrifugal (61 lbs. or more)						
Gorman Rupp 14x225						
1 ½ in. inlet and 2 ½ in. outlet	107	150	50	—	—	—
Gorman Rupp 14x230						
2 ½ in. inlet and outlet	158	100	60	—	—	—
Gorman Rupp 14x240						
2 ½ in. inlet and outlet	260	220	—	70	—	—
Mountable centrifugal						
Berkeley B1-½ XQBS-10						
1 ½ in. inlet and outlet	196	52	44	35	18	—
Cedco or (Hale) HPZZ						
1 ½ in. inlet and outlet	171	60	30	20	15	—
Darley 1 ¼ AGE						
1 ½ in. inlet and outlet	194	72	36	24	18	—
Berkeley B1-½QBS-18						
2 ½ in. inlet and 1 ½ in. outlet	356	122	114	96	70	39
Cedco HPZF						
1 ½ in. inlet and outlet	303	87	87	77	58	37
Portable positive displacement						
Pacific WA7						
1 ½ in. inlet and outlet	130	26	25	24	21	17
Pacific WX10						
1 ½ in. inlet and outlet	170	45	43	32	24	—
Edwards L-23						
1 ½ in. inlet and outlet	154	49	47	37	28	—
Edwards 120						
1 ½ in. inlet and outlet	183	48	46	45	35	—
Mountable positive displacement						
Edwards EBE						
uses ECO pump						
1 in. inlet and outlet	45	11	11	10	—	—
Pacific BE						
uses ECO pump						
1 in. inlet and outlet	45	11	11	10	—	—
Western Fire 14x120A						
uses ECO pump						
1 in. inlet and outlet	45	11	11	10	—	—
MP Duraflex W6694B						
1 ½ in. inlet and outlet	144	40	38	31	28	—
Cedco AIOF						
1 in. inlet and outlet	124	—	10	10	10	10
John Bean 101FD						
1 in. inlet and ¾ in. outlet (High pressure)	209	at 600 psi pumps 6 gpm			6.5	6.5

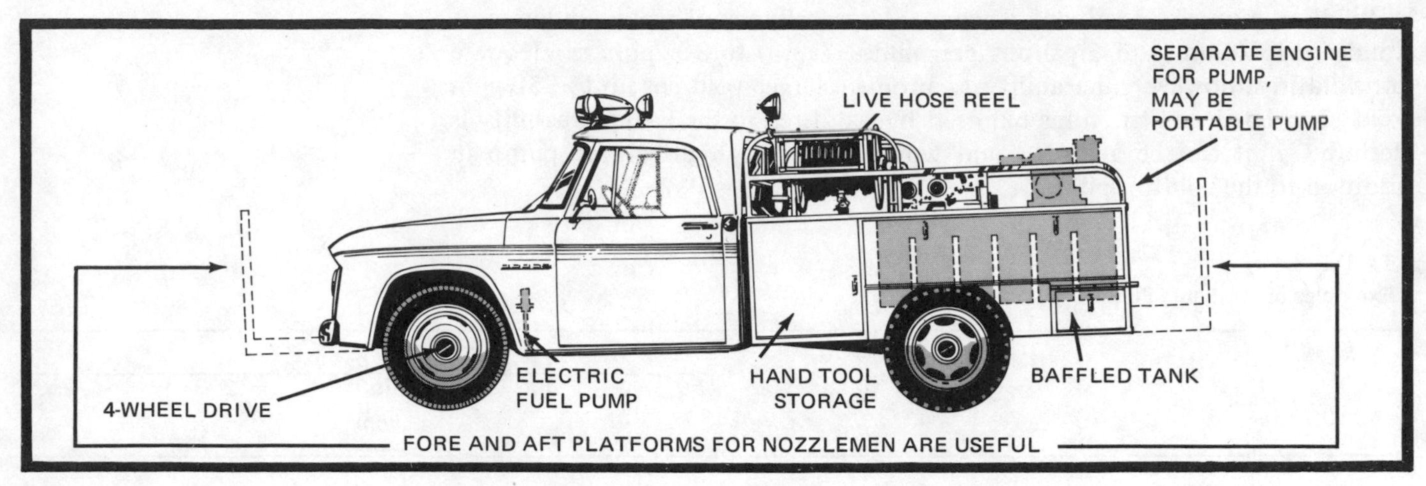

Figure 2.25. Wildfire pumpers have different characteristics from those of the regular fire service.

Wildfire pumpers, or "brush rigs" as they are often called, are operated off the road and must be constructed for this use. The all-wheel drive trucks are favorites for this reason. Excellent pumpers have been built on surplus military vehicles such as jeeps and four-by-four and 2½-ton six-by-six trucks. The commercial four-by-four trucks should be considered if the military vehicles are not available. A pumper designed for wildfire use is much less costly, better adapted to wildfire use, and more mobile than the standard fire department apparatus.

Figure 2.26. Slip-on tanker unit.

The pump, tank, and plumbing can be either a slip-on type or an integral unit. Slip-on units are self-contained and can be removed from the truck chassis without disturbing the plumbing. Usually, the removal of a few anchor bolts is all that is required. The usual size varies from 50 to 300 gallons. One-thousand gallon slip-ons are available for army six-by-six trucks. Many districts develop their own pumpers on surplus military equipment. Some pumps are mounted so that they may be quickly disconnected for portable use. It is best to have an electric starter, but some pumps have only a rope starter. Integral units vary in size from 50 gallons for jeeps to 1,000 gallons in the 2½-ton class. The tank capacity should be compatible with the size of truck chassis. Overloading should be avoided for safety as well as for maintenance reasons. The manufacturer's gross vehicle weight (gvw) recommendations should be used because these vehicles are subjected to high stresses, vibration, and other related operational treatment due to the locations where they are used. Pumps on integral units are almost always fixed, but they may be detachable so that they may be used as portables. Rural fire department tankers might have pumps operated off the engine of the truck in order to have the capability of higher gallonage for structures. However, for wildfire use they should also have a pump that can be operated by its own two- or four-cycle engine. (Refer to the appendix for weights and conversion factors.)

Tanks should be suitably baffled to prevent movement of the water during travel. They should also be treated or painted on the inside to forestall rusting. Tanks need to be of the lowest possible profile in order to keep the center of gravity as low as possible. Provision should also be provided for recirculation of the water if they are to be used during freezing weather.

Electric fuel pumps on truck engines are very desirable for wildfire pumpers to help prevent vapor lock, and, in some instances, provision has been made to introduce compressed air into the carburetor. The amount and size of hose carried will depend on the use that will be made of the unit. If the hose is to be used for pump and roll operation only, two short lengths approximately 10 feet long will be preconnected to the pump. These may be ¾ inch, 1 inch, or 1½ inches in diameter, depending on the capacity of the pump and local preference. A hose reel for a booster line of ¾-inch or 1-inch size would also be advisable but is not necessarily required.

Tankers. Tankers are primarily considered water supply vehicles for pumpers. They may or may not have a pump on board. The pump may only be capable of drafting and transferring the water, or it may have some capability for wildfire attack.

Tankers maintained for fire use only vary in capacity from 500 to 5,000 gallons. The larger sizes are semitrailer tankers, stationed in strategic locations where they can reinforce several fire control companies. A large tanker used to supply several pumpers is often called a "mother tanker."

Many types of water transport are available in most communities. These include county and city tankers, street flushers, milk transporters, oil well equipment, concrete transporters, and commercial tankers. Arrangements can often be made to use this equipment under contract or by cooperative agreement. It is important to develop the arrangement in advance and to definitely establish the method of dispatching and conditions of hire.

Guardrails. Suitable guardrails are an important safety item. They should be installed on the front, sides, or rear of the truck, depending on the positions to be used during operation of nozzles from the pumpers or tanker. They should be high enough to prevent a man from falling off in rough terrain. Often, safety belts are necessary.

FIRE HOSE

The size of the discharge hose used on wildfire control varies from a ¾-inch garden hose to a 2½-inch rubber-lined hose. The most frequently used sizes are the ¾-inch booster hose and the 1-inch and 1½-inch woven jacket rubber-lined hose.

Discharge hose is divided into three groups: linen hose, rubber-lined hose, and rubber-covered hose.

Linen hose. This hose is woven of flax fibers and synthetics. There is some leakage until the fibers swell and seal the tube. Minor sweating occurs during use, but this keeps the surface damp and resists burning. It can withstand working pressures up to 300 pounds and is highly sensitive to acids, alkalies, and sunburn. When it is treated to be mildew resistant, this hose can be used indoors, outdoors, and in areas of high humidity. It is the lightest of all types and takes up the least space; it is durable, but must be thoroughly dried after each use. It is principally for indoor use or on wildfires on long hose lays because of its light weight and compact size. It should not be used for regular fire service. It is used in 1-inch and 1½-inch sizes on wildfires. It has a much higher friction loss than the rubber-lined hose.

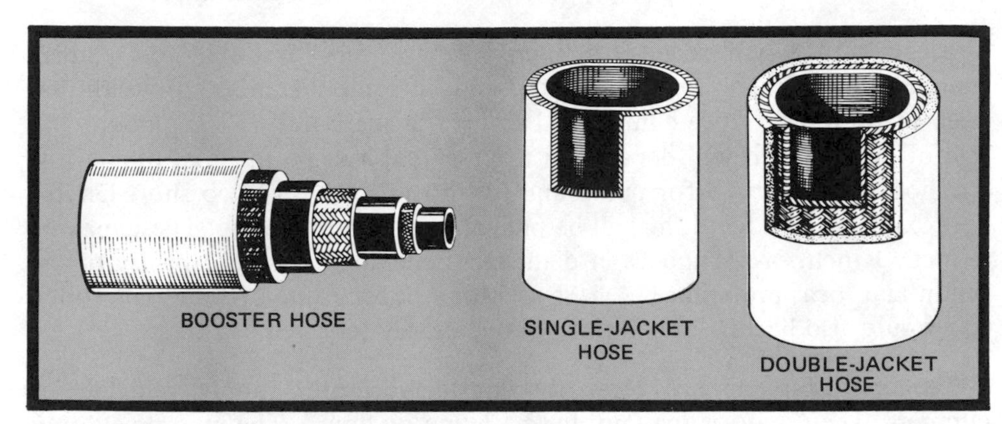

Figure 2.27. The booster hose is made up of several layers of hose material with an outer rubber covering. The single-jacket hose has a rubber interliner with a single outer protective jacket. The double-jacket hose has a rubber interliner with two outer protective jackets.

Rubber-lined hose. This hose is a circular woven jacket of cotton, polyester, or other fibers lined with a rubber tube. The single-jacket type has working pressures up to 300 pounds. The double-jacket type provides extra protection from physical injury and abrasion. It should be mildew-proofed when it is purchased and should be dried after each use. Some types are sensitive to acids, alkalies, grease, oil, and hot dry storage. Quarterly wetting and maintenance will prolong its life. It is most suitable for general use. It is used in 1-inch and 1½-inch sizes in wildfire service. It is commonly called CJRL. It has lower friction loss than the unlined hose.

Rubber-covered hose. This hose consists of a rubber tube, covered with one or more plies of woven fabric and an external rubber cover bonded together by vulcanizing. It may employ synthetic rubber compounds to resist oil, grease, acids, or other chemicals both internally and externally. The fabric may be of cotton, rayon, nylon, dacron, or similar material. It is commonly used as a booster hose. Because of its size, it has very high friction loss. It should have a working pressure of 250 pounds. It has an inside diameter of ¾-inch with a 1¼-inch outside diameter for the 1-inch size. Extra heavy duty garden hose is often used and sometimes preferred on wildfire rigs. Its use will depend on the type and extent of fuels encountered. It is a water saver in mop-up operations. Hose used in wildfire service should have a burst pressure of at least 500 pounds.

SUCTION HOSE

To withstand severe use, suction hose should be circularly woven using rayon warp cords and spring steel wire "filler" cords. If it is crushed, it can be restored by reforming the inner wire by pressure. It is provided with a smooth bare rubber tube on the inside and a tough rubber cover to resist abrasion. The 1½-inch size is most used in wildfire service. It may vary from 1 inch to 2½ inches in diameter.

CARE AND MAINTENANCE OF HOSE

Hose should be thoroughly cleaned and dried after each use, even if it has been mildew treated. It should be properly rolled or laid in trays and stored in a cool, dry location.

TRANSPORTATION OF HOSE

Fire pumpers carry hose in two ways: (1) coupled and folded so it can be pulled out in a continuous length (this is called preconnected), and (2) rolled in single lengths that are coupled as the hose is laid (a hose clamp is required and the nozzle is removed each time a length is added or the line is shut down or kinked, or a gated wye is added and one side is shut off).

Hose is usually rolled in the doughnut roll method. To make the doughnut roll, lay one length on the ground. Carry the male end back over the hose to a point 4 feet from the swiveled end. Roll from the loop end–not too tightly when starting–and tie the roll with three ropes tied with bow knots.

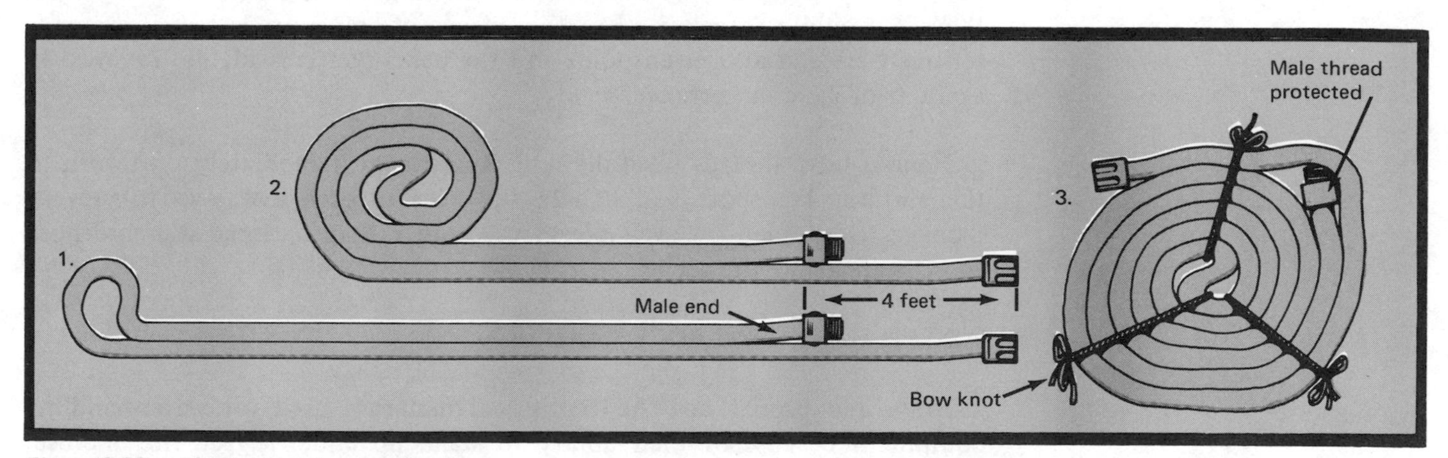

Figure 2.28. Making the doughnut roll.

A hose roller is very handy on a pumper. With a roller, the hose can be picked up after use and made into storage rolls for transport. If a roller is not available, the hose can be rolled up in a potato roll, which is simply rolling it as one would roll a ball of string.

Hose can also be transported along the line folded on packboards or in packsacks. The hose is folded in the packsack so that it plays out the top as the bearer moves forward. This method is best used with linen hose. Packboards are useful and efficient means of packing hose. Either several lengths are folded on the board and held with bungee straps, or the rolls are tied to the packboard.

Figure 2.29. Hose can be transported along the line in a packsack or a packboard as shown in the inset.

The lay can also be made by helicopter. However, if the hose catches in the tops of trees, it may take more work to get it together on the ground than it would to pack it. But this means of transport is fast if water is seriously needed at a given point and the helicopter is available. Teamwork, again, is of basic importance.

Handle hose threads carefully and repair them immediately on return to the station. Transport and handle the hose in such a way as to prevent damage. Have adapters available in case other than national standard hose threads are at the fire scene.

WATER-HANDLING ACCESSORIES

Brass and bronze are the traditional materials used for water-handling equipment because of their ability to stand up under rugged fire line use; they are a good-looking product, and their thread engagement is satisfactory. But they are heavy, and most fire-retardant chemicals are highly corrosive to brass.

Aluminum alloys are being used more and more for wildfires. They are about one-half the weight of brass, and the threads are hard-coated to prevent seizure. They tend to be less expensive than brass and bronze.

Lugs are the means of tightening the female fittings that are used intermittently, such as couplings. The rocker lug is composed of two raised knobs opposite each other on the collar of the female fitting. There are many types of lugs, all of which can be tightened or loosened with a spanner wrench. Different types include slotted, pin, pinhole, and long handle lugs.

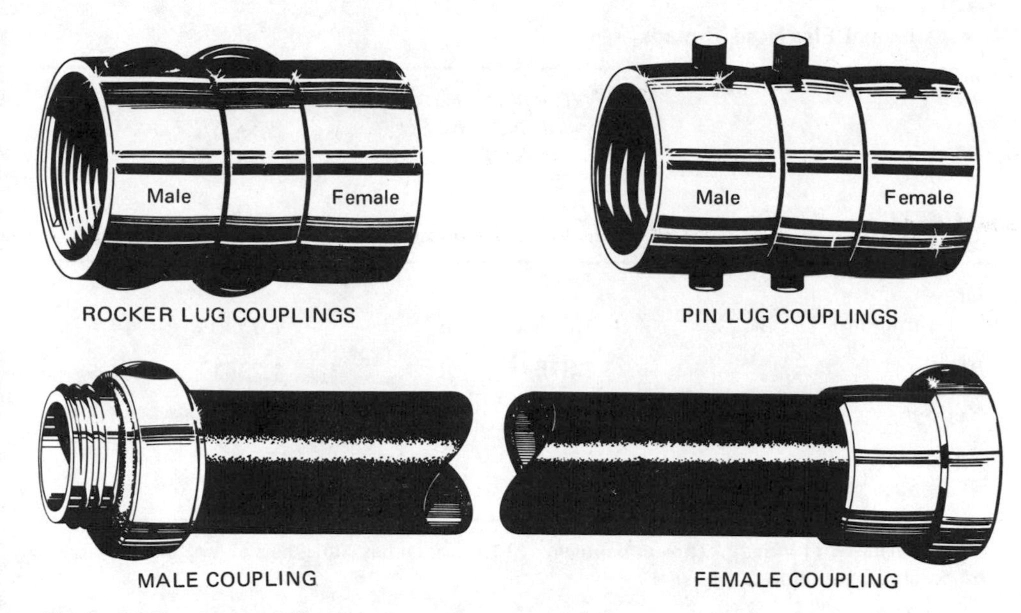

Figure 2.30. Lugs tighten couplings along a line of hose.

FIRE HOSE THREADS

Hose threads are termed straight or parallel, and a water seal is formed as the external thread lip seats against a recessed gasket in the internal thread section. In contrast, water pipe threads are tapered and seal against themselves.

Nearly all wildfire agencies have adopted the National Fire Protection Association (NFPA) standards for fire hose threads (NFPA standard number 194, 1968). At one time there were several hundred types of fire hose threads in use in the United States. Many fire services had their own special thread as a mark of distinction, but they could not hook up to any other service without special adapters. There has been much loss of property and a number of firemen have lost their lives because two outfits could not couple their hose together. The standardization program is well along in most areas. When hose or fittings are purchased, it is important to acquire the National Standard fire hose coupling threads shown in Table 2.5.

Thread standardization. United States National Standard (NS) fire hose coupling threads should be used for 1½-inch and larger size hose. Most national and state wildfire services use straight iron pipe threads (NFPA) for 1-inch hose and garden hose (GH) threads for ¾-inch hose.

A color code was established by the California region. A narrow strip of paint is brushed on the male shoulder or swivel of all fittings to designate the type of thread—chemical thread is coded blue.

The 1½-inch size hose is by far the most common size in wildfire use. The 1-inch connection is used on most nozzle bases, on 1-inch soft hose, on ¾-inch (inside diameter) hard rubber lines and reels, and on ¾-inch and 1-inch chemical (booster) hose. This size and type of hose is used in fire department practice, and the national standard thread is designated as 1-inch (eight threads per 1.375 inch outside diameter); this thread is also known as chemical or booster hose thread.

TABLE 2.5
Comparison of Fire Hose Threads

		National Standard Thread (NST, NSFH, NH)		*Straight Iron Pipe (SIPT and NPSH)*	
Thread color code	*Size (inches)*	*ODM* (inches)*	*TPI†*	*ODM* (inches)*	*TPI†*
Gray (aluminum)	¾	1.375	8	1.0353	14
Black	1	1.375	8	1.2951	11½
Yellow	1½	1.990	9	1.8788	11½
Yellow or orange	2½	3.0686	7½	2.855	8

* Outside diameter of male. †Threads per inch. ‡This gasket has a thickness of 3/16 inch, whereas others are 1/8 inch thick.

Higbee cut. The starting thread on fittings or couplings gets thinner with use, and this process can lead to thread crossing and damage. The problem is eliminated by cutting away a portion of the end of the last thread and blunting it. This makes the starting more positive and is known as the Higbee cut.

Gaskets. Gaskets should be used in each coupling and accessory. Extra gaskets should be kept with the pumper.

Expanders. Expanders for attaching couplings to hose are needed when replacing couplings or eliminating broken sections of hose. These may be available from nearby fire services.

NOZZLES

There is a wide variety of nozzles on the market. The best nozzle for wildfires should have the capability of quick change from straight to fog or spray stream and should have a quick shutoff to save water. For garden hose streams an adjustable garden hose nozzle or tip used with a quick shutoff is most often used on patrol tankers.

A forester fog stream nozzle is one of the best for use on ¾-inch, 1-inch, or 1½-inch lines. It has a control valve lever to activate either the straight stream or the fog spray stream. This nozzle has two parts, one for straight stream and one for fog spray, with several different size tips for each. It also has a quick shutoff that is operated with the one control valve lever. The control lever can also be used to adjust below tip discharge rate.

The six-shooter nozzle provides instant control with an on-and-off lever and six straight streams and six fog streams built into the head. Fog streams range from 3 to 18 gpm, and the straight streams range from ⅛ to ⅜ inch. It has a 1-inch inlet and is made of aluminum alloy.

There are several nozzles available that have a quick shutoff and are adjustable from straight stream to fog. Some have an adjustment for gallonage; others maintain a fixed gallonage. In choosing one of these, keep the gallonage low enough for wildfire use so that water will not be wasted.

TABLE 2.5 (continued)
Comparison of Fire Hose Threads

Garden Hose (GHT)		*Chemical Hose (CHT)*			
ODM (inches)*	*TPI†*	*ODM* (inches)*	*TPI†*	*Diameter of outside gaskets (inches)*	*Diameter of inside gaskets (inches)*
1.625	11½	1.375	8	$1^7/_{16}$	$^{13}/_{16}$
		1.375	8	$1^7/_{16}$	$1^1/_{16}$
				$2^1/_{16}$	$1^9/_{16}$
				$3^3/_{16}$‡	$2^9/_{16}$

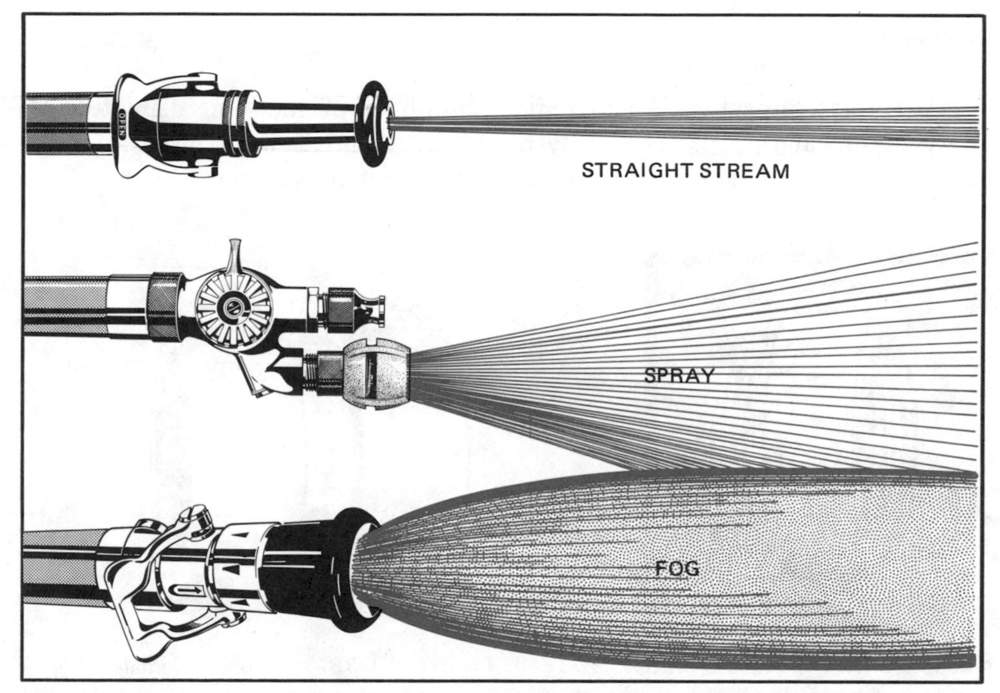

Figure 2.31. Different nozzles provide different types of streams of water. (1) Straight stream, ball shut off. (2) Combination fog-spray, straight stream, shut-off type. Different size tips available. (3) Adjustable, fog to straight stream and adjustable for gallonage.

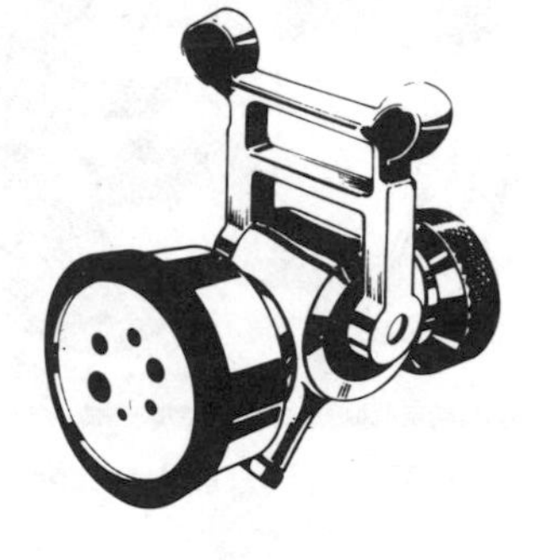

Figure 2.34. Six-shooter nozzle.

Figure 2.32. The garden hose nozzle has a quick shutoff.

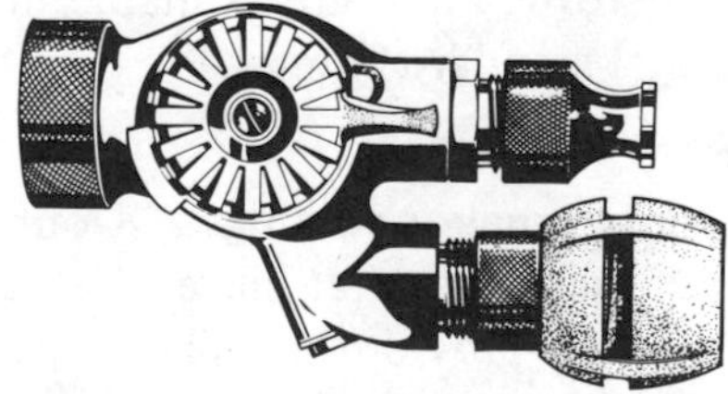

Figure 2.33. The forester fog stream nozzle has a control lever to activate either the straight stream or fog spray stream.

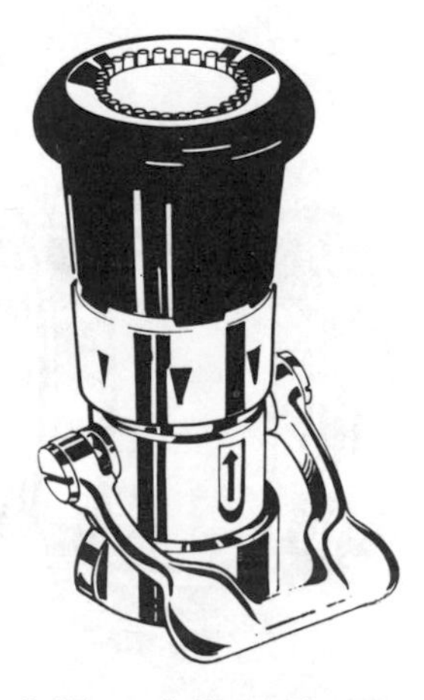

Figure 2.35. Adjustable straight stream to fog stream nozzle with quick shutoff.

In-line shutoffs can be used between couplings or in back of a nozzle for 1½-inch lines. A ring in the middle revolves to shut or open the device. Other shutoffs are available that are activated by levers.

Applicators are available in 48-inch lengths, or you can make your own. An applicator is simply a piece of lightweight pipe bent on the discharge end at a 45-degree angle. On the bent end it should have threads to receive different tips, and it should have a female coupling on the other end. Applicators are invaluable for direct attack on light fuels in order to get the water in at the base of the flames and in mop-up to reach in under logs, roots, and stumps and to explore in deep duff to search out hot spots.

FIRE HOSE SHUTOFF CLAMP

The shutoff clamp is a pocket-sized tool that enables one man to instantly shut off the water flow in charged lines up to 2 inches in diameter. It is convenient for extending working lines, installing accessories, and replacing nozzles without shutting down the line or pump. If hose lines burst during service, the clamp will permit continuous operation or facilitate replacement of hose.

Figure 2.36. In-line shutoff.

ADAPTERS

Adapters connect male and female couplings of the same diameter that have dissimilar threads. They may have lugs for intermittant use or notches on the inside to attach with epoxy for permanent conversion.

Figure 2.37. (*Left*) Permanent male adapter. (*Right*) Adapter with lugs.

Figure 2.38. Double female coupling.

Double female couplings. Double female couplings are used to connect two male sections that have the same diameters and threads. Both swivels should turn. They are available in 1-inch iron pipe and 1½-inch and 2½-inch national standard thread.

Figure 2.39. Double male coupling.

Double male couplings. Double male couplings are used to connect two female sections that have the same diameter and threads. The center is formed to provide lugs. They are available in 1-inch iron pipe and 1½-inch and 2½-inch national standard thread.

Reducers. Reducers connect the outlet end of any service to the inlet side of a smaller fitting. The male side is the outlet side and is smaller than the female side, which is the inlet side and is larger with the reduction in the direction of flow. They are usually 1½-inch national standard thread to 1-inch iron pipe.

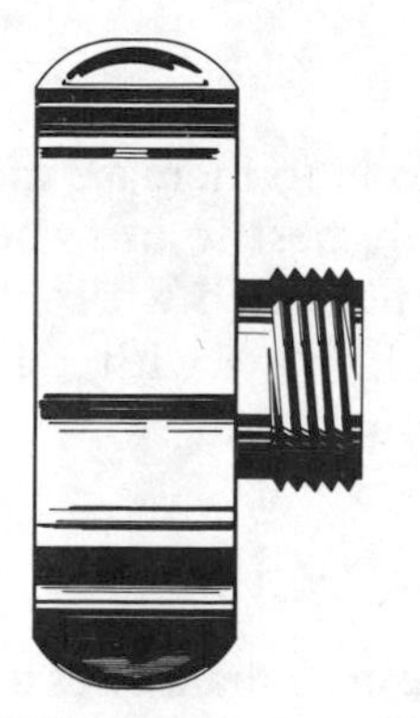

Figure 2.40. Reducers connect the outlet of any service to the inlet of a smaller fitting.

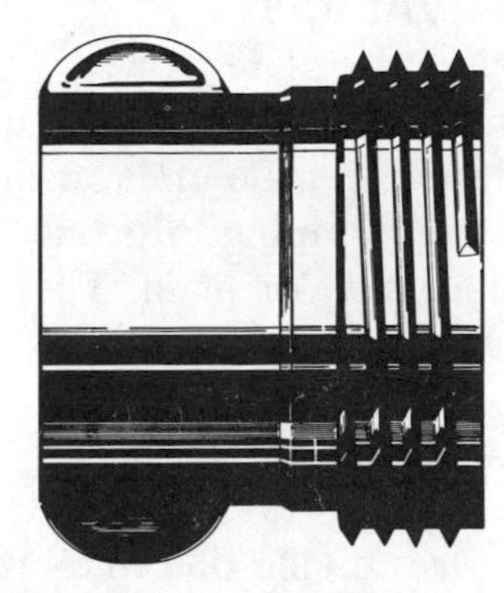

Figure 2.41. Increasers connect a male outlet to a female inlet of a larger diameter.

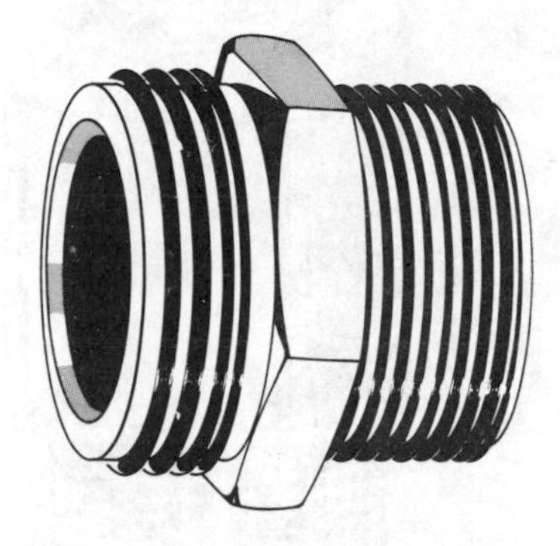

Figure 2.42. Male hose nipple.

Increasers. Increasers connect a male outlet end to a female inlet of larger diameter of the same thread. The male end is always larger as the increase is in the direction of flow. The female is available in 1-inch iron pipe to 1½-inch national standard thread.

Hose thread nipples. Hose thread nipples are designed for permanent attachment to pipelines, hose valves, hydrants, pump discharge, and similar fixed installations. They have male threads on both ends and a hexagonal wrench grip at the center. One end is always furnished with standard pipe thread (tapered). The other end is furnished with hose thread, usually national standard unless otherwise specified. They are the same size on both sides or can be furnished for reducing or increasing the hose thread side.

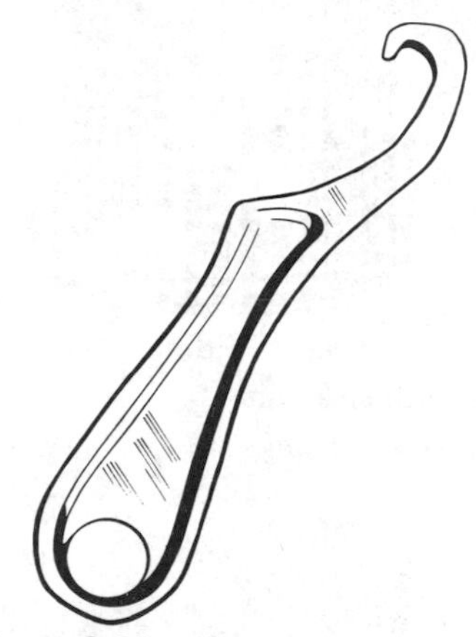

Figure 2.43. Spanner wrench.

SPANNER WRENCH

A spanner wrench is a small pocket wrench for use on rocker lugs or slotted fittings. It may also be made to use on hydrant stems. There are several different kinds in use.

HOSE CAPS

Hose caps are used to prevent damage to male hose threads on hydrants, pumper outlets, and some valves and connections. Most are not designed to withstand pressure. They are available in all sizes.

Figure 2.44. Hose cap.

STRAINERS

Strainers are used on the submerged intake end of the suction line of a pumper or a portable pump. They are designed to prevent debris and most abrasive material from entering the pump where damage can easily result. They are available in 1½-inch and 2½-inch sizes. It's a good idea to tie a bucket over any strainer to keep it out of mud and sand.

FOOT VALVE WITH STRAINER

The foot valve with strainer is a combination used on suction hose for centrifugal pumps and on ejectors. It should be capable of holding a water head equivalent to 500 psi, since the water in a line can flow past centrifugal pump impellers when the pump is not running. It is available in 1-inch through 3-inch sizes.

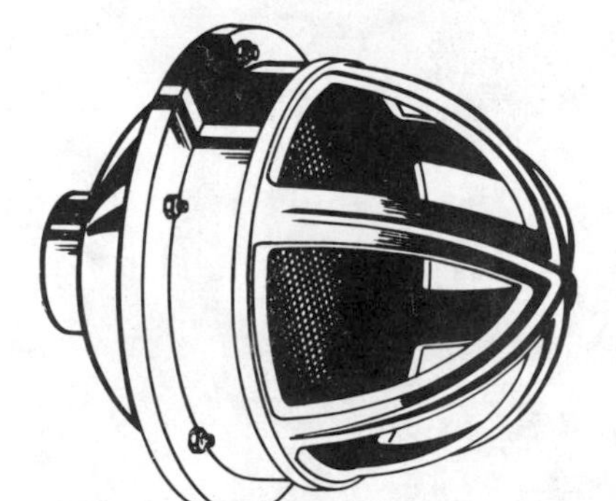

Figure 2.45. Strainer.

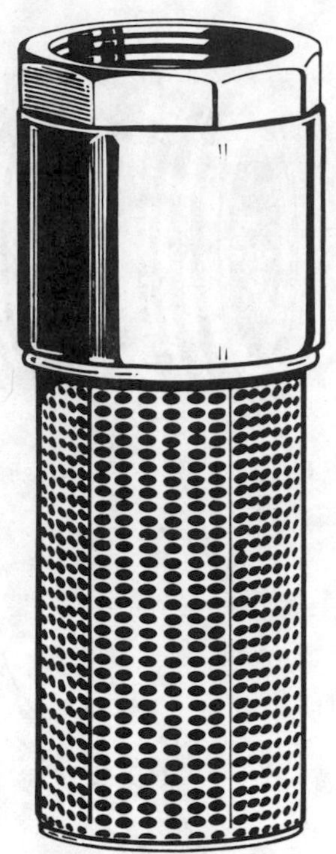

Figure 2.46. Foot valve with strainer.

SIAMESE VALVES

A siamese valve combines or unites two lines into one to increase the flow. There are two female ends to one male end. They are best to use when two pumpers are pumping into one line. They can be either gated with valves on each female side or plain. They usually have 1½-inch inlets with a 1½-inch outlet but are also available in 1-inch size–these are not gated.

WYE VALVES

Wye valves divide one line into two, so they have two male ends and one female. They may be gated or plain. They are available in 1-inch and 1½-inch sizes. They are used in hose lays to allow for extension and branch lines.

HOSE LINE TEE

A hose line tee is a mainline connection to allow for a branch outlet. It is usually for 1½-inch to 1-inch branches and is also available for 1-inch mainline to 1-inch branches. They are female on one end and male on the other with male threads on the branch outlet that is fitted with a cap with lugs.

HOSE LINE TEE VALVE

The hose line tee valve, or "water thief," is placed intermittently in 1½-inch main hose line. The male outlet for the branch line has a valve to open or close the branch line.

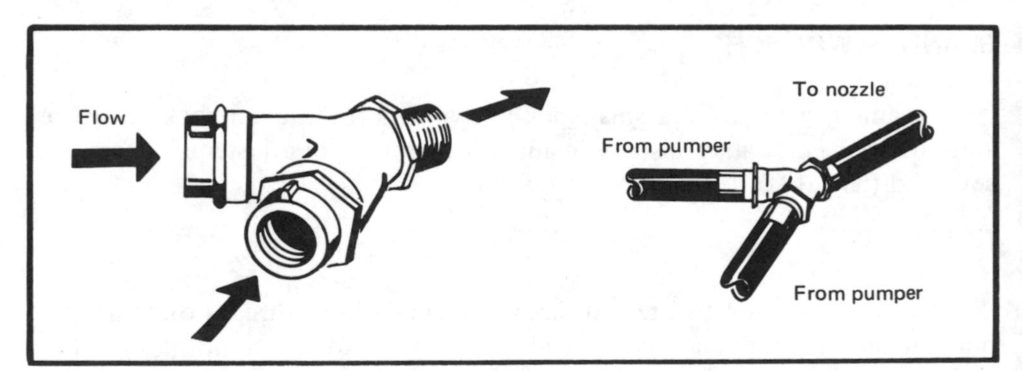

Figure 2.47. Plain siamese valve.

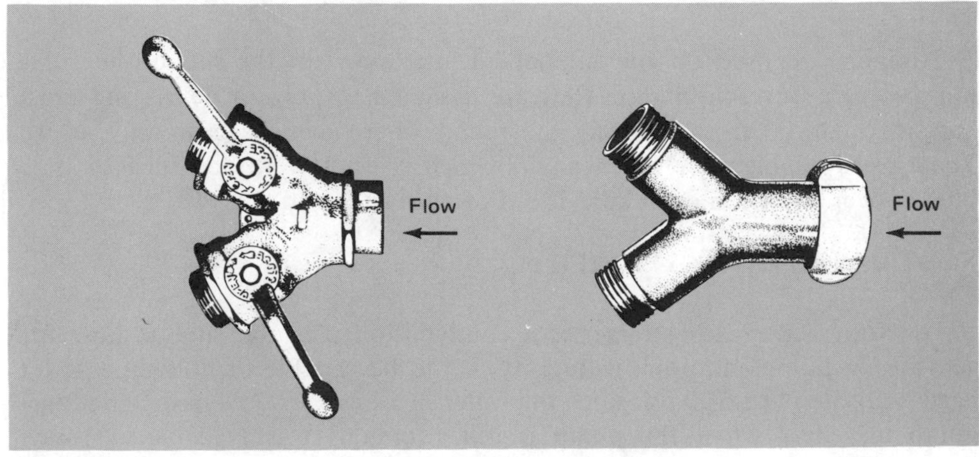

Figure 2.48. Gated wye valve (*left*) and plain wye valve (*right*).

BLEEDER VALVE

The bleeder valve is installed in 1½-inch hose lines to allow for filling of backpack pumps and for canteens if the water is potable. It is female on one end and male on the other. The small water valve is operated with an open end wrench chained to the casting. There is no interruption to the flow in the line.

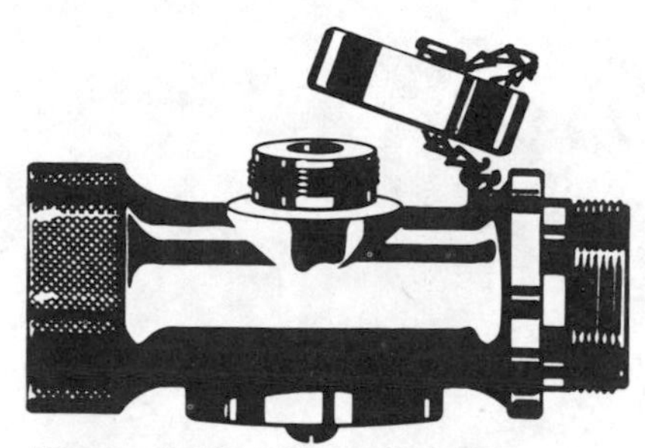

Figure 2.49. Hose line tee valve.

CHECK AND BLEEDER VALVE

The check and bleeder valve is a combination valve used at the pumper in 1½-inch lines where there is a vertical lift of 200 feet or more. A clapper valve holds the water in the line when the pump is stopped—this is the check valve. Before the engine is started (with the line full), the bleeder valve is opened to facilitate starting the pump; it is then closed when the pump has gained sufficient head to pump against the load in the hose line. A length of hose can be attached to the 1-inch bleeder valve outlet and discharged back into a booster tank.

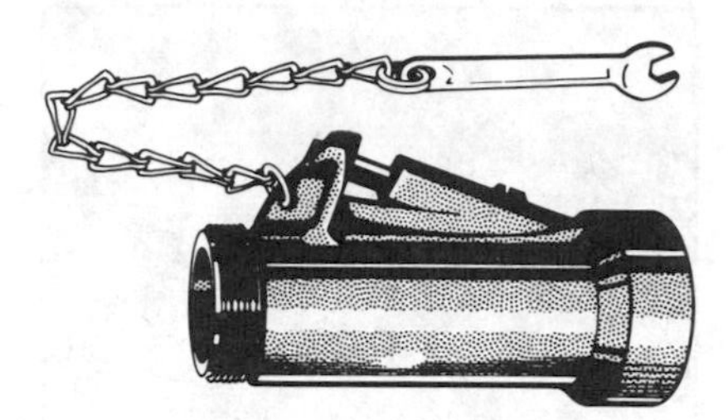

Figure 2.50. In-line bleeder valve.

AUTOMATIC PRESSURE RELIEF VALVE

The automatic pressure relief valve is used with positive displacement pumps to relieve excessive pressures and line surges. If all nozzles on the line are closed, it will bypass water from the main hose line by a spring-loaded release. A manually operated valve wheel adjusts the relief pressure from 50 to 250 psi, and a spoked locknut holds the adjustment. It is used on 1½-inch lines near the pumper. It is used with the automatic check and bleeder valve and placed after it in the line.

Figure 2.51. Check and bleeder valve.

Because of the weight of these two valves, it is best to have a pigtail, about 2 feet long, of 1½-inch hose between them and the pump. This takes the stress off the male outlet on the pump and prevents it from breaking.

GRAVITY SOCK

When water is available from a stream or spring above the fire, a gravity sock can be used as a water pickup point. This point should be at least 100 feet in vertical elevation above the point of use on the fire to obtain 50 psi at the nozzle (not considering friction loss). The sock is secured by stakes, rocks, or ropes, and the line then runs down to the fire. The sock opening is from 8 to 15 inches in diameter, and the sock is 3 or 4 feet long with a 1½-inch male connection.

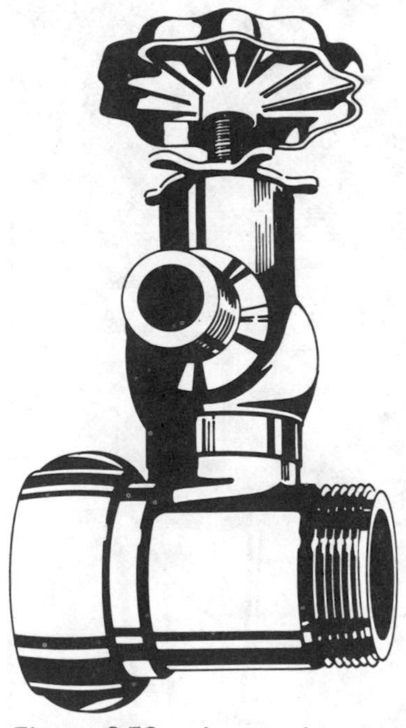

Figure 2.52. Automatic pressure relief valve.

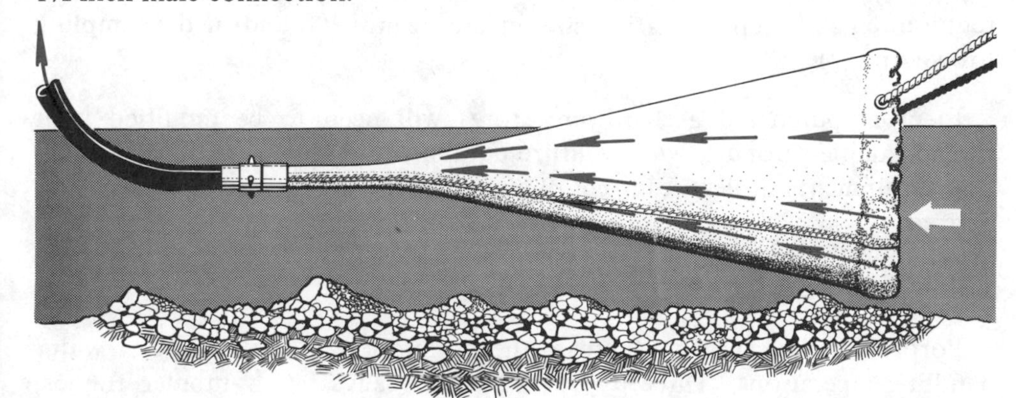

Figure 2.53. Gravity sock in use in a stream.

HYDRAULIC EJECTOR

A hydraulic ejector is very useful if the water source for drafting is 18 feet or more vertically below the pumper or if the water source is further away than the length of the draft hose. With the use of an ejector, water can be drafted and raised vertically from 40 feet to 250 feet, depending on the type of pump and ejector used. Or the ejector can be used up to several hundred feet from the pump where the pump cannot be spotted within the length of the drafting hose to the water source.

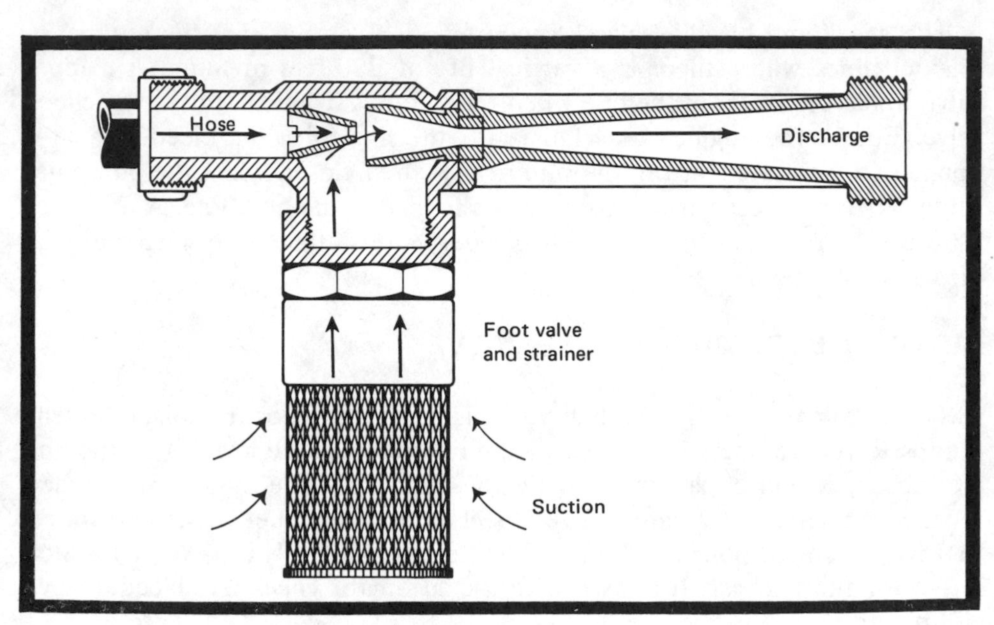

Figure 2.54. Hydraulic water ejector.

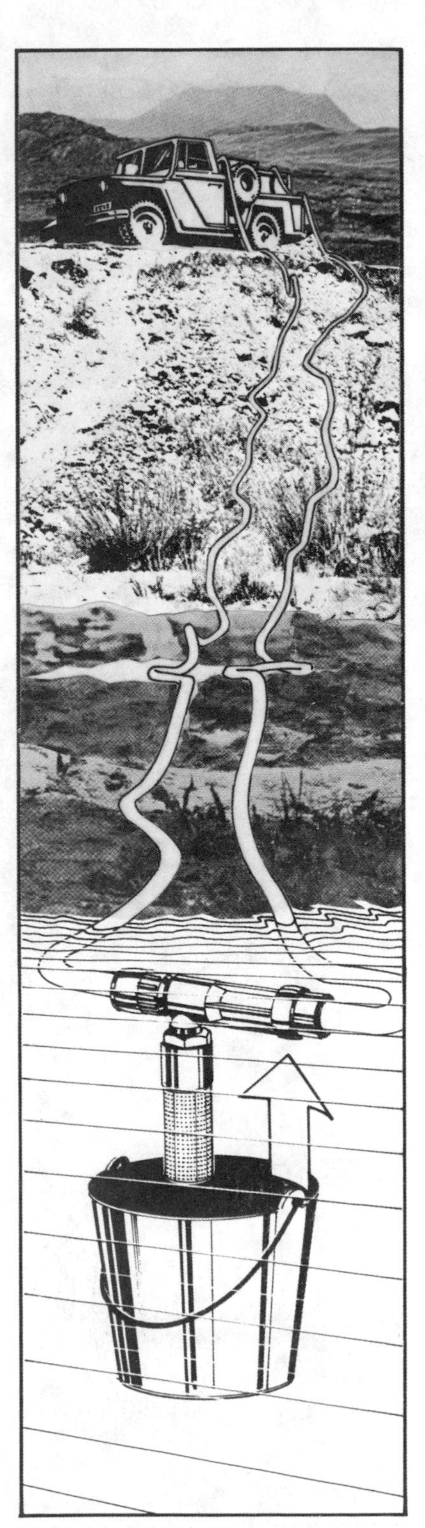

Figure 2.55. Hydraulic water ejector in use.

Water is pumped in the conventional manner from a tanker with the booster line to the water inlet of the ejector and it passes through a nozzle restriction as a high-velocity stream. This jet stream then picks up additional water through the suction part and delivers the combined flow through the diffuse chamber and out the discharge port under lower pressure through a 1½-inch line.

A few small leaks in the hose are not critical, since the entire hose lay to and from the ejector is under pressure.

The ejector should be selected to match the pressure and flow output of the pump to be used. Consideration must also be given to the usual head requirements. A representative size ejector can be found in the sampling shown in Table 2.6.

Ejectors purchased at hardware stores will need to be modified with thread adapters for use with wildfire equipment. Those purchased through fire equipment dealers can be obtained with the thread specifications desired.

TANKS

Portable tanks can often increase the production and efficiency of water handling operations. Three types are used: canvas with tubular frames, pillow shaped pods, and self-supporting pyramidal shaped tanks.

TABLE 2.6
Examples of Ejectors

Pump	Ejector model (Penberthy)	Maximum (foot)	Lift (psi)*	Ejector at ¾ maximum lift (gpm)	Pickup at maximum flow (gpm)	Ejector return maximum combined flow (gpm)
Briggs Eco	62A	101	44	7	11	20
Berkley B1-½ XQBBS-10	64A	118	51	15	28	52
Hale or Cedco APZZ	63A	122	53	10	20	36
Edwards L-23	64A	127	55	16	28	53
Pacific WA-7	63A	155	67	12	20	38
Pacific WX-10	64A	125	54	16	28	53
Cedco ALO-F (02F-1-GR 200 psi)	62A	127	55	8	11	21
Gorman Rupp	65A	127	55	23	45	85
Cedco HPZF XQBS-18	65A	148	64	30	45	88
Berkley B-1 ½	65A	152	66	31	45	88
Hale CBPD (400 psi)	65A	250	107	44	44	99

*There is no point in operating with the ejector discharge at or above the maximum lift pressure because there is no gain over the pumper flow rate (ejector pickup flow is zero).
There are a number of ejector manufacturers such as Berkley, Western Fire Equipment Company, and Penberthy.

Canvas with tubular frame tanks are usually square and can be folded for storage and transport. They have a lightweight tubular frame and are open at the top. They are available in 600-, 1,000-, and 1,500-gallon sizes.

Pillow-shaped pods are made with heavy canvas or neoprene-treated canvas in 50-, 100-, and 150-gallon sizes and are self-supporting. They can be used to transport water on pickups or flatbed trucks or for booster pump locations.

Pyramidal shaped tanks may be either self-supporting (Harodike) or open-topped. They are supported by stakes and rope anchors. They are available in sizes from 75 gallons to 250 and 500 gallons.

HOSE ROLLER GUIDES

A four-way roller guide for booster line hose keeps the hose and couplings away from the pumper body and makes its use simpler for one or two men. The line can be used in any direction from the pumper, and the guides tend to keep loose hose from getting under the truck wheels.

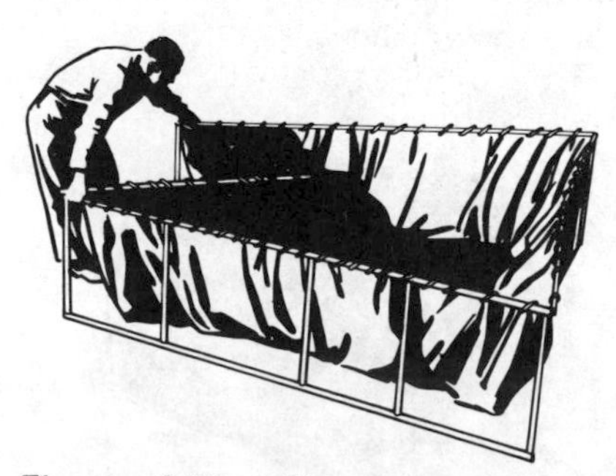

Figure 2.56. This canvas tank with a tubular frame can support 1,000 gallons.

HOSE WASHERS

There are a number of commercial hose washers, or they can be made from discarded 2½-inch couplings for 1½-inch lines. One type that is reasonable uses a jet action on the inside, and the hose is drawn back and forth in the washer until it is clean.

Figure 2.57. Pillow-shaped staging pads are made of neoprene-treated canvas.

TRAINING

Training of firemen is a continuing necessity in any fire service. Training includes both the practice required to perform operations efficiently and the learning of necessary background information. Training is a prerequisite to adequate fire service performance in prevention, presuppression, and suppression of all types of fires. The objective of training is to develop each fireman and the group as a unit to be able to develop a public that is attuned to fire prevention, to maintain adequate organization with ready equipment, and to control fires with a minimum of loss to life and property.

Rule of thumb: If the learner hasn't learned, the teacher hasn't taught. Good performance is the result of carefully planned and effective instruction.

METHODS

Classroom instruction. Classroom instruction using appropriate visuals helps to develop an optimum of understanding and a maximum of recall by the students and is the basic method of teaching. This method introduces the subject material and is useful in reinforcing points of learning, but, to be most effective, it must be followed up by other methods. The learning process is not complete until the student can demonstrate what he has learned.

Figure 2.58. One type of pyramidal shaped tank is the Harodike bag shown above.

Classroom teaching may vary from a formal lecture to a lecture with questions, a training conference to which the students contribute, a seminar that is a clearinghouse of individual studies, or a pure conference in which the conclusions are not predetermined. For most fire service training the formal lecture or lecture with questions will be the method used.

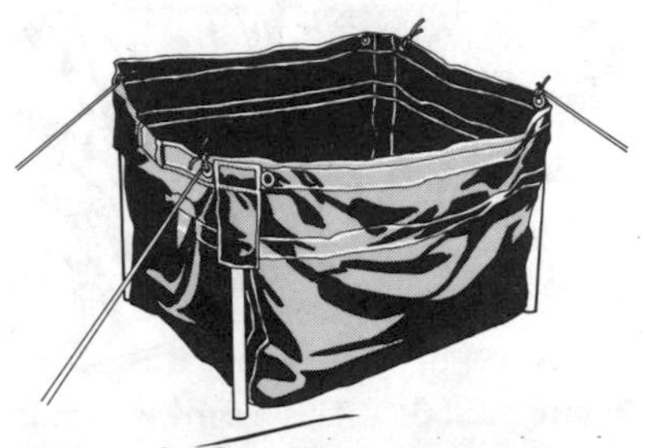

Figure 2.59. This canvas bag is supported by stakes and ropes.

Demonstration and practice. This method shows the students how to do specific operations and it allows for their participation. It should include practice sessions where the firemen practice the various evolutions, jobs, and actions with the tools and equipment they will use on wildfires.

Simulators. The simulator is realistic practice in control of wildfires. There are several different simulators in use, but basically they use the principle of developing a realistic situation in which an appointed fire boss or team solves the problems of a given fire situation by issuing instructions as to where, when, and how the men and equipment will be employed to control the wildfire.

This is a method of training where the students do and say in the process of learning. After the exercise in the simulator, one of the students leads a discussion or critique of what was done right, what might have been done wrong, and how the operations could be improved in future wildfires. This method is most useful in reinforcing classroom instructions, and good recall is developed.

Critique. A critique is simply a discussion by the group, preferably right after each fire. It is led by either the chief or the training officer to discuss the performance of the group. The discussion should bring out what was done right, actions that need to be changed, and how overall performance can be improved.

The critique should be conducted as soon after the fire as possible, while everything is clearly remembered. It should be objective and not critical of individuals; rather, it should suggest improvements. This is one of the most effective methods of training, and its use is recommended for all types of units.

Regular meetings. Regular meetings provide for upkeep of the equipment and maintenance of plans, and they give continuity to the training program. They should be held at least every one or two months, and attendance by all firemen should be required.

SOURCES OF TRAINING

Lesson plans are available for the chapters in this text. They are a guide for the officer in developing training programs for wildfire control. Overhead transparencies are also available to use with the training outlines, and they provide one of the best methods of teaching (contact the Robert J. Brady Company, Bowie, Maryland 20715).

Each state maintains a training organization for the fire services. The training authority may be located in the vocational education program or the continuing education program of your state, which is usually located at one of the universities. Arrangements can usually be made with the training authority for training of your group at your location. Contact should be made with them in developing your training program.

Most states maintain a state fire school that functions either year-round or at a selected time. These are very worthwhile, and firemen should be encouraged to attend them. Many of them now have courses in wildfire control.

Your state forestry agency has good information and material available on wildfire control for your use. Many state forestry agencies also maintain a training program.

There are many other courses that should be programed to support your unit's training for wildfire control. Two of the most important are first aid and driver training. Basic knowledge for the control of structures, flammable liquids, highway fires, and rescue operations is also very desirable.

The training officer's job is to develop and schedule the training required for his unit. He may conduct some of the classes himself, but there are many courses where he can obtain training programs presented by specialists. His principal action with them would be scheduling, making arrangements, and accommodating the visiting instructor.

WILDFIRE REPORTS

Reporting each wildfire to the proper authorities is an important function of a fire control unit. The State and Private Forestry Division of the U.S. Forest Service compiles and publishes a report each year on the numbers, causes, sizes, classes, and acreages of wildfires in the United States. The report is initially compiled by each state forester and is sent to the State and Private Forestry Division for final compilation and publication as *Wildfire Statistics* each year. Each fire-fighting organization that is engaged in wildfire suppression should contact their state forester for the reporting method to use and the location to which to send its fire reports. Standard forms are usually provided for this purpose. (Each state fire control authority also requires reports. In some states a report is needed only when the damage from a fire exceeds a specified amount.)

It is important to compile each report as soon as possible after each fire while the information is fresh in everyone's minds. Most fire services appoint one individual (perhaps the secretary for the unit) to the task of compiling the report.

Basic information needed for a wildfire report includes, but may not be limited to, (1) the name of the fire or the district number in which the fire occurs, (2) the location of the fire (township, range, and section), (3) the size of the fire, (4) the cause of the fire, (5) the estimate of damage in dollars, and (6) the date and time of the fire.

Wildfires are classified by size according to the following scheme: Class A, ¼ acre or less; Class B, at least ¼ acre but less than 10 acres; Class C, at least 10 acres but less than 100 acres; Class D, at least 100 acres but less than 300 acres; Class E, at least 300 acres but less than 1,000 acres; Class F, at least 1,000 acres but less than 5,000 acres; Class G, 5,000 acres or more.

The U.S. Forest Service classifies fires by cause according to the following definitions:

- Lightning
- Campfire (a wildfire resulting from a fire started for cooking, heating, or providing light or warmth)
- Smoking (a wildfire caused by matches, lighters, tobacco, or other smoking material)
- Debris burning (a wildfire spreading from clearing land, burning trash, burning slash, or other prescribed burning)
- Incendiary (a wildfire willfully set by anyone to burn or to spread

to vegetation or property not owned or controlled by him and without consent of the owner or his agent)

- Equipment use (a wildfire caused by railroad operations, including the burning of rights-of-way and railroad ties)
- Children (a wildfire started by children less than 12 years old)
- Miscellaneous

Other information may also be required in the wildfire report, such as time of discovery, man power and equipment needed for control, fuel type, and a narrative report of the fire control operation. Fire reports provide a realistic and factual basis for prevention planning, provide support for funding requests, and aid in organizational development.

3 FIRE BEHAVIOR

THE NATURE OF FIRE

To understand the way in which fires burn, we first need to understand the phenomenon of fire. What is it? Why and how does it burn? Why are there flames?

Fire, or the process of combustion, is a chemical reaction called *rapid oxidation*, and is accompanied by the production of *heat* and *light*. The process is similar to that of the rusting of a piece of iron or the rotting of wood, only it is tremendously speeded up.

When heat is applied to a substance, either liquid or solid, the molecules move more rapidly within that substance. When enough heat is applied, some of the molecules break the surface to form a vapor, or gas. If enough heat is present, this vapor bursts into flames.

PHASES OF COMBUSTION

The three phases of combustion are preheating, gaseous, and charcoal. Sufficient heat must be present to attain the ignition temperature of the particular fuel.

In the preheating phase the fuels are dried, heated, and partially distilled, but no flame exists. In this phase, the fuel is being raised to the ignition temperature, which is roughly between 500°F and 800°F for most wildfire fuels. The ignition temperature will be reached quickly or slowly, depending on the size of the fuel, its moisture content, and its curing stage (whether green or dormant). You should have the ability to judge when a fuel is reaching its ignition temperature.

In the gaseous phase the distilled gases actually flare and burn. The fuel has been brought to its ignition temperature, and, if an ignition source is present, flames appear above the fuel. When you observe a wildfire, you will notice that the flames are above the top of the fuel. At this stage the gases are burning, but the fuel itself is not yet glowing. This is the phase in which most fires are fought. For this reason, water should be applied at the base of the flames. If incomplete burning occurs, the unburned gases and carbon are given off as smoke, and charcoal is left.

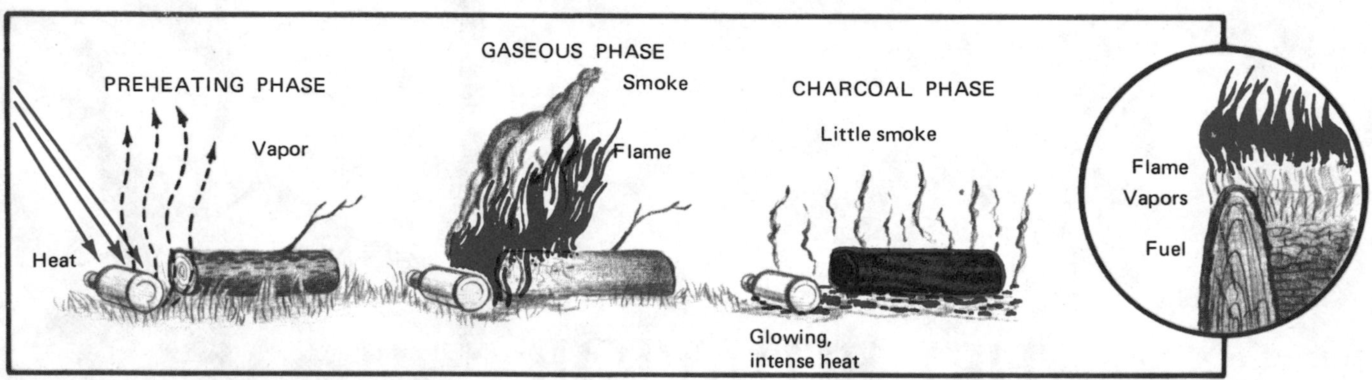

Figure 3.1. The three phases of combustion.

In the charcoal phase the fuel is consumed, and ashes are left. There is a lot of heat but little flame and smoke. Wildfire fuels in this phase can cause spot fires if wind causes mass transport or if the burning fuels break apart and roll downhill. City firemen derogatorily claim that one is "saving the foundation" if a structure fire is being controlled in this phase.

HEAT ENERGY SOURCES

Heat is a form of energy. Heat is produced by mechanical means, electrical functions, chemical reactions, and nuclear energy releases.

Mechanical means include the friction produced from a slipping clutch, grinding wheel, slipping belt, bearings, mufflers, etc., or from compression, such as in a diesel motor.

Electrical functions that generate heat include electric arcing that occurs in poor wiring or the natural arcing of lightning and resistance heating as in a light bulb, electric heater, or overloaded wiring. Static electricity is of short duration and low intensity, but it will ignite gases, vapors, and dusts. A common example is the static electricity that results from sliding shoe soles along a carpet.

Chemical reactions often cause heat and result in fire. Fire itself is a chemical reaction that results from the combination of a fuel with oxygen at the ignition temperature of the fuel. Spontaneous combustion is a condition in which sufficient air exists for oxidation of the fuel but there is insufficient movement of the air to carry the generated heat away from the fuel. This condition can occur in animal oils, vegetable oils, hay, grain, sawdust, compost, and silage.

Nuclear energy release is the most powerful of all heat energy sources. Volume per volume of the fuel, the energy released is one million times greater than the energy released from other ordinary chemical reactions. Presently this is not a source of heat to be reckoned with in wildfire control, and hopefully it never will be a source of heat for wildfires.

Heat application to fuel can be caused (1) by direct application of flames, such as a match or other fire, and (2) by direct application of embers, such as ashes, glowing fuel, or resistance heating.

HEAT TRANSFER

Once the heat source has created fire, the heat must be transferred to other fuel in order to advance or spread the fire. This is done by conduction, radiation, convection, and mass transport.

Conduction is the transfer of heat by direct contact with the heat source. An example is a frying pan on a stove. Wood is ordinarily a poor conductor, but metal is a good conductor. Think of conduction as heat being carried along a conductor as electricity is carried. Although this method of heat transfer is important in structures and flammable liquids, it has little relation to wildfires.

Figure 3.2. One form of heat transfer is conduction, the transfer of heat by direct contact with the heat source.

Radiation is the transfer of heat through space in any direction at the speed of light. Examples are sun rays, the radiant heater, or a grass fire that causes the drapes in a house to catch fire when the heat is radiated through a glass window. The angle at which radiation strikes a surface affects the amount of heat the surface receives. A beam of heat striking at an angle must cover a larger surface than one striking perpendicularly. Thus the latter radiates more heat. The intensity of the heat, however, decreases by the square of the distance from the object to the source. Radiation is important in all fires, but it is a primary method of heat transfer in wildfires. Think of radiation as a ray or wave.

Figure 3.3. Two forms of heat transfer, radiation and convection, are shown. Heat is *radiated* in a straight line in all directions; the flames produced on the tree are caused by radiant heat. Heat also is *convected* upward in any direction depending on the prevailing air current.

Convection is a circulating upward movement in a gas or a liquid because heated air rises. Examples are a hot air furnace or a teapot. Convection forms a column of molecules of heated air and gases that spread heat in any direction, depending on the movement of the air.

Owing to the principle of convection, fire creates its own draft by pulling in additional oxygen from the sides of the fire and, in the case of wildfires, throwing up the heated air above the fire. This draft accounts for the roar often heard in fast-burning wildfires. In large fires this phenomenon only reaches its climax when the amount of air is sufficient for oxidation of the amount of fuel available. Think of convection heat as a mass or a cloud.

Figure 3.4. Fire creates its own draft. An oxygen supply of cool air feeds the base and sides of the fire as heated air is thrown above the fire.

Mass transport is a principal method of heat transport in wildfires. It is primarily a function of convection, but it may be caused by rolling or falling of ignited fuels and embers. When the convection column is pushed by wind and its own draft, burning embers may be picked up in the moving column of hot air and dropped ahead of the main fire. Burning cones have been known to be deposited as much as a mile and a half ahead of the main fire. The mass transport of embers may occur in any type of wildfire, depending on the type of fuel and the action of the wind.

Figure 3.5. In mass transport of heat, burning embers are carried through the air and are deposited on new fuel.

Mass transport can occur when burning fuels roll downhill into new fuel or when burning fuels drop from above to ignite ground fuels after the crown fire has traveled through the aerial fuels. Think of mass transport as showers of burning bits of fuels.

THE FIRE TRIANGLE

Sufficient oxygen, fuel, and heat must be present for fire to exist. This interrelationship is called the fire triangle. Eliminate any one of the three, and the fire goes out.

Figure 3.6. The fire triangle is composed of sufficient oxygen, fuel, and heat to maintain a fire.

OXYGEN

Twenty-one percent of the air is oxygen. A reduction in oxygen to 15 percent extinguishes the fire. In the early gaseous stages, even a momentary reduction in oxygen puts out the fire. This can be done by smothering or covering–usually with dirt in the case of wildfires or with swatters, sacks, boughs, etc.

FUEL

Wildfires are primarily controlled by working on the fuel side of the fire triangle. This is done by confining the fire to a definite amount of fuel by means of a fire line and natural barriers, if they are available. By keeping all fire inside the line, the fire is confined and controlled. Then it can be put out over part or all of the area inside the line or allowed to burn out. The fire is robbed of additional fuel by controlling any spot fires or, better yet, preventing their occurrence. The line is usually made by removing the surface fuel with hand tools or equipment so that the mineral soil is exposed or by wetting down a line of sufficient width with water.

HEAT

In order to start a fire, fuel must be brought to the ignition temperature. If the heat drops below the ignition temperature, the fire goes out. Water is the most effective agent for this reduction of heat. The finer the water particles are per unit of water, the more heat is absorbed from the fire. Application of dirt also helps to reduce the heat.

Rule of thumb: No two fires are exactly alike. No two firemen will fight the same fire in the same way.

FIRE BEHAVIOR

Knowledge of fire behavior involves understanding how a fire acts under varying conditions. Only the fundamentals of fire behavior are outlined in this text. Fire behavior is a large and complicated subject, and a knowledge of fire behavior will require additional study on your part. A clear understanding of fire behavior is the basis for all fire fighting. The need for a thorough understanding of fire behavior cannot be emphasized too strongly—it is the essence of fire control. Basic principles must be correctly applied to any fire situation to obtain control of the fire. The application of the basic principles may vary somewhat because of equipment, location, and many other factors.

In relating fire behavior to wildfire control, there are three broad areas of consideration that we need to analyze in determining how a wildfire will act and in planning for its control: fuel, weather, and topography.

Figure 3.7. Fire behavior is determined by the types of fuel, weather conditions, and topography of a particular area.

FUEL

Fuel is any organic material, either living or dead, in the ground, on the ground, or in the air that will ignite and burn. There are infinite combinations of kind, amount, size, shape, position, and arrangement, including the "urba forest," an intimate association of wildfire fuels and human dwellings. The fuels of any general area are referred to as a fuel complex; the flammability of a fuel complex is determined by the burning characteristics of individual materials and their combined effects. Wildfire fuels are both living and dead. The dead fuels ordinarily promote the spread of a wildfire.

Light fuels such as grass, needles, small branches, and most small brush pick up moisture quickly and give it off quickly; these are fast-burning fuels. By comparison, heavy fuels, such as logs, stumps, large branches, etc., take in moisture slowly but much more deeply, and these fuels give up moisture slowly except for the exposed outside layers of dead material; these are slow-burning fuels.

The amount of fuel per acre may vary from a few hundred pounds in the case of light grass to over 100 tons in the case of slash. (Slash is the term applied to limbs, tops, and stumps left after removing logs in timber harvesting.) The volume of fuel determines the total heat that can be developed during a given fire. The total heat volume plays a big part in spreading the fire.

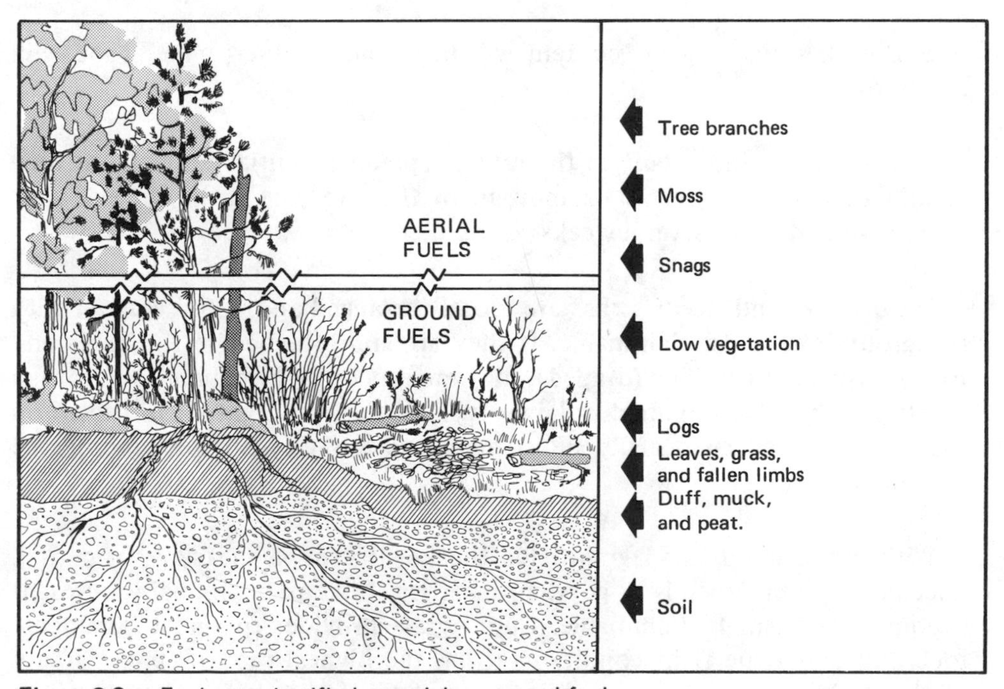

Figure 3.8. Fuels are classified as aerial or ground fuels.

Fuels are classified by location as ground or aerial fuels. Each of these classes is then evaluated for arrangement, size, compactness, continuity, volume, and moisture content. Flash fuels such as grass, moss, and draped needles are those that have the fastest rate of spread; this is because they are light and the ratio of surface area to size is very high. Flash fuels may be either aerial or ground fuels.

FUEL VOLUME

As the volume of fuel increases, the amount of heat produced by a fire in it also increases. Large volumes of fuel increase the difficulty of control not only because of the production of heat and the length of the flames but because of the operational problems of cutting through and separating the fuel. Radiation is intensified by volume, and, if the area contains large volumes of small materials, the spotting potential may be increased. Areas that contain large amounts of both small and large material, especially if part or all of it is dead, indicate a high potential for an extremely hot fire.

GROUND FUELS

Ground fuels are those lying on, immediately above, or in the ground. They may be either dead or living plant materials, including duff, roots, dead leaves and needles, weeds, field crops, sagebrush, grass, fine deadwood, down logs, stumps, large limbs, brush (includes chaparral and chamise, which are important fuels in California and elsewhere), and small trees.

Duff is partially decayed vegetative matter lying against the soil. In forests it may be several inches to several feet deep, and it creates the humus soil beneath it. It may be only a light layer of decaying vegetable matter, as in grass or sagebrush. Usually it is moist and lightly compressed; it is flammable, but it burns slowly. Duff allows fire to creep between other types of fuel, and its smoldering characteristics require specific treatment in control and mop-up.

Soil with a high organic content will burn and smolder; mineral soils will not burn.

Roots are not important to the rate of spread or initial control, but they should be carefully treated in mop-up or the fire can break out and start again several days to several weeks later.

Dead leaves and needles that are loosely arranged and not in contact with the ground are most flammable. If they are still attached to the limbs and freely exposed to the air (draped), they are especially flammable and form a highly combustible kindling for larger fuels. If they are tightly compacted on the ground, they are usually moist and have less oxygen surrounding the individual pieces.

With grass, weeds, and grainfields the key factor is the stage of curing. Succulent green grass is a good fire barrier, but, as it gradually cures, it becomes increasingly flammable. Cured grass provides the most flammable fuel, and under the right conditions it has the highest rate of spread of any of the fuels. Grass occurs in varying amounts in forest types, depending on the maturity of the stand and the amount of sunlight that reaches the ground. If the grass cover is more or less continuous, the rate of spread will be determined by the grass rather than the timber type. If it is mixed with brush, it may develop a surface fire with enough heat to trigger a crown fire under dry and windy conditions. Heavy grass fuels, when they are pushed by a strong wind, develop fires of intense heat and flame and should never be underestimated as to their rate of spread or their ability to roll over barriers. Special safety measures must be observed.

Areas that are well-covered with sage may constitute a very hot flash fuel if sufficient cured grass is intermixed in the stand.

Grainfields burn with intense heat when they are cured or curing; the rate of spread is dependent primarily on wind, because wheat and milo, for example, are very compact. However, corn, beans, and cotton may be more loosely arranged and have a higher rate of spread. Cultivated fuels usually have a large volume of fuel per acre, with resulting high heat production.

Material less than ¼ inch in diameter (such as twigs, dead needles, leaves, small limbs, bark, and rotting materials) is classed as fine deadwood. It is ignited easily and provides the kindling for larger fuels. Timber stands that are decadent, fire killed, bug killed, diseased, or heavily wind blown usually have large amounts of fine deadwood on the ground. Logging slash usually produces the greatest volume of fine fuel, and, under dry conditions, fires in slash produce strong convective currents that can cause spot fires by mass transport of embers.

Dry rotten wood is fragmented and is easily ignited by either flame or embers. It may also cause prolific spotting. Such material requires a control line around it dug down to the mineral soil (see the sections on mop-up in the chapter on tactics).

Down logs, stumps, and large limbs are heavy fuels and require long periods of dry conditions before they become highly flammable. When they do become dry, they can develop very hot fires. They are most dangerous when they contain stringers of dry wood, shaggy bark, or many large cracks and chunks. Smooth-surfaced material is less flammable because it dries out more slowly, has less surface exposed to the air, and has less fine fuel attached to it. Criss-crossed piles of logs and large limbs may develop extremely hot fires; however, the individual logs or limbs may not burn hotly unless large amounts of fine deadwood are present to support the fire.

Low brush and small trees may either slow down or accelerate the spread of fire, depending on the species and its curing stage. Normally, in the early season, when they are growing and shading other ground fuels, they act to slow fire spread, especially when they grow under a tree canopy. As these fuels dry out and cure, the other ground fuels also dry out, and they may become fire carriers. Small evergreen trees are particularly hazardous; fire in this type of understory vegetation often triggers crown fires. Fires may slowly burn through other ground fuels at night or during periods of low fire danger, and in so doing, may dry out the leaves or needles of low brush and small trees. If the fire weather changes to high or above, a reburn may occur, triggering crown fuels if they are present. Reburns usually burn intensely and cause great damage.

Sawdust is a ground fuel that produces a slow-burning, smoldering fire which has few visible signs and is very difficult to extinguish. Often it is necessary to spread the sawdust in thin layers on bare soil and completely soak it with water or allow it to be consumed. The best treatment is never to allow fire to get into sawdust in the first place. There is a definite safety hazard in walking over piles of sawdust containing fire because the burned-out areas underneath the surface may not be visible. This danger also exists for stored grain, hay, and silage, all of which produce fires that are difficult to extinguish. If such fuels are very deep, planks should be used to walk on over the piles, and a rescue rope should be attached with a bowline knot to the fire fighter working the area.

Peat or muck fires are a hazard in certain areas. The Florida Forest Service has prescribed policy and techniques for their control. These fuels are actually soils of organic composition and are classified by the amount of inorganic minerals that they contain and the stage of decomposition of the organic matter. Muck is completely decomposed, with 10 to 50 percent minerals, while peat is not completely decomposed, with 6 to 12 percent minerals. These fuels are formed by the rotting of organic or vegetable materials and are most commonly found in depressions such as poorly drained swamps, dried-up lake bottoms, or the floodplains of freshwater bodies. The depth may vary from several inches to 90 feet. Peat and muck contain oxygen and support slow-burning fire when they become dry enough to kindle. They burn to the depth at which moisture is found. This is called the moisture line and is dependent on drainage, groundwater, the number of days since the last rain, etc. These fires spread laterally at a uniform rate according to the weather factors that affect all fires. Muck fires may originate from wooded areas such as swamps or from open areas such as prairies or dried-up lake bottoms. Most of them burn in a shallow uniform zone, but they may also burn deeply in pits.

AERIAL FUELS

Aerial fuels consist of tree branches, crowns, snags, moss, and high brush. They are physically separated from the earth and from each other, and air can circulate around the fuel particles. They may be green or dead and they form the canopy of forests or tall brush.

The live needles of evergreens are highly flammable because their arrangement on the branches allows for free circulation of air and they contain oils and resins susceptible to ignition. The upper branches are more exposed to sun and wind than ground fuels. Volatile oils and resins in evergreen needles are often fast-burning fuels.

Tree crowns react quickly to relative humidity, and it is rare for a crown fire to occur with high relative humidity. Needles dry out quickly, especially when the ground becomes dry and the amount of moisture produced by transpiration is reduced.

Dead branches found in insect- or disease-killed stands are excellent fire transporters if they are close enough together for radiation and convection to take effect. Dead branches along the base of the trunks may provide an avenue for fire from the ground fuels to the crowns. Branches still containing dead needles are the most flammable and are the fastest transporters of flame to the crowns.

Broadleaf trees can, if the conditions are right, support crown fires, but such instances are rare.

Snags (dead standing trees) provide dry, easily ignited fuel and are particularly susceptible to lightning starts. Smooth, solid snags with little or no bark and few openings and cracks are not as easily ignited as those that are broken-topped, shaggy-barked, or partially decayed. Burning embers from partially decayed snags are prolific starters of spot fires. Fires may spread from one snag to another if they are close enough together to transmit intense heat by radiation. Mass transport of embers may also ignite adjacent snags.

Moss is light fuel and is extremely flashy when it is dry. It reacts quickly to changes in relative humidity, and, during dry weather, crown fires may develop easily in heavily moss-covered stands.

High brush (10 to 20 feet in height) is often classed as an aerial fuel where its crowns are distinctly separated from ground fuels. These brush fields occur in old burns or in cleared areas, and brush often forms the total cover where it is the climax type. Crown fires do not occur unless sufficient ground fuels are underneath to trigger the action or unless the area is close enough to another type that can furnish enough heat to start combustion and there is sufficient wind to maintain the crown fire. In some stands, a sufficient amount of dead stems and branches may be present to allow for fast-spreading crown fires. The chaparral type may appear green but may actually be dry and dormant. In this condition it will burn fiercely when weather conditions are favorable for burning.

Although grass may be as high as 6 feet or more in some places and may have many of the characteristics of aerial fuels, it is arbitrarily classed as a ground fuel.

FUEL CONTINUITY

Continuity is the distribution of fuels over an area or the relation of fuel particles to each other, which may or may not allow heat transfer. Two broad classes of continuity are *patchy* or *uniform*.

Patchy fuels have definite breaks in the ground cover, such as bare ground, rocky outcrops, or plots of vegetation with lower flammability than the general area. With this type of continuity the area of burn and rate of spread will be reduced. Patchy continuity may reduce the possibility of the satisfactory use of fighting fire with fire because the set fire may die out before consuming a sufficient area of fuel.

Uniform fuels are distributed evenly and continuously over the area, thus permitting rapid and uninterrupted spread to occur. A network of stringers, or blocks of fuel that connect with each other and provide a continuous path for the spread of the fire, would be considered uniform fuel. Stringers are narrow areas of fuel that may connect several areas of fuel.

Solid uniformity, as found in wheat fields, may slow the rate of spread, because the insulating effect of the closeness of stems retards radiation. However, the intensity of the heat developed is extremely high and may override the first effect somewhat.

FUEL COMPACTNESS

Compactness refers to the number of individual fuel particles per unit of volume, that is, the proximity of fuel particles to one another in respect to the free movement of air around the particles. Continuity may produce faster and greater heat spread, while compactness could allow less heat and, consequently, less combustion because of lack of oxygen.

Loosely compacted fuels, such as ponderosa pine needles or oak leaves on the ground, will burn and spread fire more rapidly than fir, juniper, or chamise needles, which lie close together and are compact. If the arrangement is such that air freely circulates around the fuel particles, the rate of burning and rate of spread will be faster than if the fuel particles are so close together that their exposure to oxygen in the air is restricted.

Compactness refers mostly to accumulation of leaves and needles on the ground or in piles. It also refers to growing ground fuels.

FUEL MOISTURE

Fuel moisture is a prime factor in judging the burning capability of fuel. It is a product of past and present weather events. When fuel moisture is high, fires are difficult to start, and, when fuel moisture is low, fires start easily and spread rapidly. Temperature, humidity, wind, precipitation, season, time of day, and topographic location all have either direct or indirect bearing on fuel moisture at a given time. Fuel moisture changes more rapidly in dead fuels than in living fuels.

Fuel moisture is not only affected by the several atmospheric conditions but is also affected by the curing stage – whether green, curing, or cured – and by the shade or wind protection received or not received by other vegetation.

Figure 3.9. Fuel moisture, a prime factor in the burning capacity of fuel, is affected by weather conditions.

Green grass feels cool and moist when it is crushed in the hand; it does not burn and is a fire barrier. As dry, hot weather prevails, grasses ordinarily progress through a period of gradual curing. Cheatgrass, as it begins to cure, is typified by a lavender tinge, commonly called the purple stage. When it is cured, it has a flash rate of spread. For most other grasses, only the tops of the blades become tan or brown in color, or only individual blades appear cured. Often a grass stand will appear green in the early curing stages, and only close inspection will determine its point in curing. In the later stages of curing, the stand appears tan or brownish in color, and close inspection is necessary to detect any green in the leaf bases.

Cured grass is completely dried, has a tan or brownish color, and feels dry or crackly when it is crushed in the hand. All grasses have high rates of fire spread when they are cured.

The progress of grass curing varies considerably with the type of grass, the weather conditions, and the soil. In the deep moist soils bordering creeks, curing is much slower than in the thin soils on slopes. Some grasses do not ordinarily reach a dangerous cured condition. Bear grass, found at high elevations, and sedges of high mountain meadows do not cure like common bunch grasses.

Shade and protection from the wind provided by forests and heavy tall brush keep ground fuels from drying out. Large variations in fuel moisture occur because of aspect, altitude, weather conditions, time of day, and time of year.

Fuel temperatures cannot readily be measured, but some rules of thumb are helpful:

- The rate-of-spread factor is doubled long before the fuel temperature is doubled.
- Fine fuels are quickly heated by direct sunlight and air temperature.
- During the hottest part of the day, all fuels on the south and west slopes will have higher fuel temperatures than those on the north and east slopes.
- Heavy fuels usually have a lower fuel temperature than surrounding fine fuels in the daytime, but the reverse is true at night.
- Ground fuels will usually have higher temperatures than aerial fuels when they are exposed to sunlight.

WEATHER

The local fire weather elements are wind, moisture, and temperature. Each of these elements contributes to the atmospheric stability of an area. They are constantly changing owing to varying pressure systems and the changing properties of huge masses of air that move in generally predictable patterns over the earth's surface. The changes in the pressure systems account for many of the short-term changes in weather as well as long-term trends and seasonal changes.

There is a net transfer of heat energy from equatorial to polar regions, which causes a circulation of air between the two regions. The movement of the earth adds another force (Coriolis force) to the movement of air. The earth is moving on its axis toward the east and turns underneath the moving

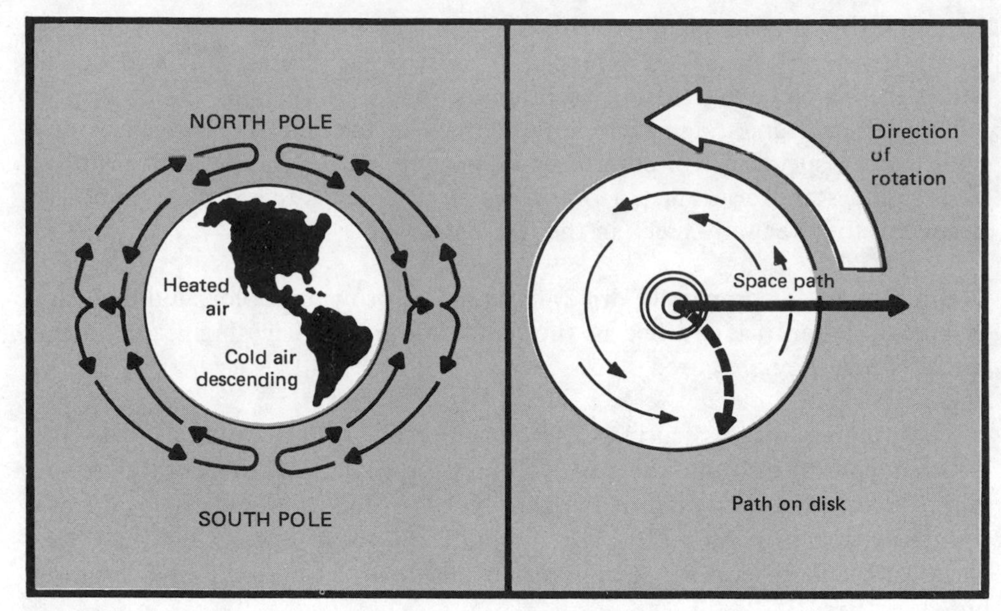

Figure 3.10. (*Left*) If general circulation could be represented by a simple convective circulation, heated air at the equator would rise to near the tropopause and would then spread out both north and south. The air would cool toward the poles and finally would descend and move back to the equator. (*Right*) A ball tossed horizontally from the center on a large disk rotating counterclockwise will travel in a straight line in space. However, because of the Coriolis force, the path in which the ball will travel will be deflected to the right.

air from the equator to the poles. The basic circulation system results in the formation of several well-defined major circulation patterns, or wind belts.

Atmospheric pressure, which determines wind patterns, is the weight of the atmosphere, or air mass, against the surface of the earth due to the pull of gravity. It is measured by balancing the weight of the atmosphere against that of a column of mercury. This measurement can be made with either a mercurial or an aneroid barometer. Atmospheric pressure decreases with altitude. At sea level it is about 15 pounds per square inch. When the barometric pressure is reported as 30.12 inches, it simply means that the pressure is sufficient to balance a column of mercury 30.12 inches tall. The layer of air over the surface of the earth weighs about 56 trillion tons and is equal to a layer of water 34 feet deep over the entire globe.

SYNOPTIC WEATHER CHARTS

There are several well-defined major circulation patterns, or wind belts, in the atmosphere, generally above 2,000 feet, and these are due to pressure patterns. To study pressure distribution, we need pressure measurements taken simultaneously at a number of stations. Since the stations are at different altitudes, a correction is made to a common level, usually sea level. The corrected readings are collected at a central point and are plotted on a weather map or chart. Lines, called isobars, are drawn through points of equal pressure; they are similar to contours on a topographic map.

A synoptic chart will show areas of lower pressure than the surrounding region. These areas are called lows or cyclones because the air flows around them in a cyclonic direction, which is counterclockwise in the northern hemisphere. Lows are characterized by convergence, which is inward and

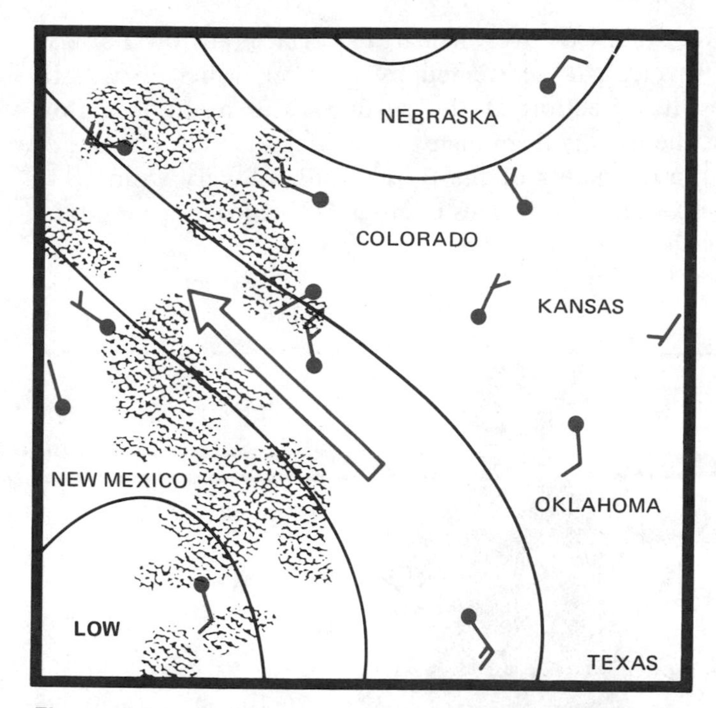

Figure 3.11. This is a typical surface weather map. Surface wind direction is indicated on weather maps by arrows flying with the wind. At the top of the friction layer the wind blows parallel to the isobars, as shown by the large arrow.

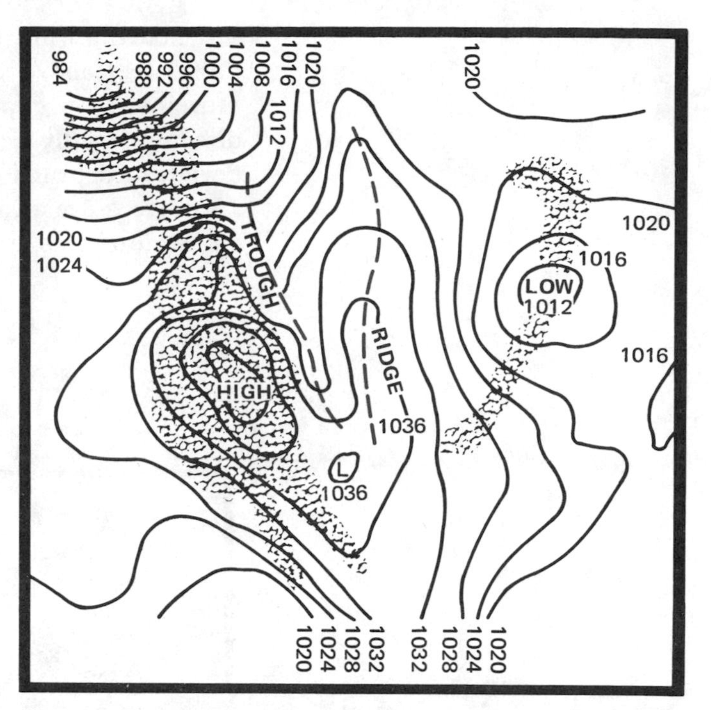

Figure 3.12. This synoptic weather map shows a line of low pressure (trough) and a line of high pressure (ridge). The lines are drawn through points of equal pressure and are called isobars.

rising air motion that results in cooling and increased relative humidity. If the air mass is lifted high enough and if it contains adequate moisture, condensation of the water vapor will produce clouds and may result in precipitation.

A line of low pressure is referred to as a trough. The pressure along the line is lower than the pressure on either side, and storm centers commonly develop in such places.

A high is an anticyclone: the winds move in a clockwise direction in the northern hemisphere, and they are surrounded on all sides by lower pressure. The airflow is generally outward and descending (horizontal divergence). Highs are areas of minimum cloudiness and little or no precipitation. If the air descends from very high altitudes, it may be quite warm and extremely dry (subsidence). When the wind speeds are high, this condition can provide extreme fire weather conditions. Subsidence is associated with foehn winds. It is much more frequent in the western half of the country than in the eastern half.

Ridges are lines of high pressure. The pressure is higher along the ridge than on either side. Ridges show characteristics similar to those of highs: descending air and a minimum of cloudiness and precipitation.

Airflow is usually parallel to the isobars and is termed gradient flow. Usually there are much higher wind speeds in the lows than in the highs. The direction of flow is always clockwise around a high pressure center and counterclockwise around a low.

Rule of thumb: With your back to the wind, high pressure is on the right.

When the wind blows against or over the earth's surface (airflow 1,500 to 2,000 feet above the surface), it is affected by friction, which retards its movement or changes its direction at the surface. The result is a flow directed slightly across the isobars from high to low pressure. The amount of deviation depends on the roughness of the terrain and will vary from 10 to 15 degrees over water to 25 to 45 degrees over land. Hurricanes, tornadoes, and waterspouts are special forms of low-pressure systems.

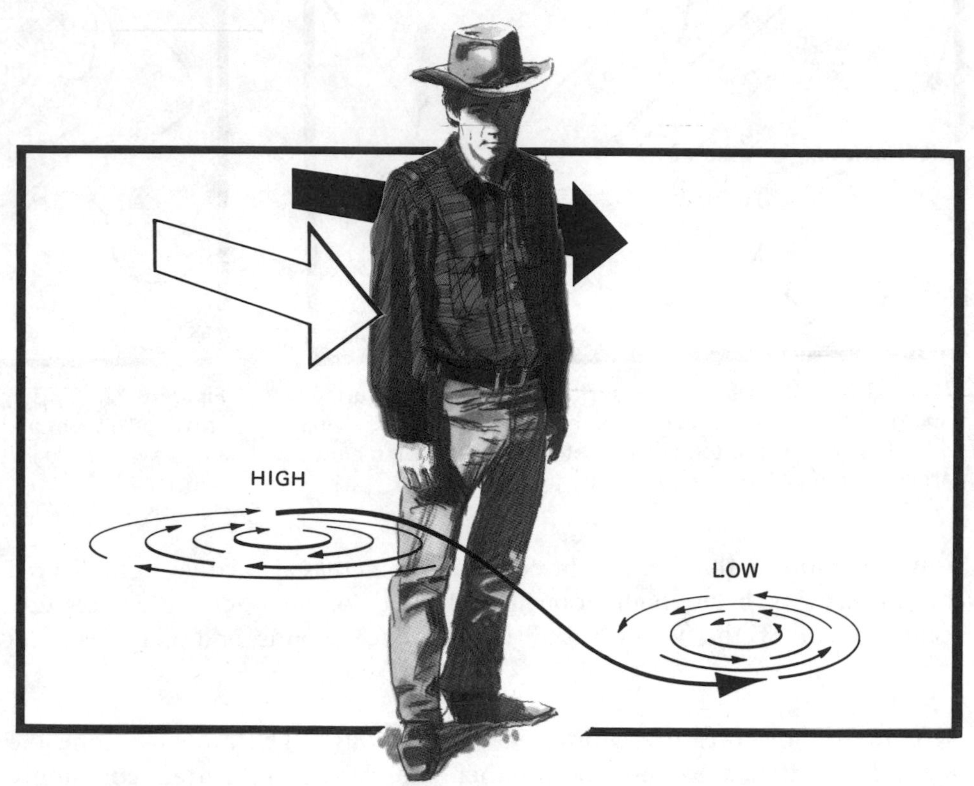

Figure 3.13. With your back to the wind, high pressure is on your right and low pressure is on your left.

WIND

Wind is the most variable and least predictable fire weather element. However, its behavior can be estimated by close study. Winds near the earth's surface are affected by the shape of the topography and by local heating and cooling. There is no substitute for an adequate understanding of local wind behavior.

Wind affects wildfires in several ways:

- Wind carries away moisture-laden air and speeds up the drying of fuels.
- Light winds aid certain firebrands in igniting fire.
- Once a fire is started, winds aid combustion by increasing the oxygen supply.
- Wind spreads fire by carrying heat and burning embers to new fuels and by bending the flames closer to the unburned fuels ahead of the fire.
- The direction of spread of fire is determined mostly by the wind.

WIND CLASS	TERMS	EFFECTS OF WIND
Less than 1 mph	CALM	Smoke rises vertically; no movement of leaves of bushes or trees.
1 to 3 mph	VERY LIGHT	Leaves of quaking aspen in constant motion; small branches of bushes sway; slender branchlets and twigs of trees move gently; tall grasses and weeds sway and bend with wind; wind vane barely moves.
4 to 7 mph	LIGHT	Trees of pole size in the open sway gently; wind felt distinctly on face, loose scraps of paper move; wind flutters small flag.
8 to 12 mph	GENTLE	Trees of pole size in the open sway very noticeably; large branches of pole-size trees in the open toss; tops of trees in dense stands sway; wind extends small flag; a few crested waves form on lakes.
13 to 18 mph	MODERATE	Trees of pole size in the open sway violently; whole trees in dense stands sway noticeably; dust is raised in road.
19 to 24 mph	FRESH	Branches are broken from trees; inconvenience is felt in walking against wind.
25 to 38 mph	STRONG	Trees are severely damaged by breaking of tops and branches; progress is impeded when walking against wind; structural damage; shingles are blown off.

Figure 3.14. The Beaufort scale of wind velocity.

Fire itself may affect the wind by modifying its characteristics and behavior.

General winds. General winds as considered here are the local winds that are produced by the broad-scale pressure gradients shown on synoptic weather charts; but they may be modified considerably by friction or other topographic effects. Local winds produced by local temperature differences will be considered convective winds. These cannot ordinarily be detected on synoptic weather charts.

Wind is the movement of air across the land surface, both horizontally and vertically. We tend to think of its action only as horizontal, but the up-and-down effects are just as important in wildfire control. Wind (moving air) is a fluid that can be compressed under pressure, expanded and contracted with heat and cold, and made moist or dry. It may pause, unmoving, and then may spring in any direction with violent gusts. Wind behavior is directly affected by known meteorological influences and by friction with the earth's surface.

Wind has three qualities fire fighters must determine at the scene of a fire. These are:

1. Velocity—the speed of the wind expressed in miles per hour (mph).
2. Direction—expressed as the direction from which it is blowing (southwest, east, etc.). Exceptions are the designations upslope, downslope, upvalley, downvalley, onshore, and offshore, which state the direction to which the wind is blowing.
3. Gustiness (turbulence)—short variations in velocity and direction. Gustincss is irregular motion of air; it can cause appreciable vertical movement.

Wind speed is measured by an anemometer. The standard height for measurement is 20 feet above the ground. Belt weather kits are also available with wind speed indicators and a small sling psychrometer for measuring weather factors in the field.

Wind direction can be determined by a wind vane. The broad tail on the vane makes it point into the wind. Wind direction may also be observed by the direction in which smoke or dust is blowing.

At regular weather stations, the direction of winds aloft is determined by tracking a gas-filled balloon from the surface up through the atmosphere or by use of a radiosonde unit that transmits temperature, moisture, and pressure data back to the ground. The rawinsonde unit is similar but is much more refined and sophisticated and transmits additional information.

A steady even flow is called laminar wind, and it affects the pattern of the fire by producing a sustained spread in one direction. The fire will form a narrow pattern if the wind speed is high or a triangular shape if the wind speed is moderate. These patterns will also be affected by topography. Laminar winds flow smoothly along with negligible vertical mixing. They are most likely to occur at night and are frequently observed over plains and rolling topography.

Gusts or turbulent winds have a great deal of vertical mixing–they cause erratic fire behavior and spread the fire in several directions. Rough topography aids turbulence.

Turbulence is caused by mechanical obstructions and thermal action. Mechanical turbulence is caused by surface friction of the wind with the ground. As the wind speed increases and the wind finds its way around and over hills, ridges,.trees, and other obstacles, eddies are set up in all directions. Wind speed and roughness of the surface increase mechanical turbulence.

Thermal turbulence is generated by instability of the atmosphere and convection activity. It is similar to mechanical turbulence in its effects on surface winds, but it extends higher in the atmosphere. It is increased by the intensity of surface heating and the instability of the atmosphere. It is affected by the diurnal change; that is, it is most pronounced in the early afternoon and is at a minimum during the night and early morning.

Figure 3.15. Turbulent wind flow is caused either by surface roughness (mechanical turbulence) or intense surface heating (thermal turbulence).

Mechanical and thermal turbulence often occur together and tend to increase the effects of each other. Consequently, surface winds are often stronger in the afternoon than at night and usually produce spurts and gusts.

Eddies are commonly produced by mechanical and thermal turbulence on the lee side of all solid objects. The size, shape, and motion of eddies are determined by the size and shape of the obstacle, the speed and direction of the wind, and the stability of the atmosphere next to the land surface. Eddies may be either vertical or horizontal. A whirlwind, or "dust devil," is a vertical eddy. Eddies may occur at the mouths of canyons and around buildings. Large cylindrical eddies that roll along the surface like a barrel are horizontal eddies.

Rule of thumb: The effect of the eddy is eight to ten times the height of the obstruction.

Frontal winds. A front is the boundary between two air masses of differing temperature and moisture conditions. The passage of a front is usually accompanied by a shift in wind direction, and this change is most important in fire control. Winds shift in a clockwise direction.

If a cold air mass is replacing a warm air mass, the boundary between the two masses is a cold front. The wind change in a cold front is usually sharp and distinct, and the air may be so dry that few, if any, clouds accompany

Figure 3.16. Eddies form on the lee side of solid objects. They vary with the size and shape of the object, the speed and direction of the wind, and the stability of the lower atmosphere.

the front. Ahead of the cold front the wind direction is usually from the south or southwest.

As the front approaches, the wind increases in speed, is often gusty, and may become violently turbulent. The wind shift is abrupt and may be less than 45 degrees or as much as 180 degrees. After passage of the front, the wind direction is usually west, northwest, or north. The gustiness may continue for some time if conditions remain unstable. A cold front is the worst type for wildfires because the winds push against the ground, holding the fire front firmly against new fuel. Usually fronts generate precipitation and thunderstorms. On occasion they will cause neither, and in these instances the winds caused by the fronts can be very significant to fire behavior.

A warm front occurs where a warm air mass replaces a cold air mass; these are more prevalent east of the Rockies. Warm fronts are fewer in the mountainous West, and they are more erratic and tend to scatter and spread. The surface wind ahead of a warm front usually blows from a southeasterly or southerly direction. The shifts are gradual as compared to the abrupt shifts of cold fronts. As the front passes, the wind change is from 45 degrees to 90 degrees and after passage it usually blows from the southwest. Winds tend to be steady rather than gusty, because warm front conditions are generally stable.

With an occluded front the wind shift is usually 90 degrees or more and changes from the south to the west or northwest. The wind change may resemble either a cold or warm front, depending on whether the air behind the occlusion is warmer or colder than the air ahead. The violent turbulence that may accompany a cold front is usually absent.

The knowledge of the existence of a front can dictate the tactics to be used in control of a fire or to avoid endangering crews on the fire line. A flanking attack may be the only logical tactic.

Squall lines often precede cold fronts in the area east of the Rockies; they are most common in spring and summer. They are narrow zones of

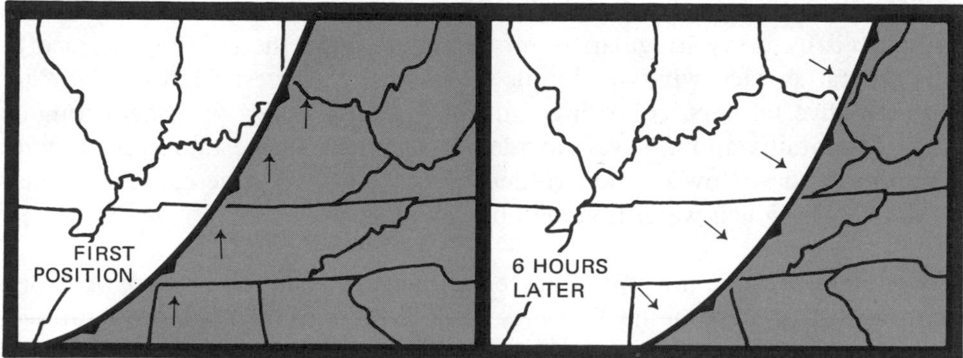

COLD FRONT

Figure 3.17. When a cold front passes through an area, the wind direction abruptly shifts, usually from south or southwest (*left*) to north, northwest, or west (*right*).

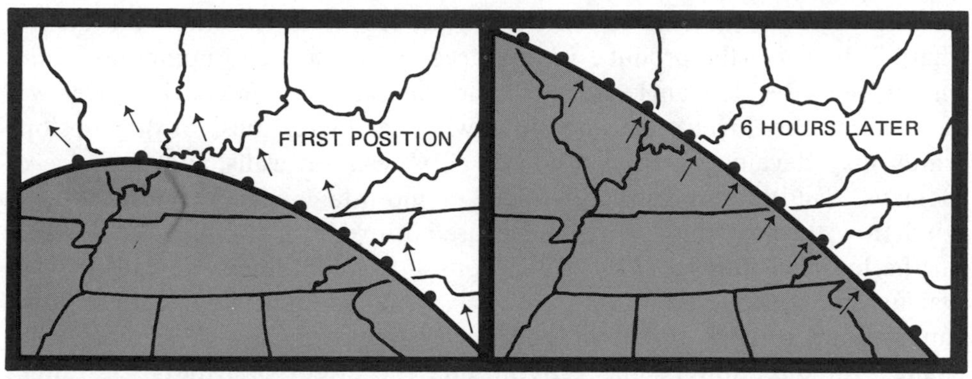

WARM FRONT

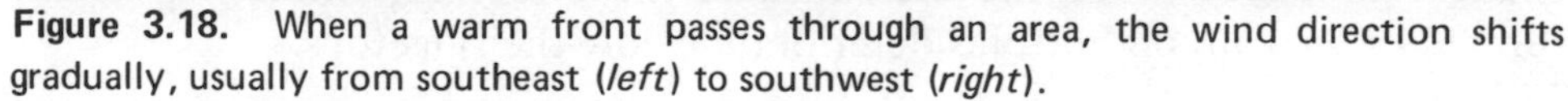

Figure 3.18. When a warm front passes through an area, the wind direction shifts gradually, usually from southeast (*left*) to southwest (*right*).

OCCLUDED FRONT

Figure 3.19. When an occluded front passes through an area, the wind direction shifts 90 degrees or more, usually from south (left) to west or northwest (right).

instability that form ahead of and parallel to a cold front and may have extremely violent surface winds. They often occur with severe lightning storms in the Midwest. Squall lines develop quickly in the late afternoon or evening and they move rapidly, dying out by late evening or early morning. Winds are from the south ahead of the line, shift to the west or northwest, and become extremely gusty with speeds of 30, 40, or even 60 mph. The winds do not last long and soon revert to their original speed and direction. Squall lines are usually accompanied by thunderstorms and rain, which may be so scattered along the line that a local area may not receive the benefit of the rain and yet may experience the strong gusty winds and the resultant increase in fire activity. About forty-five minutes or so after the passage of the squall line the winds return to about what they were ahead of the line. This feature distinguishes a squall line from a front.

Effects of topography on wind. Winds blowing over the land surface are influenced by every irregularity. Mountains provide the roughest surface to the general surface winds and thus produce the greatest friction. They are also effective barriers, collecting and holding cool heavy air or deflecting its flow. Mountains and valleys provide the channels that establish local wind direction, and airflow is often guided by principal drainage channels similar to the way in which water is guided by channels.

General winds are most pronounced at the surface in the absence of strong heating; strong daytime heating may replace the general wind at the surface. Ridgetop winds tend to be somewhat stronger than winds in the free air at the same level.

Round top ridges tend to distort surface flow the least. In light to moderate winds, there is often little evidence of any marked turbulence. Sharp ridges usually produce pronounced turbulence and numerous eddies on the lee side. The angle at which the general wind strikes the ridges will also affect the turbulence and downwind characteristics. Strong upslope winds may develop on the lee side of cliffs, canyon walls, and the ridge of plateaus. Saddles and passes in ridges or mountain chains concentrate the wind flow through these vents and cause both vertical and horizontal eddies on the lee side. Careful analysis and constant watchfulness are called for in placing and working crews in such areas. General winds blowing up or down canyons are usually turbulent because of the contact with canyon walls, changes in direction of the canyon, and the effect of tributaries. Eddies formed in the canyon bottoms deteriorate below the ridgetops.

Mountain waves. Moderate to strong stable winds form wave characteristics after passing over high mountain ranges. These waves become smaller and smaller downstream until they disappear. But they may extend as high as 40,000 feet, as in the case of the Bishop wave in California. Large-scale waves also occur in the Rocky Mountains and to a lesser extent in the Appalachians and elsewhere.

Within and beneath each wave downstream there will be a roll eddy, with its axis parallel to the mountain range. These become smaller as they form farther from the mountain range, but they affect surface wind speed and sometimes direction.

Figure 3.20. Sharp ridges in mountains cause turbulence in general surface winds and eddies on the lee side of the mountain.

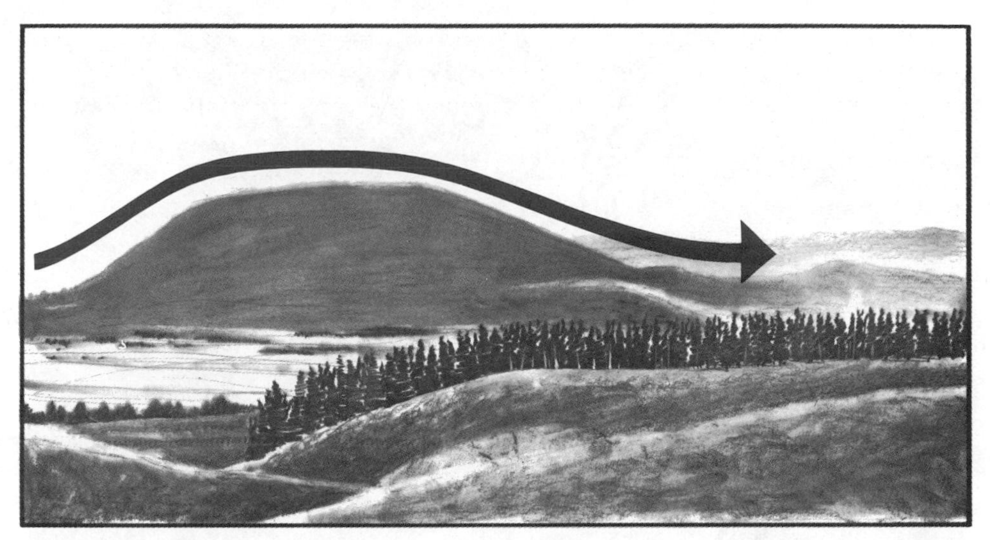

Figure 3.21. Rounded hills do not greatly disturb surface airflow, and, in light to moderate wind, there may be no marked turbulence.

Figure 3.22. Cool dense air settles in the bottom of canyons and valleys creating an inversion. It increases in strength and depth during the night. Cool air flows outward over the valley bottom.

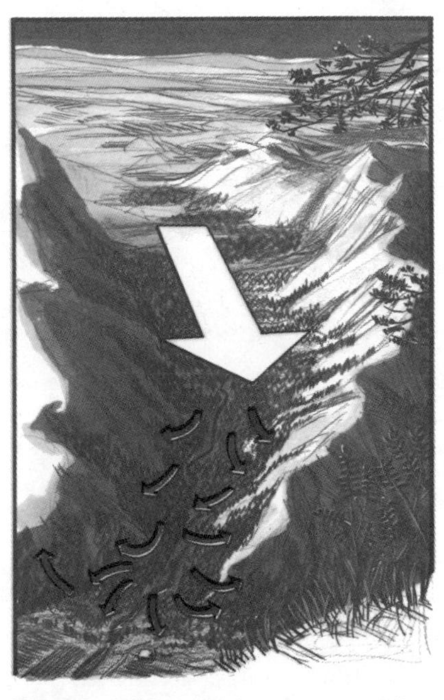

Figure 3.23. Ridge top saddles and mountain passes form important channels for general wind flow. The flow is compressed and the wind speed increases in the passes. Horizontal and vertical eddies form on the lee side of saddles.

If sufficient moisture is present, cap clouds will form over the crest of the mountains, roll clouds will form on top of the roll eddies, and wave clouds will form on top of the waves.

Foehn winds. Foehn winds are a special type of local wind formed by mountain ranges and downflowing wind that is warm and dry. These winds flow from a high-pressure area on the windward side of mountains to a low-pressure or trough area on the leeward side. They cause severe fire weather.

Surface wind speeds of 40 to 60 mph are common, and speeds of 90 mph and more have been reported. These winds often last for three days or more and usually stop abruptly. The temperature increases rapidly and may change 30 to 40 degrees in a few minutes. The relative humidity often drops

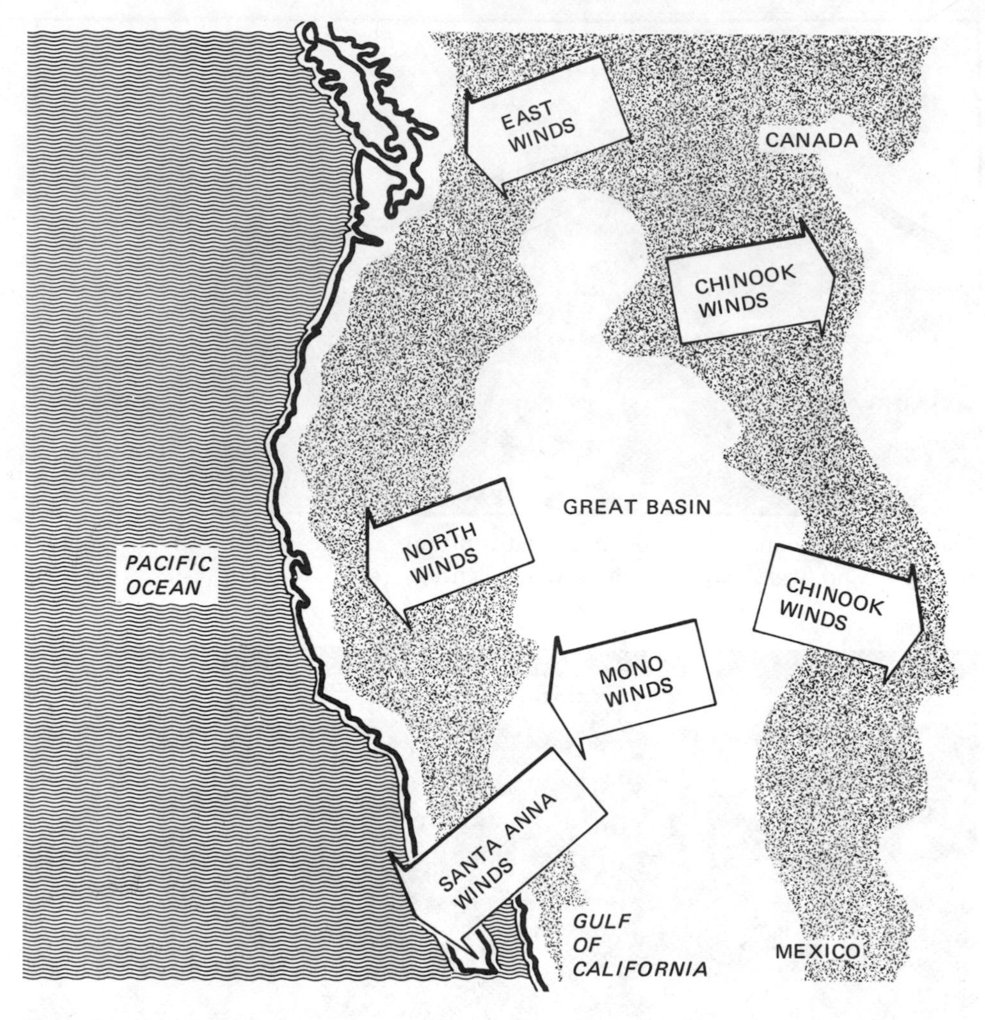

Figure 3.24. Foehn winds are known by different names in different parts of the mountainous west. In each case, air is flowing from a high pressure area on the windward side of the mountains to a low pressure or trough area on the leeward side.

quickly to 5 percent or less with foehn winds. This type of wind can either override the cooler air on the lee side of the mountains, thus affecting only the upper elevations, or it may press down all the way to the surface or alternately dip down to the surface.

The Santa Anna of southern California are notorious winds, and, because of the warm temperatures, low humidities, and sustained flow in a region of flashy fuels, they produce serious fire weather and disastrous conflagrations.

Slope and valley winds. Slope winds are local winds caused by the temperature changes brought about by day and night. The flow is up the slope during the day because of surface heating and down the slope at night because of surface cooling. Ravines and draws facing the sun act like chimneys. Upslope winds are shallow but increase in depth toward the upper portion of the ridge. At the top of the ridge turbulence can be expected because the flow from both slopes converges at the top. Downslope winds are shallow and of a slower speed than upslope winds, and they tend to be laminar.

Cool dense air accumulates in the bottom of canyons and valleys, creating a night inversion that increases in depth and strength during the night hours. Downslope winds from above the inversion continue downward until they

Figure 3.25. A strong Foehn wind may flow down the leeward side of the mountains, bringing warm and extremely dry air to lower elevations. The Foehn wind may hug close to the surface, be pushed aloft near the valley floors, or dip down alternately.

reach air of their own density. There they fan out horizontally over the canyon or valley. This may be either near the top of the inversion or some distance below the top.

Valley winds caused by diurnal changes (day and night differences) can drastically affect fire behavior. Valley winds flow upvalley by day and downvalley by night. They are the result of local pressure gradients caused by differences in temperature between air in the valley and air at the same elevation over adjacent plains.

Aspect (orientation to the sun) is an important factor that governs the strength and time of day of slope and valley winds.

Figure 3.26. The direction of slope winds changes with daytime and nighttime temperature changes. Upslope winds occur during the day. They are shallow near the base but increase in depth and speed near the top of the slope. Downslope winds occur at night. They are shallow and they tend to be laminar.

Figure 3.27. Valley winds reverse direction from day to night as a result of pressure gradients caused by temperature differences.

South and southwest slopes receive the greatest amount of heat and have the strongest upslope winds. South slopes have maximum wind speeds about midday, while southwest slopes reach their maximum about midafternoon. South upslope winds may be several times greater than those on north slopes.

Bare slopes and grass-covered slopes heat up more readily than those covered with trees or brush. Consequently, the upslope wind is greater on grass slopes than on tree- or brush-covered slopes. Often, because the surface is shaded by trees or brush, there is a very shallow downslope flow at the surface, while the upslope flow overrides the treetops.

Where an open space exists between the crowns and the ground in a forest stand, the downslope flow will be confined to this space while calm prevails in the crown region. A forest with a dense understory is an effective barrier to downslope winds. The flow may thus be diverted around dense areas through stream channels, roadways, and other openings in the forest cover.

Effects of vegetation on winds. Vegetation forms part of the rough surface that contributes to turbulence, eddies, etc. There is some air movement through its composition as well as over and around the vegetation.

Wind speed is zero at the very surface of open level ground but increases very fast in the first 20 feet above ground. When the surface is covered with dense vegetation, such as grass or brush, the effective friction surface is considered as the average height of the vegetation because the air flowing through it is negligible. However, airflow within and below the tops of trees in a forest is important. The crowns of trees (branches and leaves or needles) are very effective in slowing down wind movement because of their large friction area. If the branches extend to the ground or if the understory is full of vegetation, the wind speed is nearly constant from just above the ground to the tops of the stand, where the wind speed will increase much the same

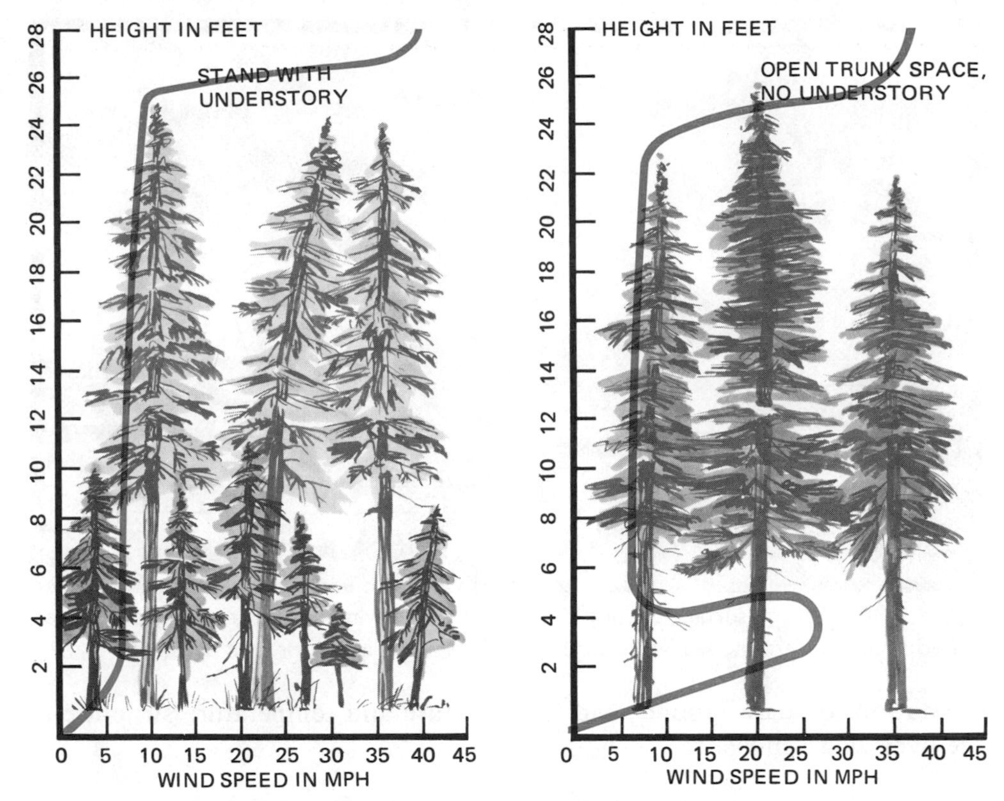

Figure 3.28. Vertical wind profiles in forest stands show that the crown canopy slows wind movement. Stands with an understory will have wind speeds nearly constant from just above the surface to the tops of the crowns. Above the crowns the wind speed increases much the same as above level ground. In stands without an understory and with open trunk space (clear of moss and limbs), wind speed will increase in the trunk space and be at a minimum in the crown space.

as it does over level ground. If the forest is open below the crowns, the air speed will increase above the surface to a point midway on the average trunk height and will then decrease in the area of the crowns or tops.

The drag of any friction surface is relatively much greater at high wind speeds than it is at low wind speeds. Although trees that have shed their leaves have a significant effect on reducing wind speeds, the reduction is much greater when they are in full leaf.

Thermal turbulence is an added problem in those areas that receive direct sunlight; the result is a general upflow of warm air and a gentle inflow from the shaded areas. Consequently, the surface wind direction in a stand of trees can be opposite to the direction of the wind above the treetops.

Local eddies are common in forest stands and may be found on the lee side of each tree trunk. They directly affect the behavior of surface fires. Openings in the fuel type cause larger eddies, and the edges of tree stands cause roll eddies, as in the case of sharp ridges and bluffs.

Figure 3.29. Eddies form on the lee side of tree trunks and affect the behavior of surface fires.

Convective winds. As used here, convective winds refer to all winds—up, down, and horizontal. These are the winds caused by local temperature differences and they may dominate the weather or be overriden and wiped out by the general winds. Since convective winds are dependent on temperature, all features of the environment that affect temperature are important. These include the season of the year; changes between night and day; cloud cover; nature of the terrain and its cover, such as water,

Figure 3.30. Strong surface heating develops upslope winds along heated slopes. Superheated air in flat terrain escapes upward in bubbles, whirlwinds, or dust devils.

vegetation, or bare ground; and the moisture and temperature structure of the overlying atmosphere.

Strong surface heating produces several kinds of convection systems. Upslope winds develop along heated slopes. Superheated air in flat terrain escapes upwards in bubbles or in the form of whirlwinds or dust devils. As land surfaces become warmer than adjacent water surfaces during the daytime and air over land becomes warm and buoyant, the denser sea breeze begins to flow inland from over the water and forces the less dense air upward.

At night, land surfaces cool more quickly than water surfaces. Air in contact with the land becomes cool and flows out over the water as a land

Figure 3.31. Openings in a timber stand tend to act as chimneys under conditions of strong daytime heating and light winds.

breeze, displacing the warmer air. General winds along an irregular or crooked coastline may oppose a land or sea breeze in one sector and support it in another.

Whirlwinds. Whirlwinds or dust devils are a definite indication of intense local heating, turbulent changing wind speeds and directions, and unstable weather. Whirlwinds are classed as larger than dust devils. They occur on hot days over dry surfaces when skies are clear and general winds are light. They develop an upward spiraling motion of strong convective force by pulling in hot air from the surface layer. Whirlwinds may remain stationary or move with the surface wind, depending on the triggering mechanism. Some may last a few seconds, and many last several minutes; a few have lasted for several hours. They may range in diameter from 10 feet to over 100 feet, with heights ranging from 10 feet to 4,000 feet. The speed of the wind inside is often more than 20 mph and in some cases exceeds 50 mph. Upward speed can be as high as 25 to 30 mph and can pick up fair-sized debris.

Whirlwinds are common in areas that have just burned over, and they can bring an apparently dead fire back to life. They can also trigger mass transport of embers to new fuel.

Firewhirls. The heat produced by wildfires causes extreme instability in the lower air and may trigger violent firewhirls. These are much more severe than whirlwinds and have been known to twist off trees 3 feet in diameter. They can carry large burning embers out of the main fire area and start numerous spot fires. As soon as the firewhirls move out of the fire area, the flame dies out, and they become ordinary whirlwinds. They usually occur in heavy fuels with attendant large volumes of heat in relatively small areas.

Thunderstorm winds. The updraft and downdraft winds inherent in cumulus cloud cycles are true convective winds. Thunderhead clouds (cumulonimbus)—especially the anvil head clouds in the later stages of development—are definite danger signals to wildfire fighters. The winds associated with these clouds can be disastrous to the fire control action.

There are three stages of thunderstorm cloud development. In the first, or cumulus, stage, updrafts are strong, sometimes reaching 30 mph. The air in these updrafts is drawn from the heated air near the surface, from fires, and from surrounding air. This air may increase the upslope winds around the ridges and peaks. The updraft and downdraft winds inherent in cumulus cloud cycles are true convective winds. The cloud grows visibly in size.

In the second, or mature, stage, rain begins to fall from the base of the cloud. Now downdrafts begin because of the drag caused by the raindrops pulling air with them. This occurs ten to fifteen minutes after the cloud cell has built upward beyond the freezing level. The cloud reaches a height of 20,000 to 30,000 feet and may reach 50,000 to 60,000 feet. The cloud top flattens and forms the characteristic anvil top. As the downdraft approaches the surface—especially a flat surface—the colder air in it may pile up and then spread out horizontally as a small, intense cold front. The outflow is strong and highly turbulent and is known as the "first gust." This causes a sharp change in wind direction and an increase in speed that is most pronounced on the forward side of the thunderstorm. The temperature at the ground may drop as much as 25 degrees, with an accompanying rise in

Figure 3.32. The three stages of thunderstorm cloud development are shown. In the *cumulus stage,* strong updrafts are characteristic. As the cloud moves over the fire, the convection column and the updrafts reinforce each other. The indraft to the fire is strengthened, and spotting possibility is increased. In the *mature stage,* rain begins to fall out the base of the cloud and starts a downdraft. An updraft is still occurring in part of the cloud. This causes sharp changes in wind direction and increases wind speed. The downdraft is the main feature in the *dissipating stage,* and updrafts stop. As rain ends, the cloud evaporates.

surface pressure in the first gust. Lightning frequency is at its maximum. Heavy rain and strong gusty winds at the surface are typical, but precipitation may not reach the ground. Rain that shows beneath the bottom of the cloud but does not reach the surface is called virga. This is a definite signal of erratic wind behavior and increasing wind speeds.

In the dissipating stage the downdrafts continue to develop, and the updrafts weaken. Finally, there are only the downdrafts, and the cloud either dissipates completely or changes into other forms.

Squall lines often develop ahead of or along with thunderstorms in the West. These have the same characteristics as squall lines ahead of fronts: the winds are strong and gusty, they begin and end quickly, and they may extend many miles beyond the original storm area.

As thunderheads approach a fire, always be on the lookout for increased and erratic fire behavior.

ATMOSPHERIC STABILITY

Stability of the atmosphere near the surface is indicated by steady surface winds, haze and smoke that tend to hang near the ground, lowering fog, clouds in layers (stratus type) with little or no indication of vertical movement, and smoke columns that rise short distances and then drift horizontally. Such stability is beneficial to wildfire control.

Figure 3.33. A stable atmosphere.

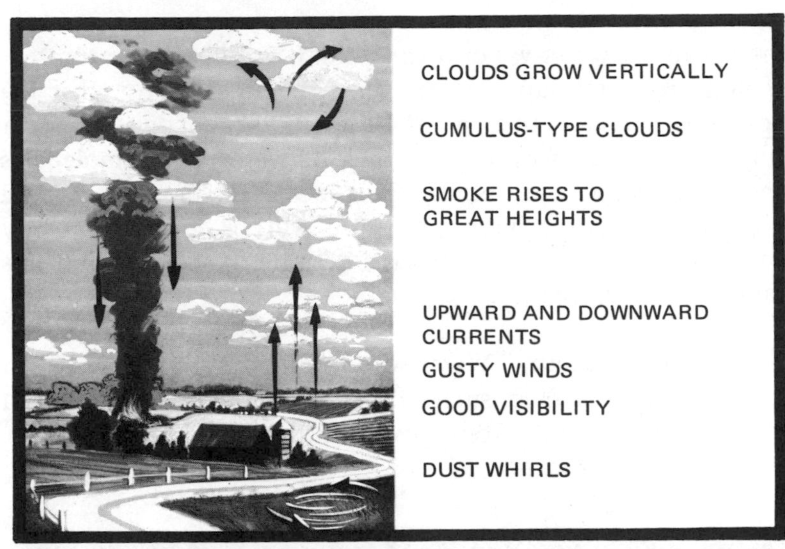

Figure 3.34. An unstable atmosphere.

Instability is indicated by turbulence (especially dust devils), mirages, cumulus-type clouds that rise vertically, good visibility, and smoke that rises vertically to great heights. It is characterized by both upward and downward currents. Instability is detrimental to wildfire control.

Lenticular, or lens-shaped, clouds that appear over ridges and to the lee side of mountain ranges are indicators of high winds aloft. These lens-shaped clouds appear over waves in the wind pattern. They may be advance warning of increased wind speed on the ground, and they should be regarded as a danger sign.

Cumuliform cloud masses in the form of turrets arranged in lines are termed castellanus. When they occur in the morning, they are often a warning of thunderstorm activity in the afternoon.

Cumulus clouds in any stage or form should be regarded as indicators of possible increase in wind speed and change of wind direction.

MOISTURE

Humidity is atmospheric moisture, or invisible water vapors in the air. Moisture in the atmosphere may be in any one of three forms: solid, in the form of ice crystals, snow, hail, or sleet; liquid, in the form of rain, dew, or drizzle; or invisible vapor, referred to as humidity. If the moisture occurs in solid or liquid forms, the control job at a fire is made easier (depending, of course, on the extent of precipitation). Moisture in the air is important because it affects the amount of moisture in the fuel, especially dead fuels. Water vapor in the air varies by time and place from near 0 to 5 percent of the air by volume. Water vapor acts as an independent gas when mixed with air.

Like air, water vapor has weight, which exerts pressure. The amount of pressure that water vapor exerts, independent of dry air pressure, is called vapor pressure. However, water vapor is lighter than dry air, so moist air is lighter than dry air; this occurrence is contrary to the fact that most things are heavier when they are moist.

When air contains all the water vapor it can hold, it is said to be saturated. The amount of pressure water vapor exerts when the air is saturated is the saturation vapor pressure. At this point, the number of water vapor molecules leaving the air balances the number of water vapor molecules entering the air. If the temperature of the air is increased, the air can hold more vapor, and the saturation vapor pressure will be greater. At any temperature the saturation vapor pressure has a definite fixed value, but this value changes rapidly with temperature.

Evaporation and transpiration. Water vapor in the air comes from evaporation from any moist surface, such as a body of water, the soil, dead plant material, or transpiration from plants. It is also produced by combustion.

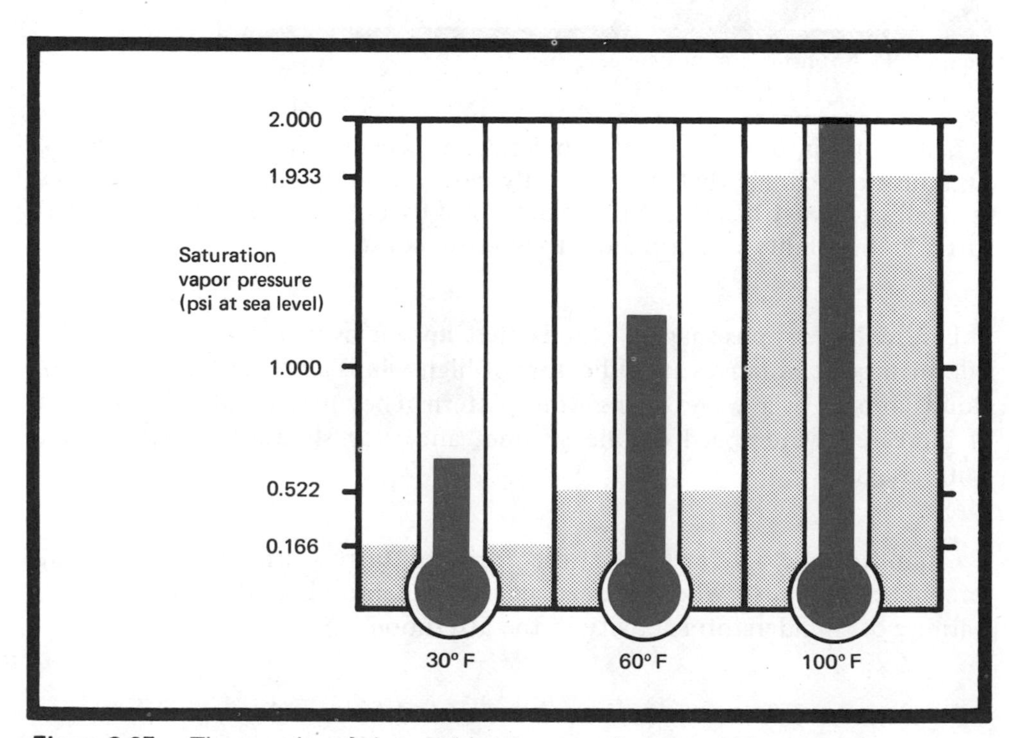

Figure 3.35. The capacity of air to hold water vapor increases with temperature.

Oceans are the most important source of water vapor, but land is also an important source. Evaporation from warm water surfaces, warm soil, and dead plant material is greater than that from cold surfaces, given the same atmospheric pressure. Wind encourages evaporation from free water by blowing away the saturated layers of air that are concentrated around the evaporating surface, replacing these layers with drier air. However, when the evaporating surface, such as wood or soil, becomes dry, wind may reverse the process by cooling the surface and thus lowering the vapor pressure of moisture that the surface contains.

Transpiration from living plants is usually highest during warm weather, but this is subject to the plants' needs. In some instances, plants may have as much as 40 square yards of surface for each square yard of ground area. Transpiration in dense vegetation can produce as much as eight times more moisture than an equal area of bare ground. Growth depends on the season and the groundwater supply. In arid areas of the West, at timberlines, and at latitudes in the far North, transpiration and evaporation may be negligible toward the end of the dry season.

Absolute humidity. The weight of the volume of water in a given volume of air is expressed as the number of pounds per 1,000 cubic feet of air and is termed absolute humidity. This expresses the actual volume of water in the air and is not as important to the fire fighter as relative humidity.

Relative humidity. Relative humidity is the amount of moisture in a volume of air compared to the total amount (saturation) that volume can hold at that temperature and atmospheric pressure. It is expressed in percentages. For an example, take a glass cylinder 10 inches high with marks at 1-inch intervals. If we fill this container to the 1-inch mark, one-tenth, or 10 percent, of the cylinder will be filled (see Figure 3.36). If filled to the top, the cylinder would be 100 percent filled. The same is true for water vapor in the air. If the water vapor in the air is one-tenth of the amount that the air can hold, the relative humidity is 10 percent. When the air has all the water vapor it can hold, it is saturated, the relative humidity is 100 percent, and the dew point has been reached.

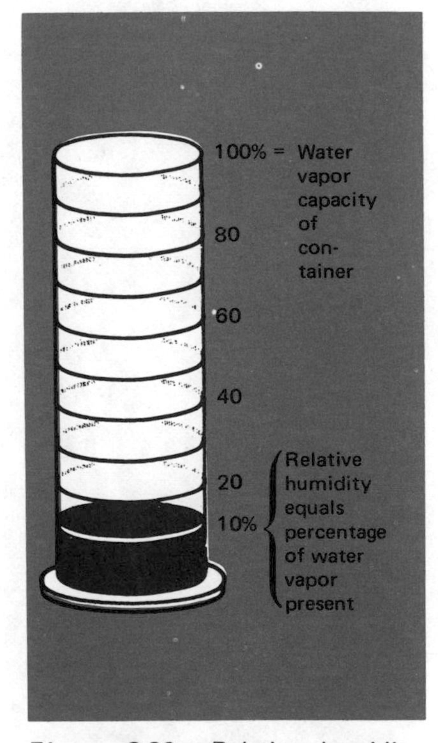

Figure 3.36. Relative humidity can be viewed in terms of a glass cylinder that is filled to 10 percent of its total capacity. If water vapor in the air is one-tenth of the amount that the air can hold, the relative humidity is 10 percent.

Relative humidity changes with temperature and pressure. If we take the same glass cylinder and cut it in half to represent a drop in temperature, the 10 percent mark now represents 20 percent on the cut-down cylinder. The same amount of liquid is present, but it represents twice the percentage of the total capacity of the cylinder. If air with 10 percent relative humidity is cooled until its saturation vapor pressure is cut in half, the relative humidity is increased to 20 percent. Therefore, relative humidity moves in the direction opposite to temperature (see Figure 3.37).

The change of relative humidity with temperature follows the diurnal change. The relative humidity is high late at night and early in the morning because the temperature is lower; the relative humidity decreases as the temperature increases during the day, and then increases as the temperature decreases at night.

Changes in elevation, and the attendant pressure changes, do affect relative humidity and therefore must be considered in its measurement. For

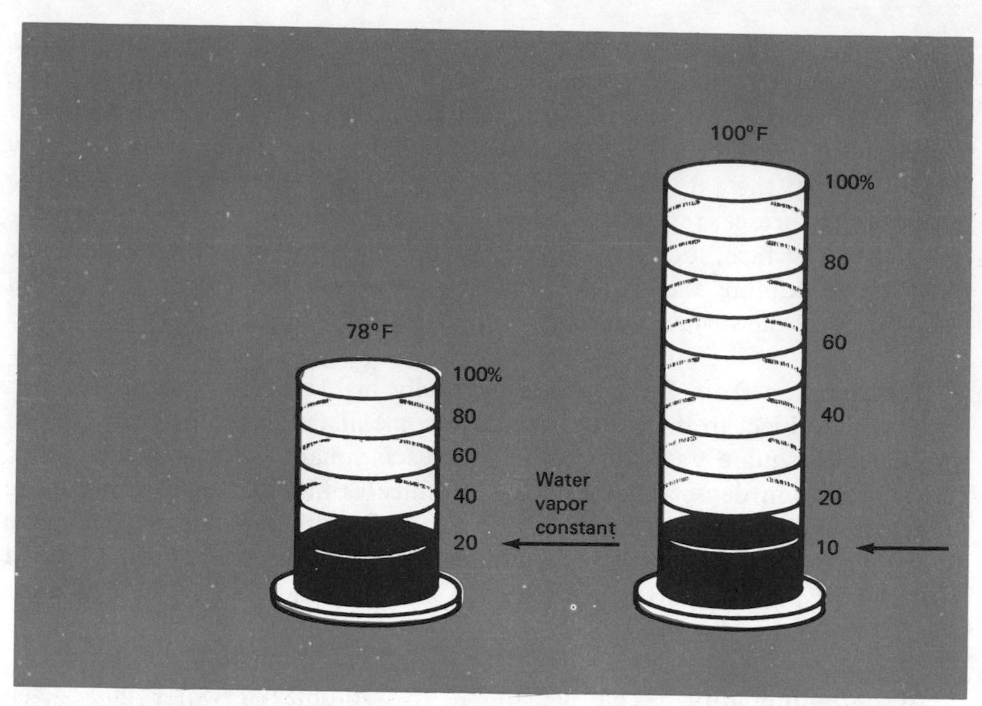

Figure 3.37. Relative humidity increases with a drop in temperature. For example, a glass cylinder that is filled to 10 percent of its capacity is cut in half to represent a drop in temperature. The amount of liquid in the cylinder remains the same but now occupies 20 percent of the volume of the cylinder.

Figure 3.38. Psychrometer.

this reason, psychrometric (relative humidity and dew point) tables are produced (Tables 3.1 and 3.2) for different elevations above sea level.

Vapor pressure is included in the psychrometric tables because by international agreement the mixing ratio is used for relative humidity calculations. Technically, relative humidity is defined as the ratio of the actual mixing ratio to the saturation mixing ratio at the same temperature. The ratio is expressed in percent. The fire fighter need only be concerned with the values determined from the psychrometric tables as developed by the wet and dry bulb readings.

A psychrometer is the most commonly used instrument for measuring humidity. Two thermometers are used: one to measure the air temperature, called the dry bulb, and the other to measure the temperature of evaporating water contained in a muslin wicking covering the thermometer bulb, called the wet bulb. The thermometers are briskly ventilated either by whirling them or by using a fan. From the wet and dry bulb measurements, the values of dew point temperature and relative humidity may be read from tables (see Figure 3.38).

Generally, as the temperature increases, the relative humidity decreases, and vice versa. But this is not always true, especially in those areas where the atmosphere normally contains a high amount of moisture.

Rule of thumb: For every 20-degree decrease in temperature, the relative humidity is doubled, and, for every 20-degree increase in temperature, the relative humidity is lowered by one half.

TABLE 3.1
Psychrometric Table for Different Elevations

Elevation above sea level (feet) Except Alaska	Alaska	Inches of Hg
0-500	0-300	30
501-1900	301-1700	29
1901-3900	1701-3600	27
3901-6100	3601-5700	25
6101-8500	5701-7900	23

TABLE 3.2
Psychrometric Table for Relative Humidity and Dew Point*

Wet bulb temperatures (°F)

Pressure 30 inches
Elevation 0–500 feet

Each cell shows the dew point (black), then the relative humidity (red).

Dry bulb temperatures (°F)	45	46	47	48	49	50	51	52	53	54	55	56	57	58	59	60
70	-14 3	0 6	+9 9	16 12	21 16	26 19	30 22	33 26	36 29	39 33	42 36	45 40	47 44	49 48	51 51	53 55
71	-26 2	-5 5	+5 8	13 11	19 14	24 17	28 20	32 23	35 27	38 30	41 34	44 37	46 41	48 45	51 48	53 52
72		-13 3	+1 6	10 9	16 12	22 15	26 18	30 21	34 24	37 29	40 31	43 35	45 38	47 42	50 45	52 49
73		-26 2	-5 4	+6 7	13 10	19 13	24 16	29 19	32 22	36 25	39 29	41 32	44 35	47 39	49 42	51 46
74			-13 3	+1 6	10 8	17 11	22 14	29 17	31 20	34 23	37 26	40 30	43 33	46 36	48 40	50 43
75			-25 2	-4 4	+6 7	14 10	20 12	25 15	29 18	33 21	36 24	39 27	42 31	45 34	47 37	49 40
76			-57	-12 3	+2 5	11 8	17 11	23 14	27 16	31 19	35 22	38 25	41 28	44 31	46 35	49 38
77				-23 2	-3 4	+7 7	15 9	21 12	25 15	30 18	38 20	37 23	40 26	43 29	45 32	48 35

*Relative humidity is shown in red. Dew point is shown in black. Values are shown where dry and wet bulb columns intersect.

Rule of thumb: Thirty percent relative humidity is about the ordinary danger point for wildfires. When humidity is above 30 percent, fires are not too difficult to handle, other factors being equal; but, below 30 percent, wildfires are more difficult to control. Twenty percent is considered to be the breaking point in arid western areas.

The amount of shade created by a given type of vegetation and its density will affect the relative humidity in the fuel area. Humidity affects wildfires primarily as it relates to moisture for fuel and secondarily as it relates to precipitation. For all intents and purposes, it does not actually affect the flame of the fire, but it does affect fuel moisture. It is most important as a fire weather factor in the layer near the ground. Near the ground, air moisture content, season, time of day, slope, aspect, elevation, clouds, and vegetation all cause important variations in relative humidity.

DEW POINT

The dew point is the point in temperature at which condensation starts. When a section of air is cooled to its dew point temperature, it is saturated with water vapor. Further cooling causes some of the vapor to condense into liquid droplets that form clouds, fog, or dew. Cooling to the dew point occurs when moist air is lifted in the atmosphere. As it is lifted, it is cooled, and, when it reaches its dew point, clouds form, as in the case of fair weather cumulus clouds. If enough condensation takes place, rain showers will result.

Dew that forms at night is another example. As air comes in contact with a cold surface, it is cooled to its dew point, and condensation forms. Materials that cool rapidly at night, such as glass and metals, are the first to show a dew deposit. This is why a parked car may be covered with dew, yet no sign of dew will be on the ground surface.

The dew point does not provide a direct measure of moisture in the air, but it does give a comparative one: the higher the dew point, the more moisture the air contains.

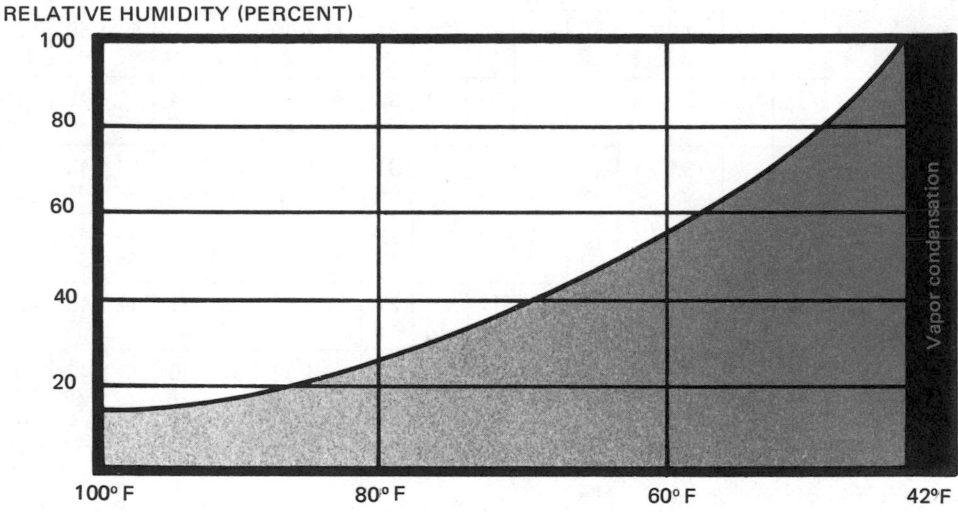

Figure 3.39. Dew point is the temperature at which condensation starts.

Small changes in dew point are common in any air mass. However, a consistent trend, either upward or downward, is a good indicator of the arrival of a new air mass. Different air masses usually have different amounts of water vapor and different dew points. By making dew point observations at intervals, the beginning of a change may be apparent before it would be otherwise detected. When the dew point temperature and the wet and dry bulb temperatures are all equal, there is 100 percent relative humidity, and the saturation point is present.

Saturation processes. In order for clouds and precipitation to form, the air must be saturated with moisture. There are two primary ways that 100 percent relative humidity, or saturation, is produced: (1) by addition of moisture to the air and (2) by a decrease in the temperature.

As cold air passes over warm water, rapid evaporation takes place, and saturation is easily reached. Lifting of air is the primary way in which air is cooled. It may be cooled by thermal, orographic, or frontal lifting.

As the ground surface is heated, the air immediately above it becomes buoyant and is cooled by being forced aloft. Thermal lifting is most pronounced in the warm seasons. It may turn morning stratus (layered and horizontal clouds) into stratocumulus (high, puffy clouds) with possible light showers. Continued heating may result in cumulus clouds that produce heavier showers and thunderstorms. In flat country, the greatest convective activity takes place over the hottest surfaces. In mountain country, it is greatest over the highest peaks and ridges.

Orographic lifted air is forced up the windward side of slopes, hills, and mountain ranges. In the West, maritime polar air from the Pacific Ocean in the winter is forced over the Coast Ranges, the Sierra Cascades, and the Rocky Mountains. As it is lifted over the western slopes, it releases its moisture, and the lee slopes receive progressively less precipitation as it moves eastward. Maritime tropical air moves into the central portion of the United States and southern Canada and is lifted over the Appalachian Mountains, producing precipitation toward the east.

Figure 3.40. Precipitation is caused as air is cooled. Warm moist air can be cooled by being forced up the slope of a warm front, causing precipitation (*top*). Warm air ahead of a cold front also results in precipitation (*bottom*).

Air forced up the slope of warm or cold fronts by frontal lifting provides for much cloudiness and precipitation in all regions in the winter and in many regions during all seasons of the year.

Humidity and fuel moisture. The amount of moisture in dead fuels is controlled in large measure by the atmospheric humidity in the absence of rain or other forms of liquid H_2O. Fire in dry fuels starts easily and spreads rapidly. When the moisture content is high, fuels are difficult to ignite, and fire spreads slowly.

Dead fuels are hygroscopic; that is, they give and take water vapor to or from the air. The amount of moisture dead fuels can absorb and hold from the air depends primarily on relative humidity.

If a piece of fuel with a lower moisture content than the air around it is exposed in that air, the moisture content of the fuel will increase, rapidly at first and then slowing down to a stop. At that point in time, the moisture content of the fuel is in equilibrium with the relative humidity. The actual moisture content of the fuel is the equilibrium moisture content. Increasing the relative humidity of the air increases the moisture content of the fuel until it is again in equilibrium with the new relative humidity.

TABLE 3.3
Equilibrium Moisture Content for Different Temperatures and Humidities

	Equilibrium moisture content		
Temperature °F	*20%**	*30%**	*40%**
40	4.6	6.3	7.9
45	4.6	6.3	7.9
50	4.6	6.3	7.8
55	4.6	6.2	7.8
60	4.6	6.2	7.7
65	4.5	6.1	7.7
70	4.5	6.1	7.6
75	4.5	6.0	7.5
80	4.4	5.9	7.4
85	4.4	5.9	7.3
90	4.3	5.8	7.2
95	4.3	5.7	7.1
100	4.2	5.6	7.0
103	4.1	5.5	6.9
110	4.0	5.4	6.8
115	3.9	5.3	6.7
120	3.8	5.2	6.5

*Relative humidity.

For every level of relative humidity, there is a corresponding equilibrium fuel moisture content. Precise equilibrium moisture content has been determined for only a few wildfire fuels. However, average values for wood are often useful for estimating fuel moisture. Table 3.3 should be used with caution because of the many differences that may occur in the location, elevation, wind conditions, diurnal changes, deep canyons as compared to exposed slopes, timbered areas as compared to open areas, the size of the fuel particles, compactness, and arrangement. However, equilibrium moisture content provides a useful tool in fire control, fire use, and the prediction of fire behavior. Fine fuels, surfaces of larger fuels, and the top layer of litter and duff are usually close to their equilibrium moisture content. These are the fuels that have the greatest effect on the ease of ignition, rate of fire spread, and the probability of fire spotting.

Time lag principle. It takes varying amounts of time for fuels to adsorb from the air or lose moisture to it. Fine fuels such as grass, leaves, and small twigs can reach equilibrium moisture content in a few minutes, but large limbs and logs take a long time.

The approach to equilibrium values from either above or below equilibrium follows a logarithmic rather than a straight-line path as long as liquid H_2O is not present on the surface of the fuels. The moisture change will be the fraction of the departure from equilibrium (I - Ie = 0.63), where e is the base of natural logarithms, or 2.7183. Under standard conditions (80°F temperature and 20 percent relative humidity), the duration of these time periods is a property of the fuel and is referred to as the time lag period.

To illustrate the moisture response, let us assume that a fuel with a moisture content of 28 percent is exposed in an environment in which the equilibrium moisture content is 5.5 percent. The difference is 22.5 percent. At the end of the first time lag period, this difference would be reduced by 0.63 times 22.5, or about 14.2 percent. The moisture content of this fuel would then be 28 minus 14.2, or 13.8 percent. Similarly, at the end of the second time lag period the moisture content would be reduced to about 8.6 percent, and so on. The moisture content at the end of five or six time lag periods very closely approximates the equilibrium moisture content.

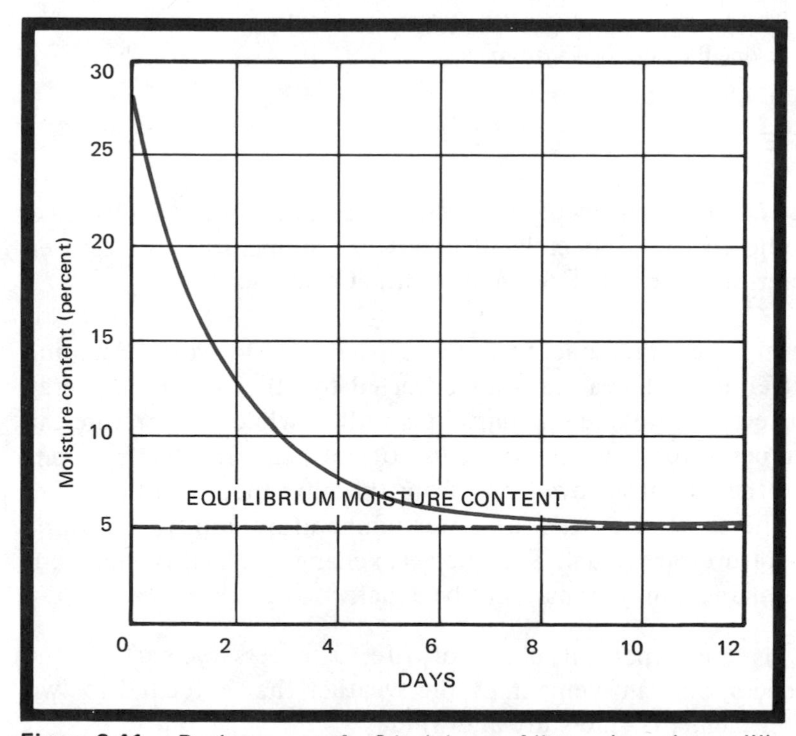

Figure 3.41. Drying curve of a 2-inch layer of litter where the equilibrium moisture content is 5.5 percent.

The average time lag period varies with the size of fuels and other factors. For extremely fine fuels the average period may be a matter of minutes, while for logs it ranges upward to many days. Using the time lag principle, it is possible to describe various fuels—irrespective of type, weight, size, shape, compactness, or other physical features—as having an average time lag period of 1 hour, 2 days, 30 days, and so on.

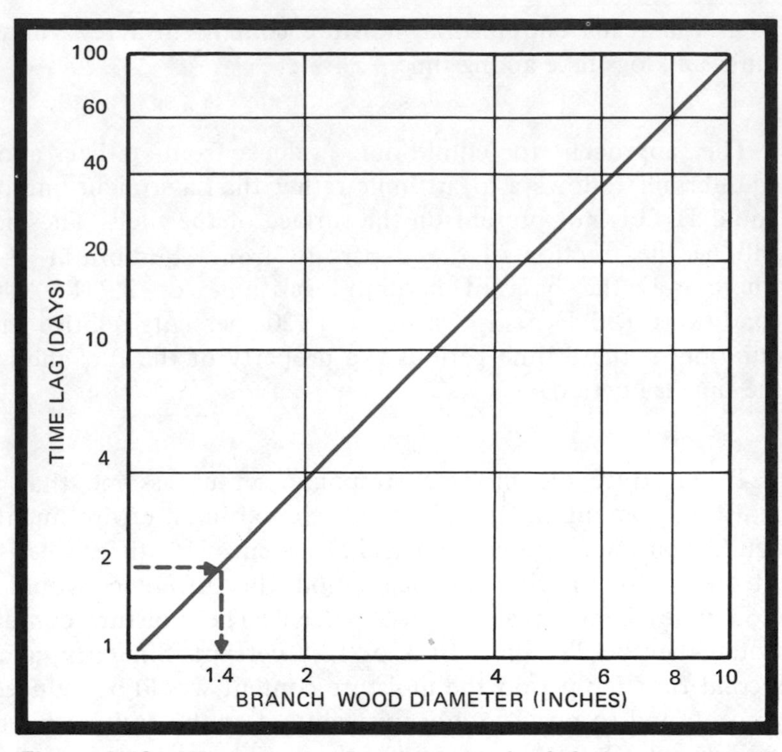

Figure 3.42. The average time lag period of fuel varies with its diameter. Branch wood is used as a point of comparison with other fuels. For example, a 2-inch litter bed with an average time lag of 2 days may be considered the equivalent of 1.4-inch dead branchwood having the same average time lag.

TEMPERATURE

Temperature is the degree of hotness or coldness of a substance, determined by the degree of its molecular activity. Temperature is measured by a thermometer on either the Fahrenheit or the Celsius scale.

Heat is a form of energy. Energy is the capacity to do work. Heat and temperature differ in that heat can be converted to other forms of energy and can be transferred from one substance to another, while temperature can do neither. Temperature shows the direction of net heat transfer from one substance to another. Heat always flows from the substance with the higher temperature to the substance with the lower temperature and stops flowing when the temperatures are equal. The energy exchange of heat is equal, but the temperature changes may or may not be equal.

Temperature is an important cause of fire, a necessary part of the combustion process, and an element of the weather that is related to two other fire weather elements–moisture and wind.

The ignition temperature of wood is between 500° F and 800° F. Wildfire fuels flame at approximately 540° F with sufficient oxygen. Openings in a forest not affected by wind can reach 170° F at ground level. Generally, the hotter the surface and the air around it, the drier the fuels.

Most extremes of climate occur nearest the ground. This rule holds true for temperature except where the ground is protected by a cover of aerial fuels. Crowns of a cover of dense aerial fuels provide effective air contact surface, and therefore the crown area has higher daytime temperatures than the air at the ground. The crown air may be 18 to 20 degrees warmer than air near the ground. The night temperatures are lower near the top of the crown canopy because this is where the principle radiation takes place.

If the vegetation is sparse, with openings and pockets, more solar heat reaches the ground. Thus, cleared areas may be 5 to 8 degrees warmer than shady areas. The openings may act as chimneys for surface fires under conditions of strong daytime heating and light winds.

Surface materials, except water and ice, show a greater range of temperature changes from solar radiation (insolation) than the air. Yet air is primarily heated and cooled by its contact with the surface.

A low sun angle results in a low surface temperature (summer versus winter). More hours of daylight cause higher surface temperatures, and more hours of darkness cause more cooling and thus lower surface temperatures.

Figure 3.43. As the sun arcs across the sky, rays that hit the surface perpendicularly transmit the greatest intensity of the heat.

Both the steepness and the aspect of slope affect surface heating and cooling. Surfaces nearly perpendicular to incoming radiation from the sun receive more heat than those parallel to the sun's rays. South-facing slopes receive more heat than north-facing slopes. Level surfaces reach their maximum temperature around noon. East-facing slopes are the first to warm, and then they very gradually cool. West-facing slopes are hottest later in the afternoon, and southwest slopes generally have the highest temperature.

Clouds, smoke, or haze in the air and trees or other objects that cause shade reduce incoming solar radiation, and thus ground surfaces are cooler. The surface temperatures respond quickly to changes in amount of shade: surface temperatures may change as much as 50 to 60 degrees within a few feet or a few minutes. Ground temperatures are fairly uniform in a broadleaf forest in the winter, while pine forests show striking differences in both summer and winter.

Figure 3.44. Clouds absorb and reflect radiation reducing surface heating.

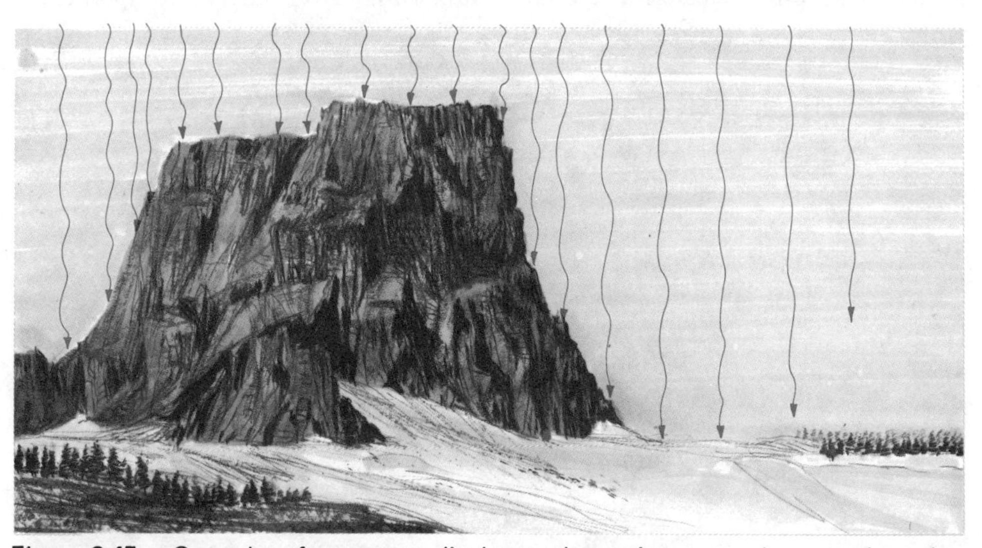

Figure 3.45. Ground surfaces perpendicular to the sun's rays receive more heat than those parallel to the incoming radiation.

The thicker and lower the clouds, the less of the sun's heat is radiated to the ground. Temperature at the ground may drop as much as 50 degrees in three minutes as a thick cloud passes overhead in clear summer weather, because the clouds both absorb and reflect much of the solar radiation.

Water vapor in the air reduces the cooling of the surface at night. It absorbs some of the radiated heat from the earth and also radiates some back to the surface. As a result, surface temperatures are much lower on clear nights than on cloudy nights. Because of the lack of water vapor in the air surface, temperatures are low in the desert at night.

A cover of smoke from wildfires can significantly lower daytime surface temperatures and raise nighttime temperatures when skies are otherwise clear. Surface temperatures are affected by the surface's capacity to absorb or reflect radiation, by its absorptivity and emmissivity, its transparency, and its conductivity.

Generally, dark materials absorb the most heat, and light materials absorb the least. Tree crowns do not absorb as much as dark soils and forest litter.

Figure 3.46. Shaded areas reduce incoming radiation and make surface temperatures cooler than those in open areas.

Tree crowns, grass, plowed land, and sand are all good radiators and give up heat rapidly at night and become cold. Absorption and emmissive capability are assumed to be the same under the same conditions of wavelength and temperature.

Transparent materials allow radiation to penetrate more deeply into them than opaque surfaces. Because radiation penetrates deeply into water, it heats a larger volume than radiation that penetrates only a surface layer of land. Thus, land surfaces become warmer during the day than water.

Wood is a poor conductor, and heat applied to it concentrates on the surface and only slowly penetrates to warm the interior. Leaf litter, other organic fuels, and dry soil are poor conductors. Rock, damp soil, and water are efficient conductors of heat but are not as good as metal.

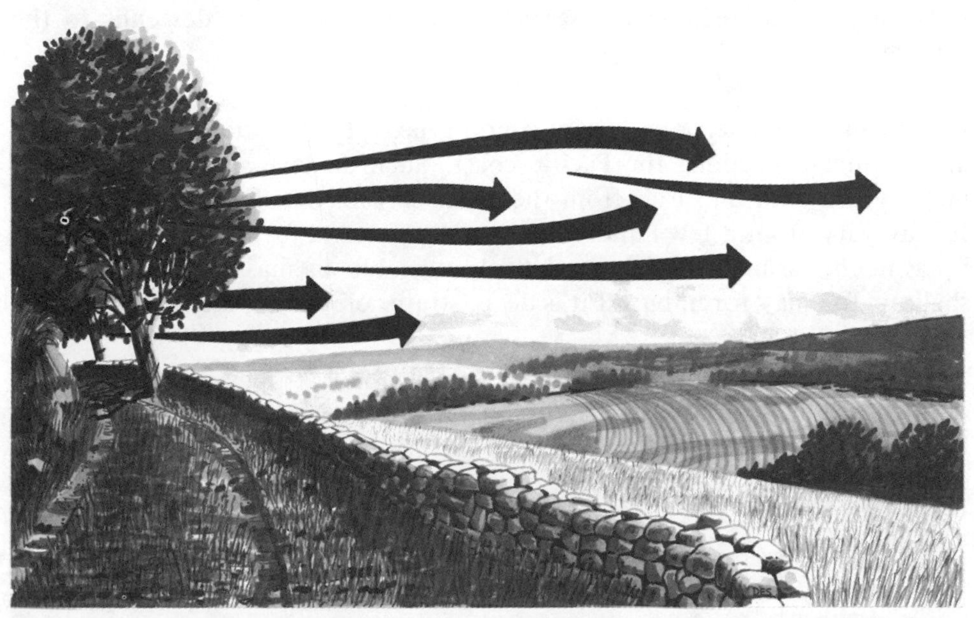

Figure 3.47. Strong daytime winds cause turbulent mixing, which carries heat away from surfaces and lowers temperatures.

Figure 3.48. In a stable air mass with little or no wind, surface temperatures increase rapidly.

Air is a very poor conductor. Porous fuels, such as duff or litter, tend to insulate the soil below because of the many air spaces within them. The surfaces of poor conductors become hotter during the day and cooler at night than the surfaces of good conductors.

Wind has a moderating influence on surface temperatures. Strong daytime winds cause turbulent mixing, which carries heat away and lowers the temperature of the surface. In a stable air mass, the temperature of the air near the surface increases rapidly. At night, wind pulls warmer air downward, where heat is transferred to the ground by conduction, thus preventing low surface temperatures.

Generally, as warm air rises, it is cooled, and as cold air settles, it is warmed. Air expands as it rises and is compressed as it descends to the surface.

Figure 3.49. Smoke released into an inversion layer will rise only until its temperature equals the temperature of the air. Then the smoke will flatten out horizontally.

Temperature inversions. A common type of temperature inversion is the marine inversion along the Pacific coast. Such an inversion is strongest at night where cool moist air from the ocean spreads over low-lying land areas. It may vary from a few hundred to several thousand feet deep and is topped by a much warmer, drier, and relatively unstable air mass. If the cold air is shallow, fog may form, but, if it is deep, stratus clouds are likely to form.

A surface inversion is formed when air is cooled at the surface by contact with cold-radiating surfaces. The temperature increases with height and may increase as much as 25 degrees in 250 vertical feet. This cold air flows down slopes and gathers in the valleys.

A night inversion is common during clear, calm, settled weather. It traps impurities in the air such as smoke and fumes, which results in a "smog" layer over valleys and depressions in the surface. Smoke rises until it reaches the bottom of the inversion and then flattens out horizontally.

Surface inversions commonly form at night and are important to fire behavior. An inversion acts as a "damper" on a wildfire, retarding its activity. When the inversion layer breaks up (usually around 10 A.M. but sometimes later or earlier), the convection columns of fires are then unhampered, and fire activity increases, sometimes very rapidly. Inversions are more common in lower mountain valleys or in basins than over flat land areas.

The average level of the inversion top is known as the "thermal belt." In mountainous areas the top of the inversion layer is usually below the top of the main ridges. Temperatures decrease both above and below the top of the inversion. The thermal belt contains the highest temperatures and the lowest nighttime relative humidity, as well as the lowest fuel moisture. Consequently, wildfires can remain quite active during the night. Although temperatures decrease above the thermal belt, this area may experience stronger winds and less stable air, thus activating the fire in that area also.

TOPOGRAPHY

It might be said that topography makes weather and determines the type of fuel in a given area. Since fire behavior is largely the result of weather and the available fuel, it might also be said that topography dictates fire behavior.

Topography is basic to general weather patterns as well as to day-to-day weather changes. As we have seen, topography profoundly affects the characteristics of winds, particularly convective winds. Topography is responsible for the location of fuel types, and it dictates their growth and flammability because of its effect on weather.

The wildfire fighter should study the effects of weather as they are produced by the topography of his area. The variations are so great that only basic generalizations can be made in this text.

Latitude (distance from the equator) and position on the continent largely dictate the weather patterns that a specific area experiences. The location of oceans and other bodies of water, as well as mountain ranges, vegetation, and elevation determine weather development, and consequently these factors relate to fire behavior.

Aspect is the position of the land surface in relation to the sun as it makes its daily rounds. It describes the direction in which a slope faces the sun. The sun shines the same on all parts of a flat land area, but, in mountains, hills, and valleys, the position of a given area in relation to the angle at which the sun's rays strike is different from that of other areas. Different aspects receive varying amounts of sunshine, wind, and precipitation.

Figure 3.50. The zone of warm nighttime temperatures near the top of the inversion is known as the thermal belt.

In general, south and southwest slopes receive more direct sunshine and, as a result, more heat, thereby increasing the fuel and air temperatures. Southeast and west slopes receive less heat, and north-facing slopes receive the least heat. The amount and direction of heat that reaches the land surface affect the kind and amount of vegetation growing there. Snow melts earlier on the south-facing slopes; summer winds are hot and dry and are generally from the south or southwest, so these slopes are drier. This dryness is reflected in the vegetation, which is drier, sparser, and more flammable than that on the north-facing slopes. Generally, the species of plants that make up the fuel type are more fire-prone on the south-facing slopes. In the summer months, temperatures are higher and humidity is lower on the south slopes. South-facing slopes have a longer fire season than north-facing slopes.

The occurrence of fires and their rate of spread will vary according to aspect, with more fires and larger fires on south slopes.

Because of the luxuriant vegetation and great volume of fuel generally present on north slopes, fires can be particularly vicious in these areas when weather conditions dry out these fuels. Ordinarily, north slopes have more moisture, especially in the ground litter, and humidities are higher.

Vegetation on broad slopes of a high mountain will probably show more response to exposure than vegetation on the sides of smaller hills, and steeper slopes have a greater variety of vegetation than more gradual slopes.

Where local dominating factors of weather influence a large area, little difference in response on any slope (exposure) may be shown in either vegetation or fire behavior. An example of such an area would be found in dry desert areas, very moist coastal timberland, areas of the Great Plains, or rolling hill country covered with the same age class and species of trees. The other extreme may be found on the Monterey coast of California, where small gulches, cut into steep mountain faces, are influenced alternately by damp fog and hot sun. Tiny exposure changes have produced moisture-loving redwood and desert cactus growing within a few feet of each other.

Where rainstorms ordinarily approach a range of mountains from the same direction, less moisture will fall on the lee side of the range. These slopes on the lee side are said to be in the "rain shadow." The plains east of the Rockies are a good example.

Prevailing winds will not only dry out slopes exposed to them but will also favor them with rainfall. Usually, flashy-type fuels grow on such exposures, so the effect of any added moisture may be short lived.

ELEVATION

Both the elevation above sea level and the elevation as compared to the surrounding countryside have their effects on fire behavior evaluation.

In the mountainous West it is usual for one to experience a hot July day along a valley bottom and view eternal snow on adjacent mountain peaks. The observer is in a temperate climatic zone, and the peaks are in a subarctic zone. Between these two extremes are a wide range of climatic conditions and resultant fire behavior problems. These climatic zones have different dates of snowmelt and grass curing, different lengths of fire seasons, different types of fuel, and different weather patterns. In the alpine zones, above the timberline, it is rare for fire to occur.

The lower elevations have a longer fire season than the higher zones. Latitude and north-south position strongly affect the length of fire season.

ELEVATION WITH RELATION TO SURROUNDING COUNTRY

Mountaintops and valley bottoms will have different burning conditions during the same twenty-four hour period.

With the stable conditions of midsummer, there is a daily interchange of air between valley bottoms and mountaintops. The air is much warmer during the day on the valley floor and this warm air tends to rise. When the sun sets, with consequent reduction in heat beamed at the surface, the colder air from the tops of mountains and ridges flows downward to the valley floors. Therefore, nighttime summer temperatures are usually lower in the valley bottoms than at the mountaintops. The mountaintop air is also drier at night. Valley bottoms have the most dangerous fire conditions during the day, but the trend is reversed at night to the mountaintops.

THE THERMAL BELT

Between the two extremes of the valley bottoms and the mountaintops is the middle of the slope area called the thermal belt. Fire conditions are usually more severe in this belt than in any other location. The thermal belt normally is the middle one third of the total slope. In the northern Rocky Mountains, this belt may vary from 3,000 to 8,000 feet above sea level. During the summer months under stable conditions, this zone does not have the range of temperature changes of the valley bottoms nor the mountaintops. The thermal belt does have higher average temperatures and lower relative humidity than any other zone.

ELEVATION AND FIRE OCCURRENCE

Generally, the mountaintops have a high fire occurrence due to lightning strikes, especially in forested areas. But fire occurrence may be greatest in areas of high human use at lower elevations. Generally, areas with the greatest use and the most burnable fuels show the greatest incidence of fires.

The combined effects of aspect and elevation will be different during various periods of the day. Fires starting at the base of slopes are often larger in size than those starting at other points on a slope, probably because, once they gain headway, they burn into the thermal belt, where they are assisted in traveling through continuous fuels to the top of the slope.

SLOPE

Slope is expressed in percentages: a 1 percent slope means a rise of 1 foot in elevation for each 100 feet in horizontal distance. Thus a slope of 45 degrees would equal a 100 percent slope.

Under the same conditions, fires will burn more rapidly on steep slopes than they will on more gradual slopes. As the steepness of the slope increases, the rate of spread increases. The fuels up the slope are closer to the fire than fuels on a level surface, allowing quicker action by radiation and convection. As a result, fires are usually larger on the steeper slopes. Difficult access for firemen is also a factor.

Fires running up a steep slope generally have a wedge shape similar to those pushed by strong winds. Usually, flames will be pulled inward from the flanks if wind direction remains steady.

Flames and the convection column are usually bent back in a vertical position or even back over the fire when the fire reaches the top of a slope, because of the natural flow of warm air up the opposite slope and the drawing action of the convection column.

Fires usually travel faster downhill at night because of the settling masses of colder air flowing toward the valley bottoms. This settling may happen in the daytime from a temperature inversion when it causes downslope winds.

In areas with unconnected aerial and ground fuels, a ground fire may creep downhill at night, drying out the aerial fuels and causing a reburn in the crowns the next day when the upslope air movement takes place.

Burning material may roll downhill, starting fires farther down the slope, which then burn back to the main fire. Such fires may seriously jeopardize or outflank crews working on the line above them.

Spot fires outside of a vertical line up a slope may also outflank crews above on the line. Planned escape routes and a lookout are a necessity, especially in heavy brush. The flanks of fires on steep slopes cannot be considered safe until the lower slope areas have been mopped up. Spotting usually occurs above fires running upslope. Slope is very much affected by wind pressure. Fires can run as much as twenty times as fast uphill as downhill.

All of the various features of topography have their effect on the direction of travel, rate of spread, and general behavior of the fire.

Narrow canyons. Wind direction normally follows the direction of the canyon. Eddies and strong upslope movements may be expected at sharp bends in a canyon. Fires spot from one side to the other because of radiant heat and eddies. Near the bottom of the slope, aspect makes little difference.

Wide canyons. Prevailing winds are normally not affected by the direction of a wide canyon. Cross-canyon spotting is not common except in high winds. Aspect has pronounced effects on general fire conditions.

Box canyons. These have the effect of a fire in a stove. Strong upslope winds are drawn in from the bottom and confined to the course of the canyon by the walls. Narrow canyons or gulches at the head of high mountain valleys cause a fire to react in the same way that fire reacts in a chimney.

Ridges. Fires burning up a ridge may change direction at the top if upslope winds are strong from a canyon ahead. The direction and course of valleys and ridges may be a dominant factor on the pattern of wind, rain, and sunshine.

The fire fighter should observe the effect of exposure upon fuel and fire behavior wherever he may be. It is very real and very important. The student of fire behavior should learn what type of fuel grows naturally in his particular area. The natural laws controlling vegetative growth are surprisingly dependable.

EFFECTS OF THE DIURNAL CHANGE

Characteristically, temperatures are lower during the night than during the day. The temperature begins to drop as the sun goes down and continues to drop until just before dawn, when it once again begins to increase as the sun rises. It reaches its peak at about 1400 hours. There are some variations due to the amount of cloud cover, season of the year, and the influence of weather changes. Temperature change is affected by aspect, so temperatures change faster on east slopes than on west slopes (see Figure 3.51).

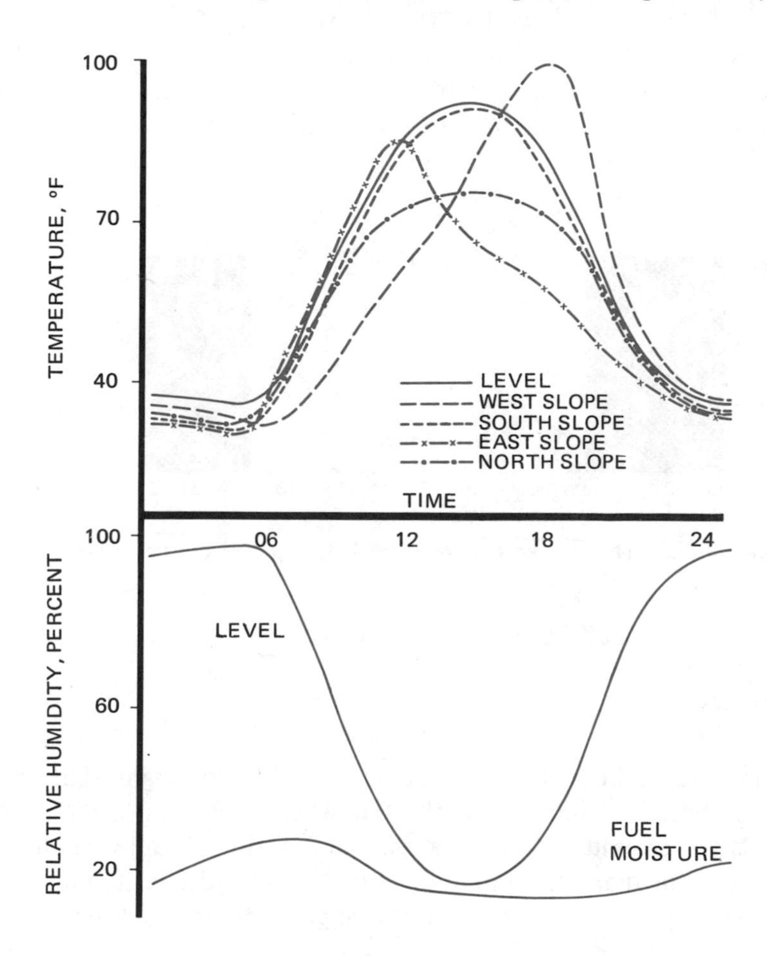

Figure 3.51. This graph shows temperature (*top*) and humidity (*bottom*) change during one diurnal cycle. Temperature changes during one diurnal cycle vary according to aspect, and humidity is almost a mirror image of the temperature change. Fuel moisture is also shown on the humidity graph.

Relative humidity is almost a mirror image of the temperature change, but it reacts in the opposite direction. As a result, fire danger increases during the heat of the day and decreases during the night. Intensity of burning and rate of spread both react to diurnal change. Therefore, most fires are more easily controlled during the late afternoon, night, and early morning—especially in the longer-burning fuels. Often, the wind dies down at dusk and remains at low speeds during the night but increases in speed during the day when thermal heating is most pronounced. In the Great Plains, and to a lesser extent elsewhere, the wind usually dies out at nightfall, and, because of the light, fast fuels on the plains, most fires die out; they can be controlled with only nominal mop-up. Therefore, fast initial attack is important because the fire will make its initial run before sundown. Occasionally, winds caused by fronts and subsidence may continue into or begin during the night. Night work on fires may not be practical because of the terrain and other hazards, but, where it can be accomplished in the heavier fuels, it may mean the difference between success and failure.

The fire day, for purposes of planning, operating, and reporting, begins at 1000 hours and ends twenty-four hours later. It is the policy of the USDA Forest Service and many other agencies to require control during the first burning period, which means that every effort must be made to have complete control of each fire by 10:00 A.M. of the day following discovery. This policy is very effective and should be followed wherever possible. Although control may not always be established during the first burning period, if the protection agency operates with this objective, the damage will be minimized.

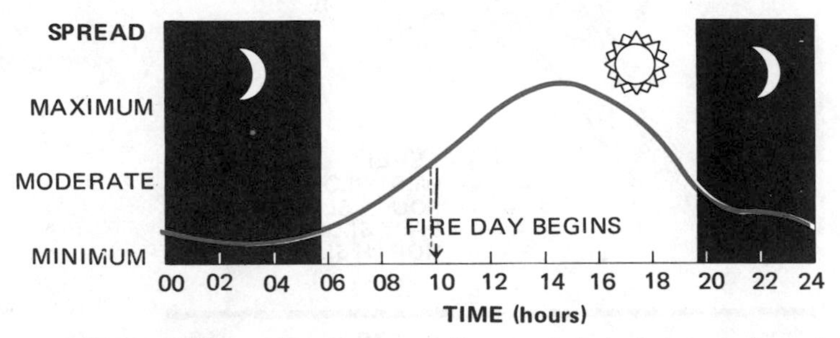

Figure 3.52. The danger of fire spread during one diurnal cycle.

CLASSIFICATION OF FIRES

GROUND FIRES

Ground fires occur in fuel that is up to 6 inches in height. The smoke height in these fires is nominal, and the flame length is from 6 inches to 4 feet. The fuels for ground fires are leaves, needles, duff, small branch wood (less than ¼ inch in diameter and lying on the ground), peat, muck, and humus soil. The type of spread is smoldering, creeping, backing, or slow running.

SURFACE FIRES

Surface fires occur in fuel that is up to 6 feet in height. The smoke height is medium to high, and the flame length is from 2 to 50 feet. The fuels for surface fires include grass, grainfields, brush and tree reproduction (up to 6 feet high), downed trees, brush, and branches, including slash. The type of spread is backing, normal, or slow to fast running. In grass and light litter, the rate of spread is high to flash.

CROWN FIRES

Crown fires occur in fuel that is above 6 feet in height. The smoke height is medium to extreme, and the flame length is from 30 to 200 feet plus. The fuels for crown fires are brush that is 6 feet or higher and all trees above 6 feet high. In some areas, 10 feet is the break point for brush. The type of spread is torch outs, spot crowning, or running crown fires. The rate of spread is fast running to extreme.

BLOWUP CONDITIONS

A blowup is a sudden increase in fire intensity, an increase in rate of spread, or both that probably nullifies existing control plans. Blowups are often the result of violent convection, and they may have other characteristics of a firestorm (see the chapter on safety).

Although only a small proportion of all fires belong in the extreme behavior group, their importance far outweighs their number.

- They are hazardous to the safety of personnel because of their extreme and erratic activity.
- They cause a disproportionate amount of heavy damage.
- Suppression costs are high.

CHARACTERISTICS OF EXTREME FIRE BEHAVIOR

Because the art of predicting extreme fire behavior has not been perfected, the characteristics listed should be considered only as a starting point.

- A rapid buildup or growth of intensity after a fire reaches a critical rate of energy output. (As the rate of fuel consumption increases, there is also a rapid increase in the size of the hot area, and a blowup fire becomes a distinct possibility.)
- A high, sustained rate of spread.
- A well-developed convection column.
- Long-distance spotting (600 feet or more).
- Fire whirlwinds.
- Horizontal flame sheets.

When fires have reached such an intensity that they exhibit the characteristics described above, the combustion chain has usually become so strong that it cannot be broken by conventional fire-fighting methods. It is then necessary to plan control for changing conditions and to try to anticipate the place and time at which the change will occur. In the meantime, only part of the perimeter may be tenable for fire control forces.

SMALL FIRES

A "small" fire is one that has not yet built up to serious proportions of intensity and spread; it can be controlled with the forces at hand. A small fire may be only a few acres—up to perhaps 40 acres, with nominal fire weather conditions and topography.

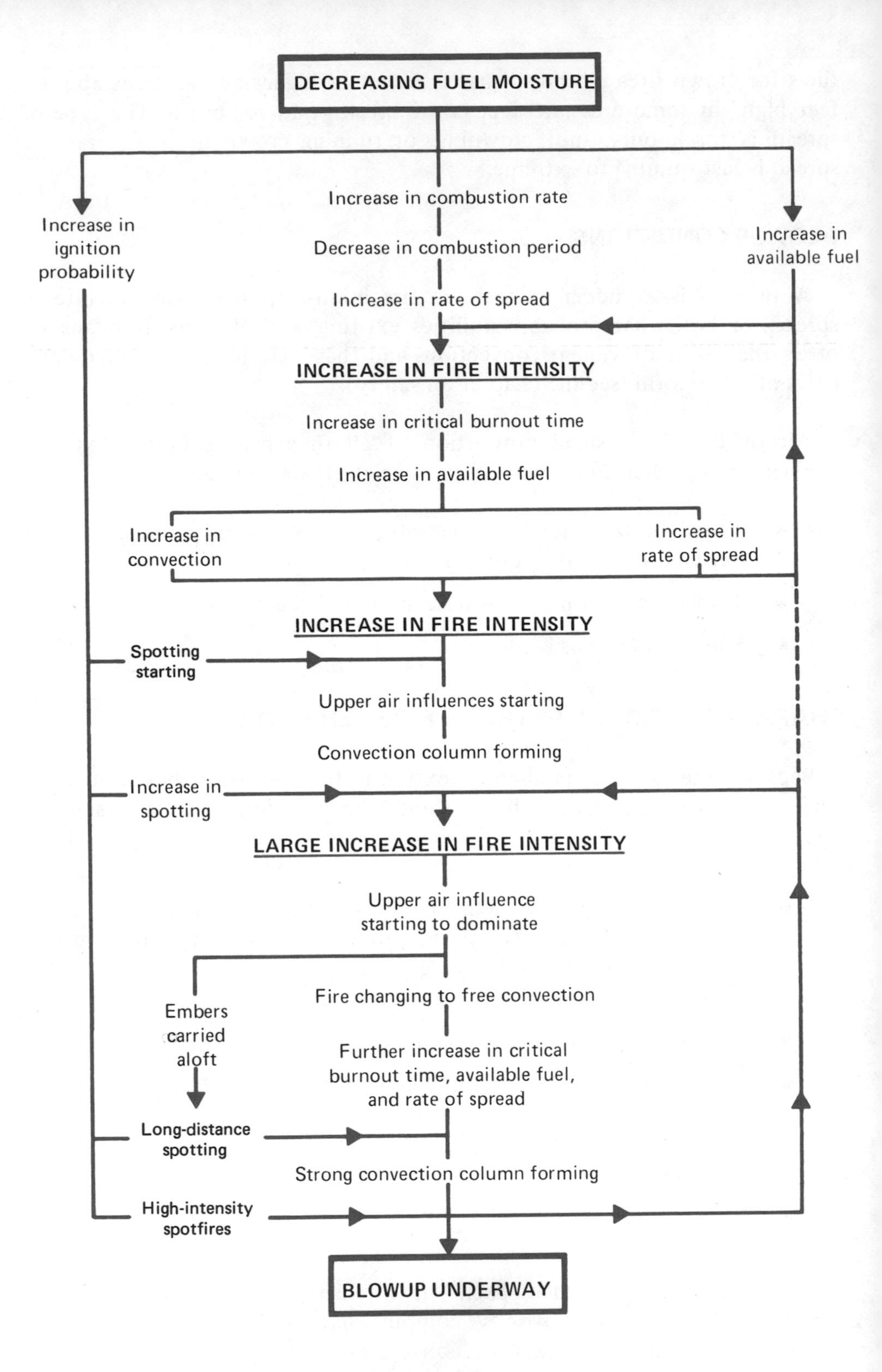

Figure 3.53. Cycle of reinforcement in forest fire behavior.

Theoretically, a fire burning in any type of fuel on flat ground would create a perimeter in the form of a circle, with the flames and smoke leaning toward the center. Since fuel varies in size, shape, arrangement, and amount of dead material, certain portions will burn more rapidly than others, thus changing the shape of the burn. A shift in wind will change the direction of the head, and a change in topography will have its effect on the shape of the area of the burn. Two or more heads may develop, according to the combination of factors present in a particular fire.

Spot fires can occur ahead of small fires as well as large fires, but they occur less frequently and at a shorter distance with small fires than with large fires. Burning embers carried by the convection column nearly always come from the underside of the heaviest smoke concentration. Spotting can be expected downwind when individual trees torch out or when large piles of dead material or slash burn briskly. When grass and weeds are burning at peak intensity, the heads and seeds are cast out ahead of the fire by the convection column.

LARGE FIRES

Fire tends to beget fire. Five to ten percent of all wildfires in a given area grow to large size. These are the fires that do the most damage, often reaching catastrophic proportions. Many large fires are the result of adverse conditions of weather and topography. Adverse may mean conditions too tough for the strength of the initial attack forces for that particular fire, or it may mean that budget and resources simply are not available to combat worse than average conditions. Planning and operations should provide for an initial attack that is strong enough to prevent a large fire. According to a rule of thumb, doubling the wind speed quadruples the spread, and reducing the fuel moisture by half (when it is already low) may cause a spread five to six times as fast.

Large fires characteristically crown and spot ahead because of the buildup of convective currents over these fires. The flame length and intensity of radiation build up as the fire increases in size. A large fire can be expected to decrease in intensity when it reaches the top of a ridge or hill because of the heavy indraft demands of a large fire running uphill.

The behavior of large fires at night is generally less predictable than in the daytime, especially if variable fuels and topography are involved.

A striking characteristic of large fires is the development of strong convection currents that rise thousands of feet into the air due to the concentrated heat mass. Atmospheric instability allows vertical motion to be speeded up and develops towering convection columns with accompanying high fire intensity on the ground. If the air mass over the fire is stable, it will resist the development of a strong convection column. Consequently, the fire will burn less intensely than it would under a strong updraft.

Long-distance spotting will most likely occur on the right flank of an advancing fire because the smoke column has a tendency, under the influence of wind, to rotate in a clockwise direction with the increase in height.

Normally, a general indraft into a convection column can be expected from all sides. If there is a prevailing wind, the speed on the lee side (advancing side) of the column will usually be less than that on the windward side. However, occasionally, pronounced outdrafts (winds moving away from the column at a greater velocity than the free air movement) have been observed on the lee side. Winds in advance of a fire have been observed, in some instances, that were two to four times the velocity of winds along the flanks or rear. Fire fighters should be aware of and alert to the possibility of this type of phenomenon.

In large fires, size, distribution, and arrangement of fuel particles are relatively unimportant: it is the total fuel volume that is important. Large-size fuels will burn faster and hotter and produce higher temperatures under extreme conditions than smaller fuels. This intense heat results in large areas being rapidly consumed with extreme violence and nearly total reduction of all combustible material.

AREA IGNITION

If a number of fires are started in and around the edge of an area, an increased intensity of combustion develops due to the multiple adjacent points of ignition—this is called area ignition. In wildfires, area ignition is likely to result from a number of spot fires burning together outside the main perimeter of the main fire. The greatest amount of burning in the main fire is at the head and decreases from there back along the flank to the rear. Usually, most of the smaller fuels remain on fire, although there may be unburned islands that have not yet ignited.

Area ignition is a useful tool in managing controlled burns. Often the center of a prescribed burn is ignited first, and then the edges are ignited. In this way, the convection columns of the edge fires are drawn toward the center, and, hopefully, any spotting falls into the burned area.

If prescribed burning is contemplated in your area, check with your state forester to obtain the best methods and planning procedures.

SPECIAL FIRE BEHAVIOR

SPOT FIRES

Spot fires are small fires (at the time they start) either ahead of or along the sides of the main fire. They may occur as ground, surface, or crown fires. If they are allowed to burn, they may burn together or may individually advance the main fire. Often they are responsible for blowups and for adding acreage to the final burn area. They are started by either flaming material or embers carried out of the main fire by the convection column and/or wind. If it is not prevented or controlled, mass transport of such burning material can be disastrous. Spot fires in the crown or tops of trees are usually called tree or snag fires. They may be caused by sparks landing on and starting fire in the tops of snags, by lightning, or by a tree being set on fire to smoke out bees or game.

Topographic and weather factors have their effects on spotting, but the type of fuel in the main fire and the ignition capability of fuels ahead of the fire are also important. Large volumes of fine surface fuels, such as slash, shaggy-barked and broken-topped snags, and rotten punky wood in the main fire produce spotting material. Crown fires frequently produce spot fires. It is not uncommon for pine cones to be cast out as much as a mile or more ahead of the flame front. Torch outs, or the crowning of individual trees or groups of trees, often cause spot fires a short distance ahead of the main fire.

When crown fires are widespread, burning embers are thrown far ahead of the main fire. Grass fires pushed by winds over 10 mph usually create a spotting problem.

In their order of susceptibility, fuels that readily ignite from burning embers are:

- Rotten wood, either on the ground, on logs, or in snags
- Moss and lichens in treetops
- Slash, particularly when it is compacted in tight piles
- Duff, peat, dried muck
- Cured grass and grain

Lowered fuel moisture and wind increase spotting. Rotten wood often ignites from sparks, while larger embers are necessary to start fires in compacted slash, duff, and cured grass. Of course, sufficient wind is required and is ordinarily present in mass transport.

CROWN FIRES

As we have seen in discussing fuels, many fuel types have the combination of ground and aerial fuels that favor crown fires. Fuel conditions that influence the probability and character of crown fires include:

- Volume and arrangement of the canopy as related to species of vegetation
- Volume and arrangement of fine dry aerial fuels such as moss and dead twigs
- Position of aerial fuels above ground fuels
- Character of ground fuels

Racing crown fires require fuel types with rather dense and continuous crown canopies to maintain their spread.

Fine, light aerial fuels have the greatest effect on crown fires. If the crown canopy is located well above the ground, sufficient ground fuels must be present to create heat intense enough to start and maintain a crown fire. Crown fires rarely develop above light ground fuels unless the trunks of the trees have enough fine fuels continuing to within 2 to 3 feet of the ground.

Where ground or surface fires dry and scorch aerial fuels without triggering a crown fire, conditions may later change, causing a crown fire to sweep back through the stand (a reburn). Because of the predrying of fuels in this situation, crown fires may start easily and burn intensely, resulting in a much more devastating burn.

NATIONAL FIRE DANGER RATING SYSTEM

HISTORICAL BACKGROUND

Several systems for uniformly describing the cumulative effects of weather in its relation to wildfire behavior were developed by the USDA Forest

Service regions in the early 1930's. By 1940 the need for a nationally uniform fire danger rating system was emphasized at a fire control conference called by the USDA Forest Service at Ogden, Utah.

In the 1950's, there were eight different fire danger rating systems in use throughout the United States. Improved transportation and communications made mutual aid agreements practical between fire control organizations. This was especially true in the case of state compacts and interagency and interregional agreements between federal agencies and states. Because of these developments, the need for a national system of rating fire danger and fire behavior was necessary to improve and simplify communications among all people concerned with wildfires.

Although a system was developed in 1961 and put into use in 1964 using a spread index and a buildup index, it was not universally adopted immediately; only with modifications was it later adopted for the several geographical areas.

In 1973 the present National Fire Danger Rating (NFDR) system was put into universal operation after having been tested by all federal agencies having primary fire control responsibility and many state agencies in the same category. Some 150 stations operated the system from Maine to California and from Florida to Alaska before it was perfected in its present form. The present system was developed by a National Fire Danger Rating Research Unit established at the Rocky Mountain Forest and Range Experiment Station, USDA Forest Service, maintained at Fort Collins, Colorado, in cooperation with Colorado State University. Assistance was also furnished by the USDA Forest Fire Laboratory at Riverside, California, by the fuel science project and the Fire Physics Project of the Northern Forest Fire Laboratory at Missoula, Montana, and the Southeastern Forest Experimental Station at Atlanta, Georgia.

OPERATION OF NFDR

The system is operated by the National Fire Weather Service. Fire weather forecasters are located in the principle weather service regional offices. Data for operation are furnished by a considerable number of fire weather stations established on national forests, national parks, Bureau of Land Management units, and national wildlife refuges and by state forestry agencies, often utilizing regular weather service stations. There are sufficient fire weather stations and sufficient available data from all weather service sources in the network that meaningful data are available for any part of the United States when the system is activated during the usual fire season for a particular region of the country.

For particular information for any section, contact your state forestry agency or the National Weather Service of the U.S. Department of Commerce, as they can furnish the information for meaningful use of the system by any fire service.

PHILOSOPHY

The framework within which the NFDR system has been developed is termed the philosophy of the system and is summarized in the following six points:

1. Only the initiating fire is considered. This is defined as a fire that is not behaving erratically; it is spreading without spotting through fuels that are continuous on the ground and there is no crossing in aerial fuels. Other than showing that extreme behavior is a result of increasing fire danger, it is not yet possible to exactly define specific conditions that cause fires to behave erractically.

2. The system provides a measure of that portion of the total job of containment that is due to fire behavior. By considering only containment, rather than the total job of extinguishment through mop-up, the scope of forecasting fire behavior is limited to the head fire. The part of the total control job dealing with accessibility, soil conditions, and resistance to line construction must be evaluated by other means.

3. The length of the flames at the head of the fire is assumed to be directly related to the contribution that fire behavior makes to the job of containment.

4. This system evaluates the worst conditions on a rating area by *(a)* taking the measurements when fire danger is normally the highest (usually early afternoon), *(b)* measuring the fire danger in the open (not in the protection of a stand of timber or brush), and *(c)* measuring the fire danger components on extreme aspects (southerly or westerly) where possible. This makes it necessary to lower, not raise, the fire danger values to fit areas away from the immediate vicinity of the fire danger recording station.

5. The system provides ratings that can be physically interpreted in terms of fire occurrence and behavior. These ratings can be used alone or in combination, giving the system the flexibility needed to meet all the different problems of fire control planning and dispatch.

6. Ratings are relevant not absolute. The ratings are linearly related to the consideration being evaluated. That is, when an index doubles, the activity of that index doubles. Because there are so many variables to be measured, our understanding of some relationships is not exact, and the fire danger can vary considerably within a rating area. It is not practical to try to predict exactly what will happen in a given situation.

CLASSIFICATION OF WILDFIRE FUELS

Five classes of fuels, three dead and two living, are considered by the NFDR system. Since the system is designed to measure the job of containment – not total extinguishment – of the fire, only those fuels involved in combustion within the immediate flaming front need to be considered. Because moisture content to a large extent determines the flammability of fuels, the separation of dead fuels into classes was based on the rapidity of the moisture content response of individual fuel particles to changes in relative humidity. This response is identified by time lag classes. Time lag is the time in which a fuel particle loses enough moisture to burn readily. It is defined as the time necessary for a fuel particle to lose approximately 63 percent of the difference between its initial moisture content and its equilibrium moisture content. (See the chapter on fire behavior.) Living fuels are classified according to whether they are woody or herbaceous. Accordingly, three classes of dead fuels are designated as shown in Table 3.4.

TABLE 3.4
Three Classes of Dead Fuels

	Time lag class (hours)		
Dead fuels	*1*	*10*	*100*
Time lag class interval (hours)	0-2	2-20	20-200
Approximate equivalent fuel dimensions (inches)*	¼	¼-1	1+ to 3
Litter and/or duff	surface layer only	surface to ¾ inch underground†	¾+ inch to 4 inches underground‡

* Although the time lag of an individual fuel particle can be determined exactly, correlating time lag with size is not reliable because the relationship is not exact. Time lag in natural fuels of the same size and species varies, owing to structural differences, surface weathering, density, and so on. The figures given above for roundwood are reasonable and are considered acceptable. Because physical properties of litter vary over a wider range, however, the figures are only "best guesses."

† This layer is from the bottom of the surface litter to ¾ inch in the ground. This layer dries out slowly.

‡ This layer is from ¾ inch in the ground to 4 inches in the ground. Moisture is contained in this layer for a long time.

Two classes of living fuels are considered. They are (1) grass and herbaceous plants and (2) twigs less than ¼ inch in diameter and the foliage of woody plants. One-quarter inch is considered to be the upper size limit of living woody material that can be dessicated and consumed within the flaming front of an initiating fire.

THE CONCEPT OF FUEL MODELS

In its simplest form, the rating of fire danger is the prediction of the behavior of a potential fire. The principal factors of fire behavior can be classified as being variable or constant over a period of time. The variables are weather dependent; they are (1) wind, (2) fuel moisture, and (3) fuel temperature. The factors of fuel and topography are more constant.

1. Wind affects fire behavior by increasing the flow of oxygen to the fire and by bending the flames over the unburned fuel and thus increasing the flow of hot gases from the combustion zone. Both actions contribute to the preheating of the unburned fuels.
2. Fuel moisture is governed by insolation (effects of the sun), air temperature, humidity, and precipitation. The lower the fuel moisture, the less energy required to dry and ignite the fuel particles.
3. Fuel temperature is dictated by air temperature and insolation. The higher the fuel temperature, the less energy required to raise the fuel to the ignition point.

The factors of wind, fuel moisture, and fuel temperature are evaluated daily in the NFDR system.

Slope is the one factor of topography that is accounted for in the system. Increasing slope accelerates burning. The steeper the slope, the closer the flames are to the unburned fuels; the result is that drying and preheating are more efficient.

Fuels are appraised by fuel models. These mathematical models consider such fuel bed properties as compactness (bulk density) and loading (weight per unit area) by classes of living and dead fuel particles. They also consider fuel properties such as density, heat content, mineral content, moisture content, and geometry. The fuel model is a simulated fuel complex for which all the required fuel descriptions have been determined; it consists of a complete set of the varying fuel values necessary for the solution of the mathematical model.

Each fuel model is formulated by the typical values for the loadings by fuel classes, surface area to volume ratios, and bed depths. Fuel particle properties such as density, heat content, and mineral content are assigned constant values. The moisture content by fuel classes are variables determined by normal functions and meteorological processes and therefore must be measured daily. Fuel moisture content, wind, and fuel temperature account for the short-term variations in fire danger.

For a given set of burning conditions, the spread component, energy release component, and burning index will depend on which fuel model is being used. As an example, at a given fuel moisture, wind speed, and slope, the spread component for fuel model A (grass) is higher than that for fuel model E (oak, hickory, and pine). Figures 3.54-3.57 illustrate the response of the several fuel models to changes in burning conditions.

The nine fuel models that follow are only general descriptions because they represent all wildfire fuels from Florida to Alaska and from East Coast to California.

SELECTION OF FUEL MODELS

Several options may be considered as a basis for selecting the base area. The base area may be chosen on the basis of: (1) Where the most fires occur; (2) Where the potential cost of suppression plus loss of property or man-made structures is greatest; and (3) Where the fire control personnel feel there is a key area to which they are able to relate fire danger across the protection unit.

Regardless of which of the above options is used, a careful analysis of the fire history on the protection unit is absolutely essential.

Once the base area has been chosen, the next step is to select the fuel model that best represents the fuels found there. Nine fuel models have been formulated for use with the NFDR system at this time; however, it is unlikely that more than two or three will be usable on any one protection unit. Besides the one selected for the base area (base fuel model), other applicable models should be determined for dispatch purposes for fires in fuels other than those covered by the base fuel model.

The following key and narrative descriptions should help in selecting the correct fuel model. Keep in mind that the models are not based on cover types but are based on how much fuel, by classes, is present and how it is

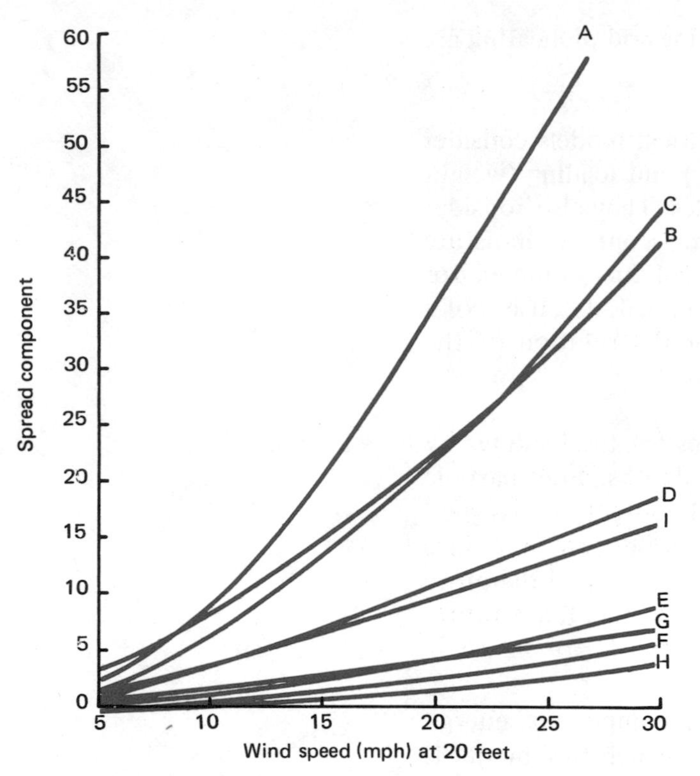

Figure 3.54. The spread component for each fuel model (A-I) is plotted against varying wind speeds.

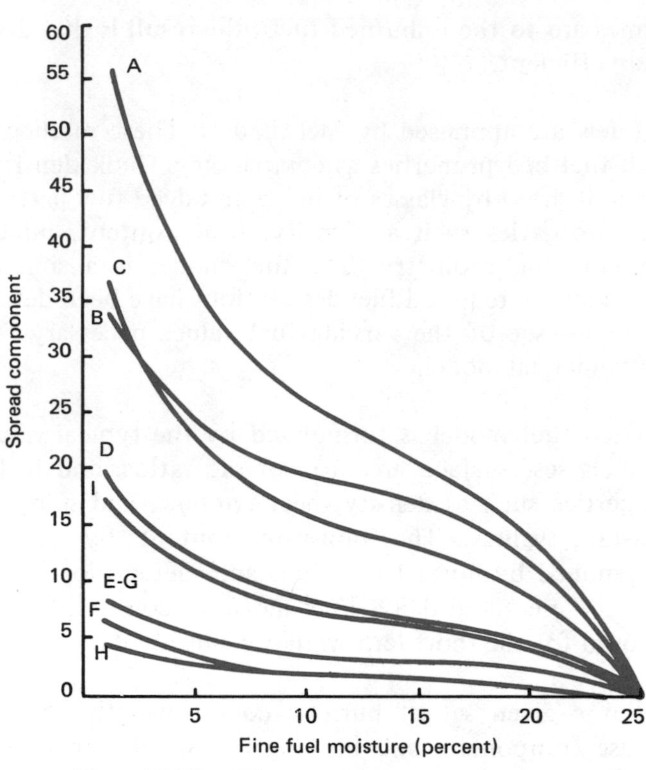

Figure 3.55. The spread component for each fuel model (A-I) is plotted against varying fine fuel moisture values.

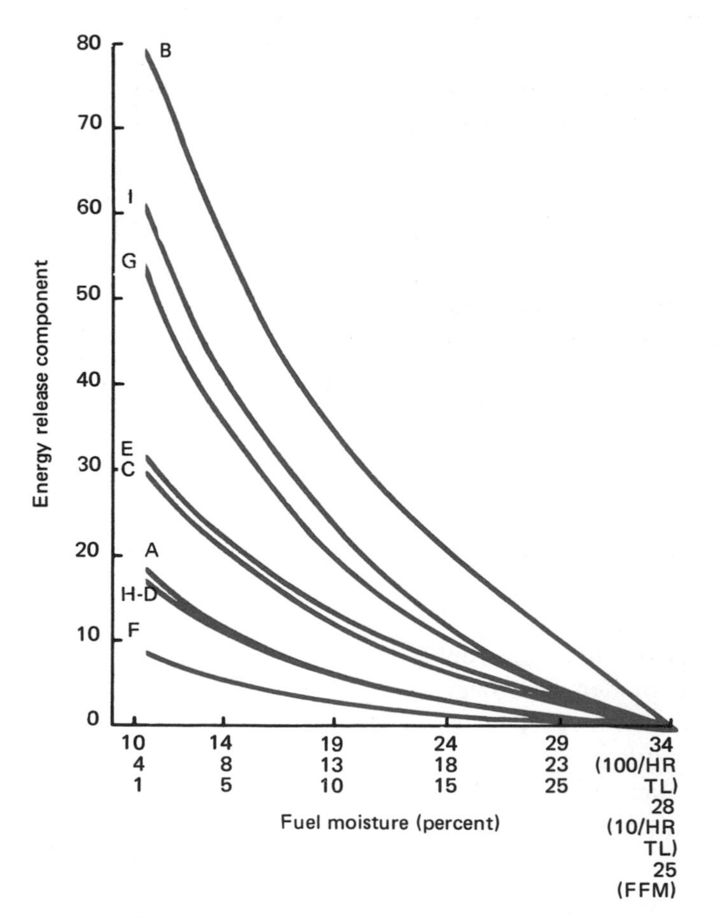

Figure 3.56. The energy release component for each fuel model (A-1) is plotted against varying fuel moisture values.

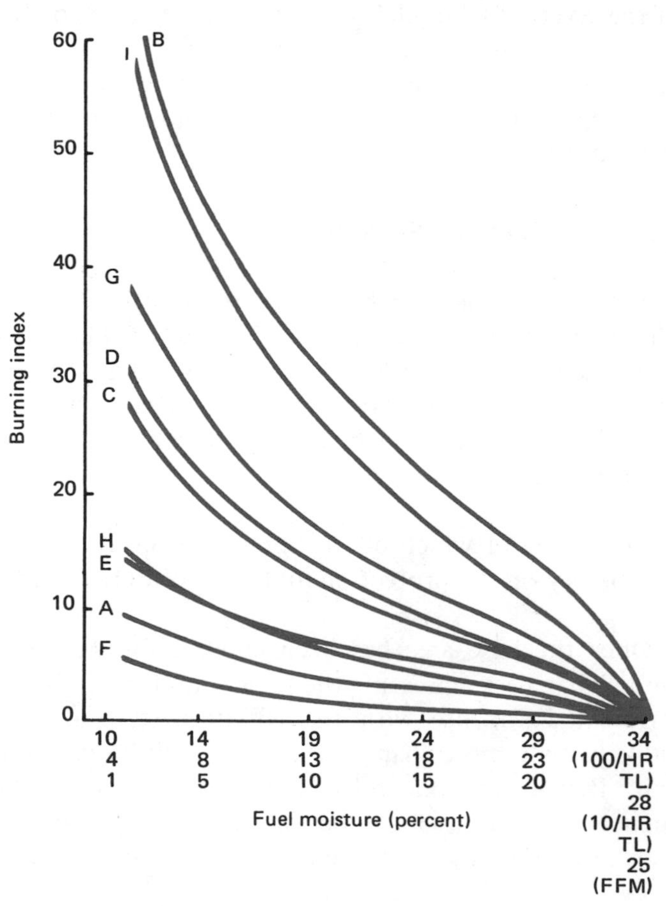

Figure 3.57. The burning index for each fuel model (A-1) is plotted against varying fuel moisture values.

arranged. Rather diverse cover types are grouped together because they have similar fuel properties.

Because data were lacking, fuel models could not be constructed for several common wildfire situations. In the following key, such situations are designated with an asterisk, and a "best fit" model is suggested.

FUEL MODEL KEY

I. The area is not timbered; less than one third of the area is occupied by trees. (Stunted tree species and conifer reproduction are grouped with shrubs and called brush. Slash is not considered brush.)
   - A. Grass and other herbaceous plants or mosses and lichens are the predominant fuel. Brush, slash, and trees together occupy less than one third of the area . . . . Model A
   - B. Grass and other herbaceous plants are not the predominant fuel.
     - 1. Brush or tree reproduction makes up the predominant plant cover and occupies one third or more of the area.
       - a) The foliage of the predominant woody cover species burns readily.
         - (1) The predominant cover occupies two thirds or more of the area.
           - (*a*) One-third or more of the woody portion of plants is dead; much of it is 2 inches or larger in diameter, or there is a duff-litter layer at least 3 inches deep. Cover must average 6 feet or more in height . . . . Model B
           - (*b*) The cover contains little dead woody fuel larger than 2 inches in diameter, and the duff-litter layer is less than 3 inches deep . . . . . . Model D
         - (2) The predominant cover occupies one third but less than two thirds of the area . . . . Model C*
       - b) The foliage of the predominant cover species does not burn easily . . . . Model F
     - 2. Slash is the predominant fuel.
       - a) The foliage is still attached to the slash.
         - (1) Coniferous slash . . . . Model I
         - (2) Hardwood slash . . . . Model D*
       - b) The foliage is no longer attached to the slash; settling is evident.
         - (1) Herbaceous plants have invaded the area . Model C*
         - (2) Brush has invaded the area . . . . Model D*

II. The area is timbered. One third or more of the area is occupied by trees.
   - A. The area has been thinned or partially cut, leaving slash as the major fuel component.
     - 1. Coniferous slash with needles attached . . . . Model B*
     - 2. All hardwood slash or coniferous slash with needles no longer attached . . . . Model G
   - B. The area has not been thinned or partially cut.
     - 1. Grass and other herbaceous plants are a common ground fuel; the canopy of the overstory is commonly "open" . . Model C
     - 2. Duff-litter, branchwood, and tree boles are the primary ground fuel; the canopy of the overstory is "closed," though openings may be common in the stand.

a) Two thirds or more of the overstory consists of deciduous species.

(1) The overstory is dormant; the leaves have fallen, and the leaf litter is not compact . . . . Model E

(2) The overstory is not dormant or the leaf litter has been compacted by rain or snow . . . . . . Model H

b) One third or more of the overstory consists of evergreen species.

(1) The overstory is mature or overmature and is often decadent. There is an exceptionally heavy accumulation of branchwood, downed trees, and duff-litter on the forest floor . . . . Model G

(2) The overstory is immature or mature. There is only a nominal accumulation of debris on the forest floor.

(*a*) Brush or reproduction occupies less than one third of the area.

(i) The ground fuel is primarily needles 2 inches or more long (most pines) . . Model E*

(ii) The ground fuel is primarily needles less than 2 inches long . . . . Model H

(*b*) Brush or reproduction occupies one third or more of the area.

(i) The foilage of the understory burns readily, or needle drape is prevalent . Model D*

(ii) The foliage of the understory does not burn readily, and there is little or no needle drape . . . . Model F

Fuel model A. Cover types represented in this model are grasslands and other lands such as tundra, where the primary carrier of fire is a continuous bed of fine fuels. Brush or trees may be present, but the crowns should not occupy more than one third of the area. Concentrations of brush or trees within the type are such that control lines need not be placed close to them. Examples of areas covered by fuel model A are western grasslands, the savannah, and the tundra. Open stands of pinyon and juniper and desert shrubs such as mesquite, creosote bush, and paloverde may be included, but fire spread will be overrated during those periods when grazing or lack of rain prevents herbaceous ground fuels from developing. (Savannah is an area of grassland without trees or containing only scattered trees occurring in a generally forested area.)

Fuel model B. This model represents chaparral or other dense brush and recently partially cut or thinned conifer stands where slash is a serious fuel problem. These situations are characterized by heavy loadings of dead woody material over 2 inches in diameter, or duff-litter layers 3 or more inches deep. In the brush, the primary cover plants must average 6 or more feet in height. Foliage of brush plants typically becomes easily involved in the fire. Individual plants in these associations almost always form a dense, continuous fuel bed occupying two thirds or more of the area. The typical cover included in this model are mature California and some other southwest chaparral, the pine barrens of New Jersey, and the high pocosins of the central Atlantic coast. (Pocosins are swamps and low flats from southern New Jersey to northern Florida and central Alabama. They are the usual habitat of pond and pitch pines.)

Fuel model C. Apply this model where grass or other herbaceous plants are the primary carrier of the fire. The overstory is typically open, occupying one third but not more than two thirds of the area. Usually, enough branch material is on the ground to contribute significantly to fire intensity. In the more open areas, concentrations of brush or trees are such that it is difficult to avoid placing control lines close to them. Types included may be young conifer plantations where the trees occupy less than two thirds of the area; open areas of ponderosa, sugar, longleaf, slash, and sand pines; and areas of wire grass-scrub oak and timber-sagebrush-grass associations. Desert shrubs and pinyon-juniper stands are included in this model for areas where a continuous ground fuel develops most years and where pinyon makes up more than one third of the stand.

Fuel model D. Use this model where there is a heavy loading of fuels 1 inch or less in diameter and little or no material greater than 2 inches in diameter. Usually, the living foliage burns readily. For example, this model applies to the low pocosins of the Atlantic coast, palmetto-gallberry areas, sagebrush and conifer plantations, and other situations where the woody plants occupy two thirds or more of the area. Also covered here are black spruce and the black spruce-aspen-poplar areas of Alaska; but the latter area is included only where the spruce makes up one third or more of the overstory. This model covers those chaparral situations that are not dense enough or heavy enough to qualify for model B.

Fuel model E. This model applies to hardwood and mixed conifer-hardwood stands when the hardwoods are dormant and before the leaf litter has been compacted by rain or snow (when hardwoods are leafed out, use model H). This model was constructed primarily for the oak-hickory type of area where the litter consists of large, coarse leaves that do not compress readily, as do the leaves of such species as maple, tulip poplar, aspen, and similar species. Under very high wind conditions, this model will underrate fire danger since it cannot account for the increase in spread due to rolling or blowing leaves. This model also covers closed stands of conifers with needles 2 inches or more in length (most pines). (The short-needled conifers produce a much denser, more compact litter bed, which is better covered by model H.)

Fuel model F. This model represents situations where there is a fairly continuous cover of young brush or shrub species that contain little or no dead material and where the foliage does not burn readily. Types covered by this fuel model are laurel, salal, vine maple, alder, and mountain mahogany. Also included are young stands of chamise and manzanita. Grass, ferns, and other herbaceous plants may be present, but, if there is a continuous cover of ground fuels, model A should be used.

Fuel model G. This model applies primarily to dense conifer stands where a heavy buildup of downed tree material has accumulated. Natural breakup of overmature stands, insect and disease damage, wind or ice storms, and thinnings or partial cuts are typical events that create the heavy amounts of fuel that typify this model. The canopies of these stands are usually closed, but large openings, the result of the downing of timber, are common. Deep litter and a very high loading of dead fuels larger than 1 inch in diameter are also characteristic. The amount of undergrowth may be quite varied. Types covered by this fuel model are hemlock and Sitka spruce, coastal Douglas fir, or wind-thrown or bug-killed lodgepole pine and spruce as well as thinned or partially cut conifer stands where there is heavy slash after the fines have dropped off and the slash has settled. When the slash is fresh, fuel model B should be used.

Fuel model H. Most closed short-needled conifer types and hardwoods, when in leaf or after compaction of the leaf litter by rain or snow, are represented by this model. These associations contain a variable amount of undergrowth, but initiating fires seldom burn other than on the ground through a shallow, dense litter layer that contains only a small amount of dead branch wood. Pine and pine-hardwood associations (where pines make up one third or more of the overstory) and dormant hardwoods (before the litter has been compacted) are best covered by model E.

Fuel model I. This model was designed to satisfy the need for determining fire danger ratings for clear-cut conifer areas. The loadings used are highest of all fuel models, but represent midrange values for most such areas in the West. Exceptions are: (1) clear-cut areas in the southern pine forests (model C is more appropriate) and (2) ponderosa and other pines of similar growth characteristics (use model D).

SLOPE CLASS

Like the fuel model, the slope class for a protection unit should be selected carefully. The basic consideration for the assignment of the slope class is the topography in that portion of the base area where initial attack is commonly initiated. Once again, knowledge of the fire history in the protection unit is essential.

Slope Class	*Slope %*
1	0 - 20
2	21 - 40
3	>40

The slope class should be selected to represent the most commonly encountered slope in the area represented by the selected fuel model.

FIRE WEATHER STATIONS

Fire danger rating is an integration of weather elements and other factors affecting fire potential. Fire weather stations are established in locations that are representative of the area protected. Standard weather stations are also adapted for fire weather use by making certain necessary modifications. The data collected each day at these stations are transmitted to central weather service offices by teletype, radio, or phone for use by the fire weather forecaster.

Fire weather stations are used to measure the principal weather elements of temperature, relative humidity, dew point, wind speed and direction, precipitation, and dead fuel moisture. Added to these observations are mental elements of risk, slope, and live fuel moisture for both woody and herbaceous vegetation. These elements are applied to the characteristics of one of the nine fuel models that best fits the type of fuel in the area for which a rating is being determined. These mental estimates are made according to a standard procedure to provide uniformity in the use of available information.

All readings taken from the instruments at the fire weather station are recorded on a standard form from the National Weather Service (Table 3.5). Then fire behavior indexes and components can be developed from the data recorded. In the following explanation of the indexes and components used to calculate fire behavior in the NFDR system, fuel model A has been used as an example, and the values for this model are calculated in Tables 3.8 to 3.18.

STRUCTURE OF THE NFDR SYSTEM

Three fire behavior components are developed from weather station instrument readings, objective estimates based on these readings, and mental estimates of observable conditions made according to standard methods to provide uniformity of such estimates. The three fire behavior components are: ignition component (IC), spread component (SC), and energy release component (ERC). Each component is evaluated on a scale of 0 to 100 for all fuel models.

From the components plus an appraisal of risk, three indexes are developed: occurrence index (OI), fire load index (FLI), and burning index (BI). (See Table 3.2.) The occurrence index is evaluated on a scale of 0 to 100 for all fuel models, but the fire load index and the burning index for each fuel model are evaluated on a scale unique for each fuel model (see Tables 3.6 and 3.7). A seasonal severity index (SSI) may also be developed.

The individual components or indexes can be used alone or in combination to determine the class of fire danger. In this way, the NFDR system can be tailored to meet the needs of the individual fire control agency. The increased flexibility with the multiple components and indexes gives a wide choice in how the system is used.

Each component will be described, followed by a description of the indexes (see Figure 3.58).

Ignition component. The ignition component (IC) is a basic part of the system, since it is derived from basic factual readings of temperature and relative humidity coupled with the state of the weather as observed, i.e., cloudy or sunny. It appraises the fuel moisture of dead fuels less than ¼ inch in diameter and live herbaceous fuels. Since these are present in all fuel models, their capability of ignition is basic to all fuel types. Therefore the rating is based on a scale of 0 to 100 for all fuel models. This component represents the ease with which fine fuels are ignited. Ignition normally takes place in the dead part of the fine fuels.

Living material in the fine fuel complex shields the dead fuel from the firebrand or otherwise reduces its efficiency. Therefore, an adjustment is made dependent on the percentage of the fine fuels that are living. This is termed the herbaceous vegetation condition. Fire behavior is very sensitive to this factor, making a rigorous evaluation necessary. It is best done with a range forage volume transect conducted on a representative site within the base area.

TABLE 3.5
Ten-Day Fire Danger and Fire Weather Record

WS FORM D–9a (1-72) U.S. DEPARTMENT OF COMMERCE NOAA NATIONAL WEATHER SERVICE	AGENCY		UNIT		STATION NAME	STATION NUMBER
10 DAY FIRE DANGER AND FIRE WEATHER RECORD	STATION ELEVATION	FUEL MODEL	SLOPE CLASS	BASIC OBSERVATION TIME *(LST)*	PERIOD OF RECORD *(Month, Day, Year)* FROM	TO

DAY OF MONTH	OCCURRENCE													BURNING INDEX								FIRE LOAD INDEX					
	IGNITION COMPONENT									RISK FACTORS				SPREAD COMPONENT				ENERGY RELEASE COMPONENT									
		TEMPERATURES												WIND													
	STATE OF WEATHER	DRY BULB	WET BULB	DEW POINT	RELATIVE HUMIDITY	1-HOUR TL FUEL MOISTURE	HERB. VEG. CONDITION	FINE FUEL MOISTURE		LIGHTNING	MAN-CAUSED	TOTAL		DIRECTION	SPEED	WOODY VEG. CONDITION		10-HOUR TL FUEL MOISTURE	100-HOUR TL FUEL MOISTURE								
1	2	3	4	5	6	7	8	9	10	11	12	13	14	15	16	17	18	19	20	21	22	23	A	B	C	D	E
1																											
2																											
3																											
4																											
5																											
6																											
7																											
8																											
9																											
0																											
31																											

DAY OF MONTH	DAILY (24–HOUR) DATA													100–HOUR TL FUEL MOISTURE							
	TEMPERATURES		RELATIVE HUMIDITIES			PRECIPITATION					LIGHTNING										
	MAXIMUM	MINIMUM	MAXIMUM	MINIMUM	AVERAGE	KIND	BEGAN	ENDED	DURATION	AMOUNT	BEGAN	ENDED	ACTIVITY LEVEL	CORRECTION FOR RELATIVE HUMIDITY	CORRECTION FOR PRECIPITATION	YESTERDAY'S 100-HOUR TL FUEL MOISTURE	TODAY'S 100-HOUR TL FUEL MOISTURE				
24	25	26	27	28	29	30	31	32	33	34	35	36	37	38	39	40	41	F	G	H	I
A	X	X	X	X	X				X	X	X	X		+/– X	X	X					
1										.					+						
2										.					+						
3										.					+						
4										.					+						
5										.					+						
6										.					+						
7										.					+						
8										.					+						
9										.					+						
0										.					+						
31										.					+						

OBSERVER
CHECKED BY
REMARKS

The moisture content of the dead fuel is the one-hour time lag fuel moisture (1 HR TL FM). This is determined by the temperature and humidity of the air immediately in contact with the fuel particle. The temperature is read directly from the dry bulb thermometer (data obtained from column 3 of Table 3.5) and the humidity (column 6 of Table 3.5) is derived from the psychrometric tables. Using Table 3.8 the 1 HR TL FM is obtained. If the sky is less than one-half cloud covered, the sunny column (Code 0-1) is used; otherwise, the cloudy column (Code 2-9) is used for the dry bulb temperature. This accounts for the insolation or the effects of the sun's radiation. The 1 HR TL FM value obtained from Table 3.8 is recorded in column 7 of Table 3.5 and is used in Table 3.9.

Herbaceous vegetation condition, or the stage of curing (column 8 of Table 3.5) and the 1 HR TL FM values (column 7 of Table 3.5) determine the fine fuel moisture (FFM) factor for both living and dead fuels. This is

TABLE 3.6
Comparison of Numerical Values for Components and Indexes of NFDR

	(1)	*(2)*	*(3)*	*(4)*	*(5)*	*(6)*	*(7)*	*(8)*	*(9)*	*(10)*
Fuel model	*1 hour TL, FM*	*Fine fuel moisture*	*IC*	*OI*	*SC*	*ERC*	*BI*	*FLI*	*Correction to 100 TL, FM for RH*	*Correction to 100 TL, FM for Precipitation*
A	1-25+	2-25+	0-100	0-100	0-100	0-19	0-12	0-12		
B	1-25+	2-25+	0-100	0-100	0-87	0-100	0-100	0-100	+9- -72	0-16
C	1-25+	2-25+	0-100	0-100	0-74	0-34	0-41	0-44		
D	1-25+	2-25+	0-100	0-100	0-33	0-31	0-51	0-53	+9- -72	0-16
E	1-25+	2-25+	0-100	0-100	0-17	0-36	0-22	0-22		
F	1-25+	2-25+	0-100	0-100	0-14	0-11	0-8	0-8		
G	1-25+	2-25+	0-100	0-100	0-13	0-85	0-58	0-59	+9- -72	0-16
H	1-25+	2-25+	0-100	0-100	0-8	0-34	0-27	0-28	+9- -72	0-16
I	1-25+	2-25+	0-100	0-100	0-28	0-96	0-90	0-90	+9- -72	0-16

TABLE 3.7
Comparative Numerical Estimates of Adjective Ratings for the Burning Indexes*

Fuel model adjective rating	*B 100 scale*	*F 8 scale*	*A 12 scale*	*E 22 scale*	*H 27 scale*	*C 41 scale*	*D 51 scale*	*G 58 scale*	*I 90 scale*
Low	0-30	0-2	0-3	0-6	0-8	0-12	0-15	0-17	0-27
Moderate	31-50	3-4	4-5	7-10	9-13	13-20	16-25	18-29	28-45
High	51-70	5-6	6-8	11-15	14-19	21-29	26-35	30-40	46-61
Very high	71-90	7	9-11	16-20	20-24	30-36	36-45	41-50	62-80
Extreme	91-100	8	12	21-22	25-27	37-41	46-51	51-58	81-90

*A scale less than 100 is used.

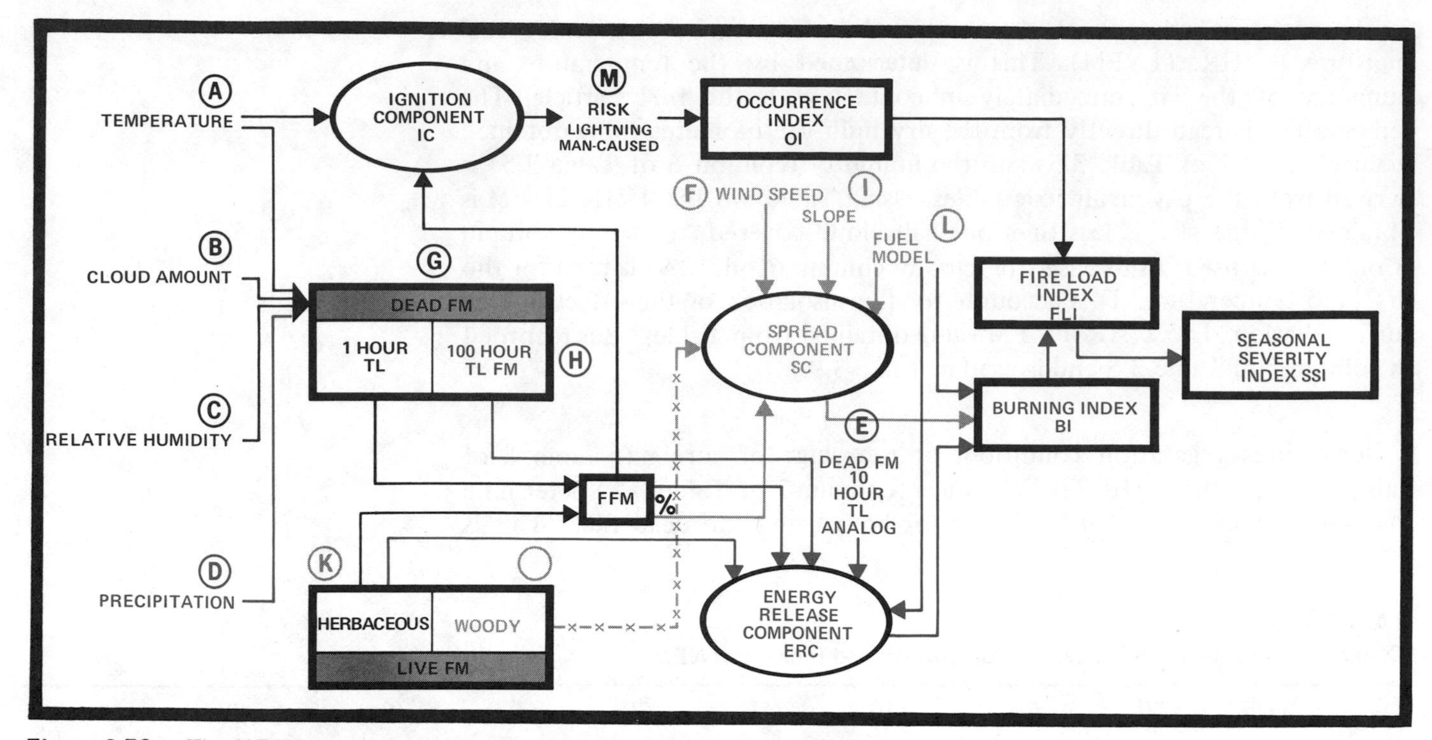

Figure 3.58. The NFDR system rates fire danger. A, B, C, D, E, and F are measurements taken at the fire weather station. G and H are objective estimates determined from measurements. I, J, K, L, and M are subjective estimates derived by a standard procedure. These data determine ignition component, occurrence index, spread component, energy release component, burning index, fire load index, and seasonal severity index.

TABLE 3.8
One-Hour Time Lag Fuel Moisture (percent)

State of weather		*Relative humidity (percent)*																				
Code 0-1	*Code 2-9*																					
Dry Bulb Temperature (°F)		0 ↓ 4	5 ↓ 9	10 ↓ 14	15 ↓ 19	20 ↓ 24	25 ↓ 29	30 ↓ 34	35 ↓ 39	40 ↓ 44	45 ↓ 49	50 ↓ 54	55 ↓ 59	60 ↓ 64	65 ↓ 69	70 ↓ 74	75 ↓ 79	80 ↓ 84	85 ↓ 89	90 ↓ 94	95 ↓ 99	100
SUNNY 10-29		1	2	2	3	4	5	5	6	7	8	8	8	9	9	10	11	12	12	13	13	14
30-49		1	2	2	3	4	5	5	6	7	7	7	8	9	9	10	10	11	12	13	13	13
50-69		1	2	2	3	4	5	5	6	6	7	7	8	8	9	9	10	11	12	12	12	13
70-89		1	1	2	2	3	4	5	5	6	7	7	8	8	8	9	10	10	11	12	12	13
90-109		1	1	2	2	3	4	4	5	6	7	7	8	8	8	9	10	10	11	12	12	13
109+		1	1	2	2	3	4	4	5	6	7	7	8	8	8	9	10	10	11	12	12	13
	CLOUDY 10-29	1	2	4	5	5	6	7	8	9	10	11	12	12	14	15	17	19	22	25	25+	25+
	30-49	1	2	3	4	5	6	7	8	9	9	11	11	12	13	14	16	18	21	24	25+	25+
	50-69	1	2	3	4	5	6	6	8	8	9	10	11	11	12	14	16	17	20	23	25+	25+
	70-89	1	2	3	4	4	5	6	7	8	9	10	10	11	12	13	15	17	20	23	25+	25+
	90-109	1	2	3	3	4	5	6	7	8	9	9	10	10	11	13	14	16	19	22	25	25+
	109+	1	2	2	3	4	5	6	6	8	8	9	9	10	11	12	14	16	19	21	24	25+

read from Table 3.9 and is recorded in column 9 of Table 3.5. The scale of 0 to 100 for this factor is the same for all fuel models.

The FFM factor and the dry bulb temperature (column 3 of Table 3.5) adjusted for the state of the weather (cloudy or sunny; column 2 of Table 3.5) determine the IC in Table 3.10; the scale of 0 to 100 is the same for all fuel models. The IC is recorded in column 10 of Table 3.5.

The adjective rating for the class of fire danger when only the IC is used is shown in Table 3.11. This has been adjusted for wind speed. It is developed for a specific regional area.

The IC is probably used in more locations than any other component or index to describe the fire weather danger class. It indicates how easily the fine fuels will ignite.

Occurrence Index. An appraisal of risk is a mental estimate of both lightning and man-caused potential for starting fires. The IC and the appraisal of risk (column 13 of Table 2) are used to determine the occurrence index (OI). This would be a number on the 0 to 100 scale that indicates the potential fire incidence within a rating area (Table 3.12). The OI is recorded in column 14 of Table 3.5.

The development of the risk factor requires a study of and analysis of the causes and occurrence of man-caused fires and the recording of and prediction of lightning activity, including an allowance for holdover fires.

Spread component. The spread component (SC) is derived from the mathematical model developed by the Northern Forest Fire Laboratory. This model integrates the effects of wind speed and slope, together with fuel bed and fuel particle properties to compute the fire spread rate.

Since the characteristics of the fuels through which the fire is burning are basic to determining the rate at which the fire front moves, a unique SC table is necessary for each of the fuel models characterizing fuel types. (Refer to Table 3.6. Note that fuel model A, which is for the grass fuels, has the largest span, 0 to 100.)

The FFM factor (column 9 of Table 3.5), as developed in the IC, is used along with slope and wind speed to determine the SC for fuel models A, C, E, G, H, and I. However, for fuel models B and F the woody vegetation condition of live fuels is also considered in the calculation of the SC because of the kind of fuels involved in these two models. Table 3.13 shows the determination of the SC for fuel model A.

The SC indicates the rate of forward spread of the fire and will give an indication of the time within which a fire must be contained to prevent it from exceeding an acceptable size. It can also be used to determine the necessary locations for fire-fighting units, since travel time and rate of line construction by the various mixes of men and machines are key factors. Table 3.14 shows a sample description of the meanings of the adjective ratings and the numerical ratings. The SC is recorded in column 18 of Table 3.5.

TABLE 3.9
Fine Fuel Moisture (FFM) Values

1-hour TL fuel moisture (percent)	*Herbaceous vegetation condition (percent)*								
	0 ↓ 4	5 ↓ 14	15 ↓ 24	25 ↓ 34	35 ↓ 44	45 ↓ 54	55 ↓ 64	65 ↓ 74	75 ↓ 75+
1	—	2	3	4	5	8	13	18	21
2	—	3	4	5	7	10	16	19	22
3	—	4	5	7	9	14	18	20	22
4	—	5	6	8	12	16	19	21	23
5	—	6	8	11	14	18	20	22	23
6	—	7	10	13	16	19	20	22	23
7-8	—	9	12	15	18	20	21	22	23
9-10	—	12	15	17	19	20	22	23	24
11-12	—	14	17	18	20	21	22	23	24
13-14	—	16	18	19	20	21	22	23	24
15-16	—	17	19	20	21	22	22	23	24
17-18	—	19	20	21	21	22	23	23	24
19-21	—	21	21	22	22	23	23	24	24
22-24	—	24	24	24	24	24	24	24	25
25-25+	—	25+	25+	25+	25+	25+	25+	25+	25+

TABLE 3.10
Ignition Component

State of weather		*Fine fuel moisture (percent)*														
Code 0-1	*Code 2-9*															
Dry Bulb Temperature (°F) SUNNY	CLOUDY	1	2	3	4	5	6	7 ↓ 8	9 ↓ 10	11 ↓ 12	13 ↓ 14	15 ↓ 16	17 ↓ 18	19 ↓ 21	22 ↓ 24	25 ↓ 25+
10-19	10-39	88	75	64	54	46	39	30	21	14	9	5	2	0	0	0
20-29	40-49	90	77	66	56	48	41	32	22	15	9	5	2	0	0	0
30-39	50-59	93	80	68	58	50	42	33	23	16	10	6	3	0	0	0
40-49	60-69	95	82	71	61	52	44	35	25	17	11	7	3	1	0	0
50-59	70-79	98	85	73	63	54	46	36	26	18	12	7	4	1	0	0
60-69	80-89	100	87	76	65	56	48	38	28	19	13	8	5	1	0	0
70-79	90-99	100	90	78	68	58	50	40	29	21	14	9	5	2	0	0
80-89	100-109	100	93	81	70	61	53	42	31	22	15	10	6	2	0	0
90-99	110-119	100	97	84	73	63	55	44	32	23	16	11	7	3	0	0
100-109	120-120+	100	100	87	76	66	57	46	34	25	18	12	8	4	0	0
110-119		100	100	90	79	69	60	49	36	27	19	13	9	4	1	0
120-120+		100	100	92	80	70	61	50	37	28	20	14	9	5	1	0

TABLE 3.11
Adjective Rating for Fuel Model A, Slope Class 1 Using the Ignition Component Only

Ignition Component	*Wind speed (mph)*	
	0-18	*19+*
91-100	E	E
71-90	VH	E
51-70	H	VH
31-50	M	H
0-30	L	M

TABLE 3.12
Occurrence Index

Ignition component	*Total risk*																			
	1 ↓ 5	6 ↓ 10	11 ↓ 15	16 ↓ 20	21 ↓ 25	26 ↓ 30	31 ↓ 35	36 ↓ 40	41 ↓ 45	46 ↓ 50	51 ↓ 55	56 ↓ 60	61 ↓ 65	66 ↓ 70	71 ↓ 75	76 ↓ 80	81 ↓ 85	86 ↓ 90	91 ↓ 95	96 ↓ 100
0	0	0	0	0	0	0	0	0	0	0	0	0	0	0	0	0	0	0	0	0
1-5	0	0	0	1	1	1	1	1	1	2	2	2	2	2	2	2	3	3	3	3
6-10	0	1	1	2	2	2	3	3	4	4	4	5	5	6	6	7	7	7	8	8
11-15	0	1	2	2	3	4	4	5	6	7	7	8	9	9	10	11	11	12	13	13
16-20	1	2	2	3	4	5	6	7	8	9	10	11	12	13	14	15	16	17	17	18
21-25	1	2	3	4	5	7	8	9	10	12	13	14	15	16	17	19	20	21	22	23
26-30	1	2	4	5	7	8	10	11	13	14	15	17	18	20	21	23	24	26	27	29
31-35	1	3	4	6	8	10	11	13	15	16	18	20	22	23	25	27	29	30	32	34
36-40	1	3	5	7	9	11	13	15	17	19	21	23	25	27	29	31	33	35	37	39
41-45	1	4	6	8	10	13	15	17	19	21	24	26	28	30	33	35	37	39	42	44
46-50	1	4	6	9	12	14	16	19	21	24	26	29	31	34	36	39	41	44	46	49
51-55	2	4	7	10	13	15	18	21	24	26	29	32	35	38	40	43	46	49	51	54
56-60	2	5	8	11	14	17	20	23	26	29	32	35	38	41	44	47	50	53	56	60
61-65	2	5	9	12	15	18	22	25	28	31	35	38	41	46	48	51	54	58	61	64
66-70	2	6	9	13	16	20	23	27	30	34	38	41	46	48	52	55	59	62	66	69
71-75	2	6	10	14	17	21	25	29	33	36	40	44	48	52	55	59	63	67	71	75
76-80	2	7	11	15	19	23	27	31	35	39	43	47	51	55	59	63	67	71	76	80
81-85	3	7	11	16	20	24	29	33	37	41	46	50	54	59	63	67	72	76	80	85
86-90	3	7	12	17	21	26	30	35	39	44	49	53	58	62	67	71	76	81	85	90
91-95	3	8	13	17	22	27	32	37	42	46	51	56	61	66	71	76	80	85	90	95
96-100	3	8	13	18	23	29	34	39	44	49	54	59	64	69	75	80	85	90	95	100

Energy release component. The energy release component (ERC) is also expressed by using a table unique to each fuel model. Thus only an appropriate part of the 0 to 100 scale is used in Table 3.14 for fuel model A (c.f. Table 3.5).

This component indicates the burning rate or intensity of burning. The burning rate is, for the most part, dependent on the same fuel properties that are used to calculate the SC.

The SC is determined primarily by the fine fuels, whereas the ERC calculation requires the fuel moisture content values for the 10-hour time lag fuels and the 100-hour time lag fuels for those fuel models where these fuels are applicable. These are the heavier fuels, and they produce larger volumes of heat for longer periods of time.

Table 3.15 shows the ERC values for fuel model A based only on the FFM values (column 9 of Table 3.5) which should be the only type of fuel involved. Note that the values are only slightly higher than the FFM values. Table 3.6 indicates the difference in the ERC values for the fine fuel models.

For fuel models B, D, G, H, and I the ERC is developed from the 100 HR TL FM, the 10 HR TL FM, and the FFM. The 100 HR TL FM is based on two computations. One is derived from the previous day's 100 HR TL FM, corrected by the twenty-four hour relative humidity to give a factor to use as

TABLE 3.13
Spread Component for Fuel Model A

Slope class			*Fine fuel moisture (percent)*														
1	*2*	*3*															
Windspeed (mph)			1	2	3	4	5	6	7 ↓ 8	9 ↓ 10	11 ↓ 12	13 ↓ 14	15 ↓ 16	17 ↓ 18	19 ↓ 21	22 ↓ 24	25 ↓ 25+
0-1			1	1	1	1	1	1	1	1	1	1	1	1	1	0	0
2			2	2	1	1	1	1	1	1	1	1	1	1	1	0	0
3			3	2	2	2	2	2	1	1	1	1	1	1	1	1	0
4	0-2		4	3	3	3	2	2	2	2	2	2	1	1	1	1	0
5	3		5	5	4	4	3	3	3	2	2	2	2	2	1	1	0
6	4-5	0-2	7	6	5	5	4	4	4	3	3	3	3	2	2	1	0
7	6	3	9	8	7	6	6	5	5	4	4	4	3	3	2	1	0
8	7	4-5	11	10	9	8	7	6	6	5	5	4	4	4	3	1	0
9-10	8-9	6-7	15	13	12	10	9	9	8	7	6	6	5	5	4	2	0
11-12	10-11	8-10	21	19	16	15	13	12	11	10	9	8	8	7	5	2	0
13-14	12-13	11-12	29	25	22	20	18	16	15	13	12	11	10	9	7	3	0
15-16	14-15	13-14	37	32	28	25	23	21	19	17	15	14	13	12	9	4	0
17-18	16-17	15-16	46	40	36	32	29	26	23	21	19	18	16	14	11	5	0
19-20	18-20	17-19	56	49	43	39	35	32	29	26	24	22	20	18	14	7	0
21-22	21-22	20-21	68	59	52	47	42	39	34	31	28	26	24	21	16	8	0
23-24	23-24	22-23	80	70	62	55	50	46	41	36	33	31	28	25	19	9	0
25-26	25-26	24-25	93	81	72	64	58	53	47	42	39	36	33	29	23	11	0
27-27+	27-27+	26-26+	100	88	77	69	62	57	51	46	42	39	36	32	24	12	0

TABLE 3.14
Sample Description of Spread Component Index Ratings

Index	*Start*	*Spread*	*Control*
Low 0-30	Not easy, precipitation may be present	Slow, smouldering, very little spotting	Patchy, irregular, easy
Moderate 31-50	Will start from accidental causes	Slow to moderate, may spot some	No special problems
High 51-70	Fairly easily from most causes	Rapid, spot easy	May be difficult, hit hard while small
Very high 71-90	Easily from most causes	Develop fast high rate of spread, much spotting	Direct attack at head is unsafe
Extreme 91-100	Easily from any cause	Burn intensely, rapid spread, many spots	Flank attack only is safe—*Blowup Potential*

TABLE 3.15
Energy Release Component (ERC) Values

Fine fuel moisture (percent)														
1	2	3	4	5	6	7 ↓ 8	9 ↓ 10	11 ↓ 12	13 ↓ 14	15 ↓ 16	17 ↓ 18	19 ↓ 21	22 ↓ 24	25 ↓ 25+
19	17	16	15	13	12	11	9	7	6	4	3	2	1	0

the current day's 100 HR TL FM. The other computation is the previous day's 100 HR TL FM, corrected by the duration of precipitation to show the current days's 100 HR TL FM. The 10 HR TL FM is determined from weighing the ½-inch fuel moisture sticks or by an optional method. For fuel models C, E, and F the ERC is developed from the FFM and the 10 HR TL FM. The 100 HR TL FM is not used because of the character of the fuels in these models.

Burning index. The burning index (BI) is derived from the SC (column 18 of Table 3.5), which portrays the rate of spread, and the ERC (column 21 of Table 3.5), which portrays the flame length.

How fast a fire spreads is indicated by the rate of combustion per unit of area within the flaming front and the width of the flaming zone. Considering these together gives a measure of the potential difficulty of the containment job.

The table for the BI for each fuel model will be different from the others (see Table 3.6). The table for model A (Table 3.16) uses only the values from 0 to 12, while the heavy fuel models B and I show a range of 0 to 100 and 0 to 90, respectively. This difference is due in large measure to the unit volume of the different kinds of fuels. Therefore, the table for each fuel model is unique for that type of fuel and may not cover the entire range from 0 to 100. Nevertheless, the range that is used will express the full range of fire danger from low to extreme for that component or index.

TABLE 3.16
Burning Index (BI) Values

Spread component	Energy release component																			
	0	*1*	*2*	*3*	*4*	*5*	*6*	*7*	*8*	*9*	*10*	*11*	*12*	*13*	*14*	*15*	*16*	*17*	*18*	*19*
0	0	0	0	0	0	0	0	0	0	0	0	0	0	0	0	0	0	0	0	0
1-3	0	0	0	0	1	1	1	1	1	1	1	1	1	1	1	1	1	1	1	1
4-9	0	1	1	1	2	2	2	2	2	2	2	3	3	3	3	3	3	3	3	3
10-15	0	1	2	2	2	2	3	3	3	3	3	3	4	4	4	4	4	4	4	4
16-21	0	1	2	2	3	3	3	3	4	4	4	4	4	4	5	5	5	5	5	5
22-27	0	2	2	3	3	3	4	4	4	4	5	5	5	5	5	5	6	6	6	6
28-33	0	2	2	3	3	4	4	4	5	5	5	5	5	6	6	6	6	6	7	7
34-39	0	2	3	3	4	4	4	5	5	5	5	6	6	6	6	7	7	7	7	7
40-45	0	2	3	3	4	4	5	5	5	6	6	6	6	7	7	7	7	7	8	8
46-51	0	2	3	4	4	5	5	5	6	6	6	7	7	7	7	8	8	8	8	8
52-57	0	2	3	4	4	5	5	6	6	6	7	7	7	7	8	8	8	8	9	9
58-63	0	2	3	4	5	5	5	6	6	7	7	7	8	8	8	8	9	9	9	9
64-69	0	3	3	4	5	5	6	6	7	7	7	8	8	8	8	9	9	9	9	10
70-75	0	3	4	4	5	5	6	6	7	7	8	8	8	8	9	9	9	10	10	10
76-81	0	3	4	4	5	6	6	7	7	7	8	8	8	9	9	9	10	10	10	10
82-87	0	3	4	5	5	6	6	7	7	8	8	8	9	9	9	10	10	10	11	11
88-92	0	3	4	5	5	6	7	7	8	8	8	9	9	9	10	10	10	11	11	11
93-99	0	3	4	5	6	6	7	7	8	8	9	9	9	10	10	10	11	11	11	11
100	0	3	4	5	6	6	7	7	8	8	9	9	10	10	10	11	11	11	12	12

The BI is a number related to the potential amount of effort needed to contain a fire in a particular fuel type within a rating area. Since the BI is made up of the SC and the ERC, a more accurate determination of the men and equipment required for prompt containment of a particular fire can be made by appraising both of these values.

If risk is not considered, the BI can be used as the basis for manning, readiness, plans, and size-up. Table 3.17 shows the relationship of the adjective classes for the BI to several functions.

Fire load index. The fire load index (FLI) is obtained by multiplying the difficulty of containing a single fire, which is the BI (column 22 of Table 3.5) by the probable number of fires, which is the OI (column 14 of Table 3.2), to give a measure of the total fire containment job on a protection unit for a particular day. It will have a unique span for each fuel model because of its relation to the BI. Table 3.18 shows the FLI for fuel model A. The FLI value is recorded in column 23 of Table 3.5.

Seasonal severity index. The seasonal severity index (SSI) will provide a yardstick for the potential fire problem of one fire protection district as compared to another or for evaluating accomplishments. It is best used for administrative decisions where several units are under one command.

The SSI is computed by summing the FLI recorded for a given period for each administrative unit so that one can be compared with the other for evaluating accomplishments, distribution of funds, changing manning levels, etc.

TABLE 3.17
Burning Index—Regular Season*
Description of Fire Danger Class Ratings†

Class	*Length of day*	*Humidity/ temperature*	*Fuel state*	*Ignition*	*Resistance to control*
Low 0-30	Long, burning period moderate	High to moderate. Temperature moderate.	Fine fuels in green stage	Difficult	Low
Moderate 31-50	Long, burning period moderate	Moderate. Temperature moderate.	Fine fuels in transition	Moderate, lightning moderate	Moderate
High 51-70	Long, burning period long	Moderate to low. Temperature high.	Fine fuels in transition	Easy, high lightning occurrence	High
Very high 71-90	Long, burning period long	Moderate to low, low at night. Temperature high, little cooling at night.	Fine fuels in curing state	Easy, lightning occurrence	High
Extreme 91-100	Long, burning period long	Very low. Temperature high, little cooling at night.	Fine fuels cured	Very easy	Extreme

*Preseason and postseason descriptions would vary with length of day, burning period, diurnal changes, and related weather and fuel factors.

†This is a sample only. Descriptions would vary with regions of the continent.

APPLYING FIRE DANGER RATING VALUES

The importance of the basic aspects of fire behavior – ease of ignition, rate of spread, and rate of combustion – cannot be over emphasized since they dictate what is needed for control. Thus the IC is basic to predicting fire danger. Additionally, the SC and the ERC provide the fire manager with a more precise estimate of the requirements for the prompt containment of a fire.

Therefore the IC, SC, ERC, and the BI are the most important ratings that will be used for suppression by most fire control agencies and for determining the adjective ratings for class of fire danger for public information. The FLI and SSI will be helpful in management decisions. As prevention is emphasized, the OI will be used more often because of the study of risk.

The NFDR system can be tailored to meet the needs of individual fire control agencies. The increased flexibility of the multiple index approach gives the fire manager a wide range of choices, allowing him to balance his needs against the capabilities of the system.

FIRE CLIMATE REGIONS

Fire weather of a particular day is a dominant factor of the fire potential on that day. Fire climate, which may be considered as the combination of the daily weather factors over a long period of time, is a principal factor in

TABLE 3.18
Fire Load Index (FLI) Values for Fuel Model A

Occurrence Index	*Burning Index*												
	0	*1*	*2*	*3*	*4*	*5*	*6*	*7*	*8*	*9*	*10*	*11*	*12*
0	0	0	0	0	0	0	0	0	0	0	0	0	0
1-5	0	0	0	0	0	0	0	0	0	0	0	0	0
6-10	0	0	0	0	0	0	0	1	1	1	1	1	1
11-15	0	0	0	0	1	1	1	1	1	1	1	1	2
16-20	0	0	0	1	1	1	1	1	1	2	2	2	2
21-25	0	0	0	1	1	1	1	2	2	2	2	3	3
26-30	0	0	1	1	1	1	2	2	2	3	3	3	3
31-35	0	0	1	1	1	2	2	2	3	3	3	4	4
36-40	0	0	1	1	2	2	2	3	3	3	4	4	5
41-45	0	0	1	1	2	2	3	3	4	4	4	5	5
46-50	0	0	1	1	2	2	3	3	4	4	5	5	6
51-55	0	1	1	2	2	3	3	4	4	5	5	6	6
56-60	0	1	1	2	2	3	4	4	5	5	6	7	7
61-65	0	1	1	2	3	3	4	5	5	6	6	7	8
66-70	0	1	1	2	3	3	4	5	6	6	7	8	8
71-75	0	1	1	2	3	4	4	5	6	7	7	8	9
76-80	0	1	2	2	3	4	5	6	6	7	8	9	10
81-85	0	1	2	3	3	4	5	6	7	8	8	9	10
86-90	0	1	2	3	4	4	5	6	7	8	9	10	11
91-95	0	1	2	3	4	5	6	7	8	9	9	10	11
96-100	0	1	2	3	4	5	6	7	8	9	10	11	12

fire control planning. Climatic differences produce important variations in the nature of fire problems in different localities and regions.

Climate generally determines the kind and amount of fuels, it dictates the pattern of the fire control job both seasonally and from year to year, and produces the background in which the current weather influences fire control operations. The fire climate of a region is the summation of the integral weather elements that affect fire behavior over a period of time. Because of the nature of the effects of the several weather elements on fire behavior, simple averages of the elements are of little control value. The seasonal distribution, the extremes, the frequency, the season when it is received, and the duration must all be considered in describing precipitation of a fire region.

Fire potential responds to the combined effects of all of the fire weather elements. For instance, strong winds are very important to fire behavior, provided they occur in dry weather. Daily fire danger rating is dependent on current fire weather, while seasonal and average fire danger ratings are dependent on the fire climate. In studying fire climate, one of the most important behavior characteristics of weather is its variation with time.

The areas of North America in which wildfires are a problem have a wide variety of fire climates. Latitude alone accounts for major changes from south to north. The shape of the continent, its topography, its location with respect to adjacent oceans, and the hemispheric air circulation patterns also contribute to the differences of climatic types.

GEOGRAPHICAL FEATURES

The extent of the North American continent permits full development of continental air masses both north and south and east and west. The continent is surrounded by water and is invaded by various maritime air masses. How both continental and maritime air masses influence climate is largely determined by the surface topography (see Figures 3.59 and 3.60 and Table 3.19).

Only about one fourth of the continent is covered by significant mountain chains, which are mostly in the West, except for the Appalachians in the East and the Sierra Madre Oriental in eastern Mexico. With the exception of the Brooks and associated ranges enclosing interior Alaska and adjoining Canada, all of the mountain systems have a north and south orientation.

TABLE 3.19
Characteristics of Summer Air Masses

Air mass	*Lapse rate*	*Temperature*	*Surface RH*	*Visibility*	*Clouds*	*Precipitation*
cP at source region	Unstable	Cool	Low	Good	None or few cumulus	None
cP over midcontinent, southeastern Canada, and eastern United States	Unstable	Moderately cool	Low	Excellent	Variable cumulus	None
mP at source region	Stable	Cool	High	Fair	Stratus, if any	None
mP over West Coast	Stable	Cool	High	Good, except poor in areas of fog	Fog or stratus	None
mP over Rockies	Unstable	Moderately cool	Moderate	Good	Cumulus	Showers at high elevations
mP over midcontinent, southeastern Canada, and eastern United States	Unstable	Warm	Low	Good	Few cumulus	Showers windside of Appalachians
mT at source region	Unstable	Warm	High	Good	Cumulus, if any	Showers
mT central and eastern continent	Unstable	Hot	Moderate	Good during day,except in showers; poor with fog in early morning	Fog in morning, cumulus or cumulonimbus in afternoon	Showers or thunderstorms
cT	Unstable	Hot	Low	Good except in dust storms	None	None

cP — Continental Polar; mP — Maritime Polar;cT — Continental Tropical; mT — Maritime Tropical

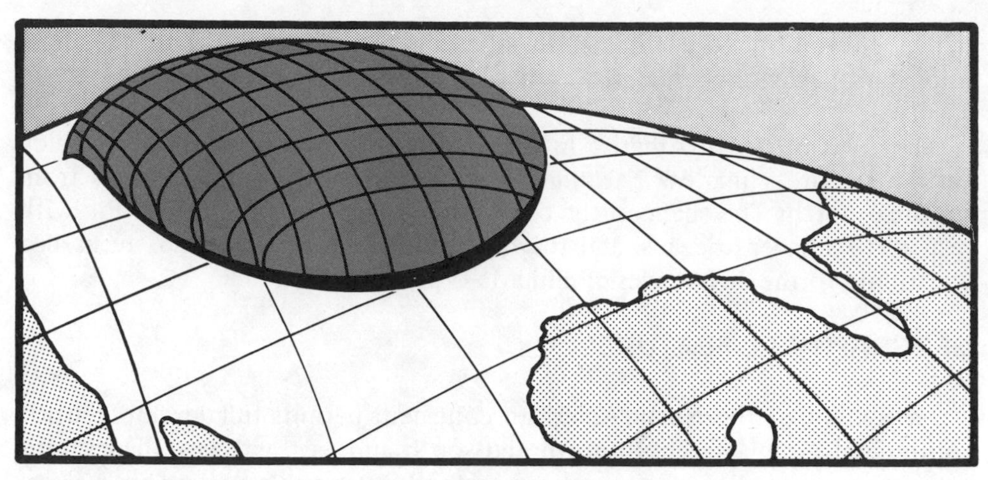

Figure 3.59. A body of air, usually 1,000 miles or more in diameter, which has assumed uniform characteristics of temperature and moisture, is called an air mass.

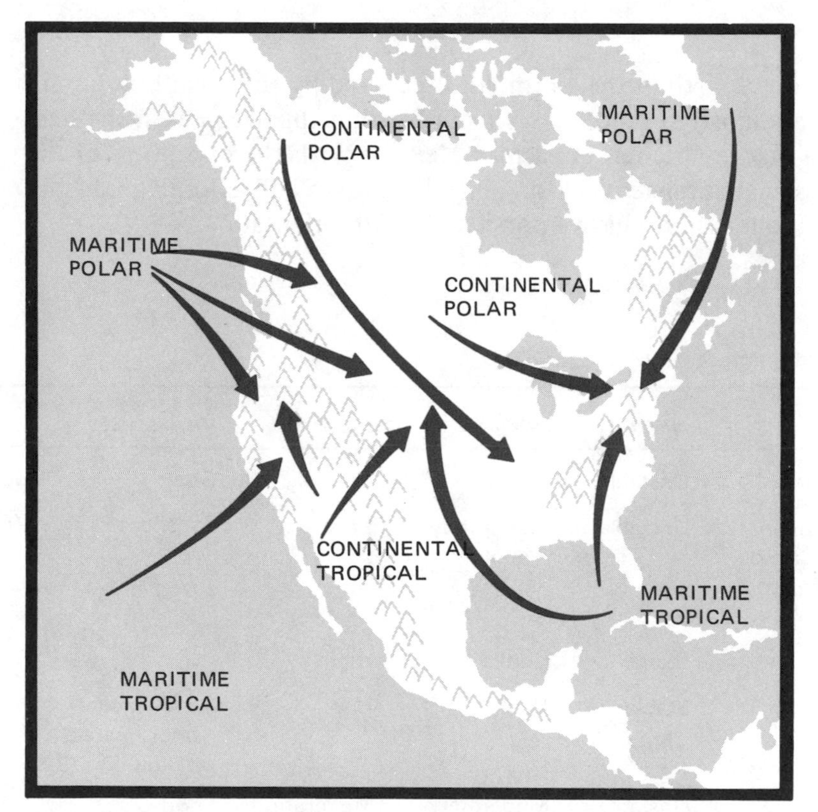

Figure 3.60. Air mass source regions as defined by Table 3.19.

The Rocky Mountain system forms the continental divide from the Arctic Ocean to northern New Mexico. The vast intermountain region west of the Rockies and northern Sierra Madre Mountains is known as the Cordillerian Highlands. A large part of it is called the Great Basin, and it is all upland country. The Appalachians are the only topographic barrier in the East that has a significant influence on general circulation. There is no such barrier between the Arctic and the Gulf of Mexico.

INFLUENCE OF OCEANS

The Pacific Ocean has a strong maritime influence on the whole length of the western shore of North America. However, this influence extends inland for only a relatively short distance because of the north and south barriers formed by the several mountain ranges.

The ocean current known as the North Pacific Drift touches the continent about Puget Sound, where it divides. The Alaska current flows northward and then westward along the Alaska coast. The California current flows to the south along the west coast. Prevailing westerly winds off these temperate waters have a strong moderating influence along the coast both summer and winter.

The Atlantic Ocean influences climate of the east coast, but the effects do not extend far inland because the prevailing wind movement is offshore.

The Great Lakes have a moderating effect in both summer and winter and contribute some moisture for precipitation in adjacent areas.

PRESSURE AND GENERAL CIRCULATION

Pressure is usually low near the equator, high around the horse latitudes (equivalent to northern New Mexico), low in the polar front zone (the latitude of the northern portion of the Canadian provinces), and high in the polar regions.

These pressures give rise to (1) the tropical northeast trade winds blowing onshore from the Atlantic and the Gulf of Mexico between the tropics and 30°N, (2) prevailing westerlies off the Pacific between 30°N and the polar front zone, and (3) polar easterlies north of the polar front zone. As seasonal heating and cooling change, these pressure and wind systems move somewhat north in summer and south again in winter.

The Pacific and Azores-Bermuda high pressure systems, with their clockwise airflow, dominate the summertime wind pattern over large portions of the continent. With the northward movement of the Pacific high during the spring, prevailing winds along the west coast gradually shift from generally southwesterly to northwest and north. The circulation around the Bermuda high is the dominant feature along the Mexican gulf coast and the central and eastern United States (see Figures 3.61 and 3.62). An intense heat low in the summer in the Southwest influences the general weather pattern in most of the southwestern United States and northern Mexico.

TEMPERATURE VARIATIONS

Temperatures vary with the intensity of solar radiation at the earth's surface among other factors. There is a close relationship between average temperatures and latitude. Another major influence on temperature patterns is the distribution of land and water surfaces. At any given latitude, mean temperatures are higher in summer and cooler in winter over land than they are over water. A third major influence on temperature is elevation because the temperature usually decreases with height in the layer of air near the earth's surface. Consequently, an area a few thousand feet above sea level may have an average maximum temperature comparable to a low elevation area many hundreds of miles farther north.

In the summer the differences in temperature between the northern and southern sections of the continent are much less than those in winter. The highest temperatures in summer are found in the lowlands of the Southwest; the lowest temperatures are found in northeastern Canada.

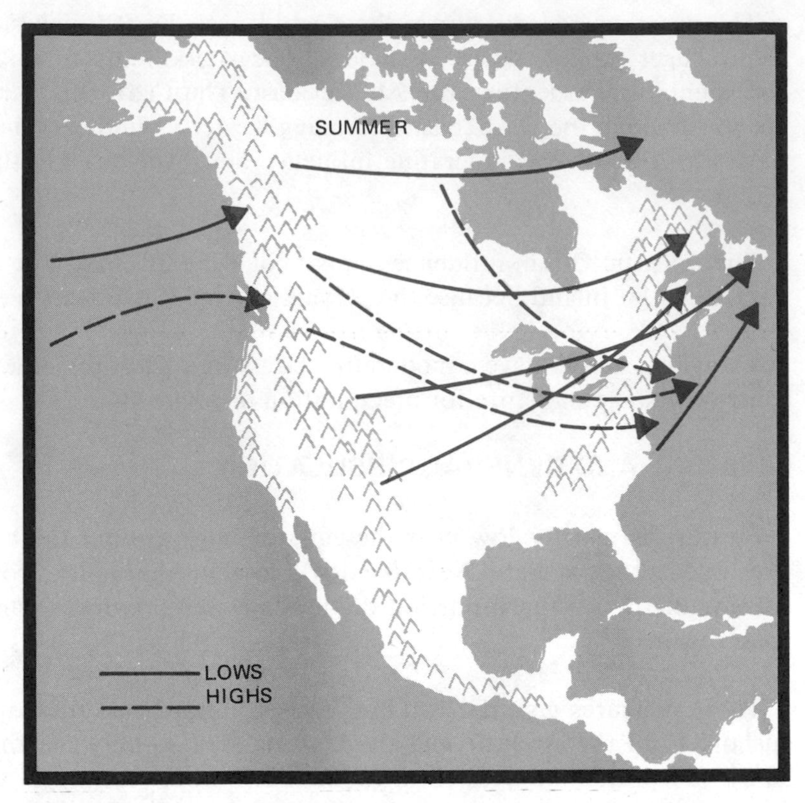

Figure 3.61. In the summer, the preferred tracks of migratory highs and lows are rather far north, mostly across southern Canada or the northern United States. A few lows travel northeastward along the Atlantic coast.

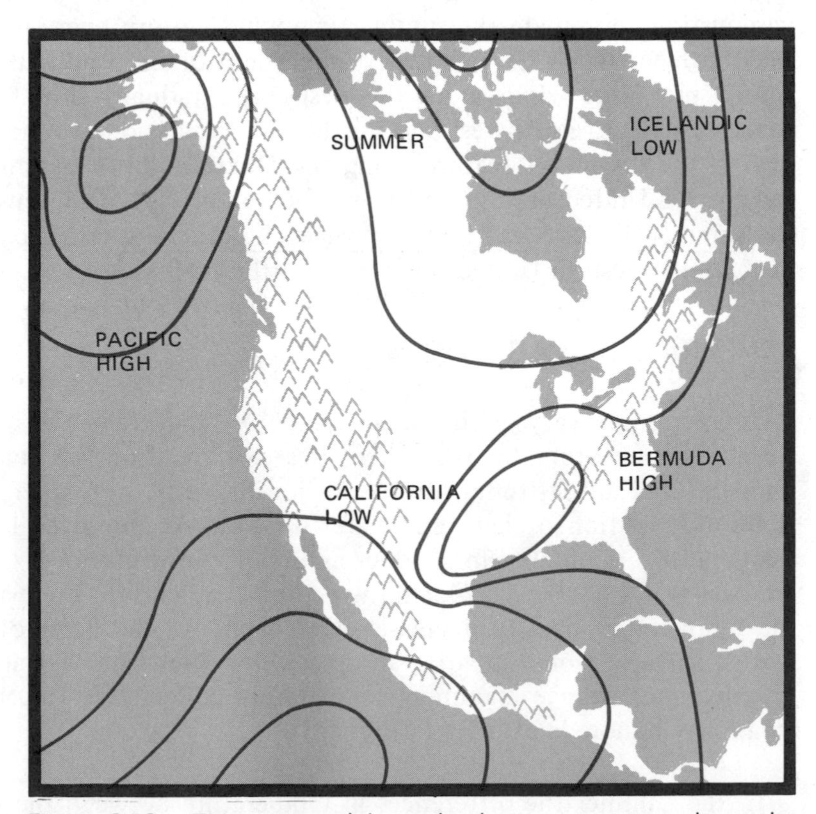

Figure 3.62. This average July sea-level pressure pattern shows the Pacific and Bermuda highs as being strong and rather far north. The Icelandic low is weak, pressure is generally low over the continent, and the intense heat in the Southwest forms the California heat low.

In general, autumn temperatures are higher than spring temperatures in North America. There are some exceptions: in Texas and the interior of British Columbia, temperatures are higher in April than in October.

PRECIPITATION PATTERNS

Both annual precipitation and seasonal distribution of precipitation depend on (1) the moisture content of the air and vertical motions associated with surface heating and cooling, (2) major pressure systems, and (3) frontal and topographic lifting (orographic lifting). This lifting has its greatest effect when the prevailing moist wind currents blow across major mountain ranges.

Moist air from the Pacific drops much of its moisture on the Pacific northwest coast with lesser amounts to the north and south of that region. As the moist air is lifted across the Sierra Cascades, more moisture is precipitated out. Finally, as this air is lifted across the Rockies, most of the rest of its moisture is extracted. As the air mass crosses the crest of the ranges, there is a marked decrease in precipitation, and there is subsidence on the leeward side, which further reduces the degree of saturation. Such leeward areas are said to be in the rain shadow because of the similarity to the shadows cast by the western mountains as the sun goes down. The Great Basin is in such a rain shadow and ranges from semidesert to desert. The area east of the Rockies has the same characteristics and has much lower annual rainfall than the western slope of those mountains. Moist air from the Gulf of Mexico often replaces air of Pacific origin east of the Rockies and often extends well into Canada (see Figures 3.63 and 3.64).

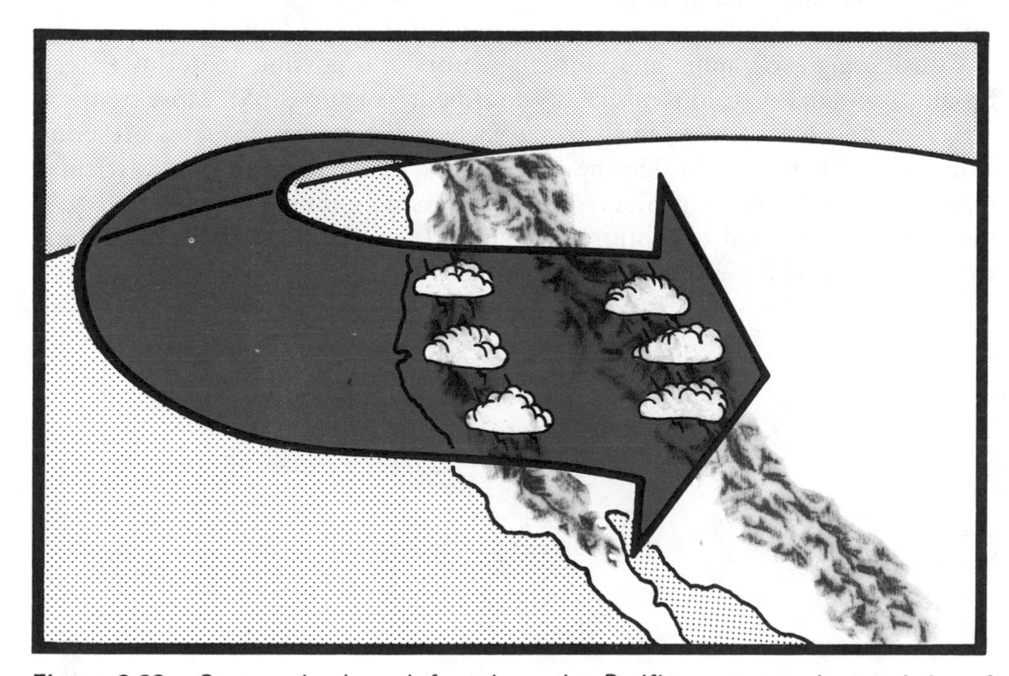

Figure 3.63. Stratus clouds and fog along the Pacific coast are characteristics of maritime polar air in summer. Heating and lifting of the air are likely to produce clouds in the Sierras and showers or thunderstorms in the Rockies if sufficient moisture is present.

In most areas of the continent there is considerable variation in annual rainfall. Wet and dry years may occur irregularly in poorly defined patterns or as wet and dry fluctuations of variable duration. Common variations are

Figure 3.64. Maritime tropical air moving onto the continent is conditionally unstable. Daytime heating and orographic lifting produce showers and thunderstorms in this warm, humid air mass.

the following: normally moist but with occasional critically dry years, typically dry with only infrequent relief, or longer-period fluctuations of alternating wet and dry years. The seasonal distribution of precipitation varies widely over the continent and is often as important in fire weather as the total amount of annual rainfall.

Considering geographic and climatic factors, it is possible to define fifteen broad fire climate regions over the continent (Figure 3.65). Most of these differ in one or more aspects, giving each a distinctive character affecting wildfire problems. In considering the climatic characteristics of a particular region, it should be remembered that generalities must be made and that there are many local exceptions. Table 3.20 outlines the more important characteristics of these fire climate regions.

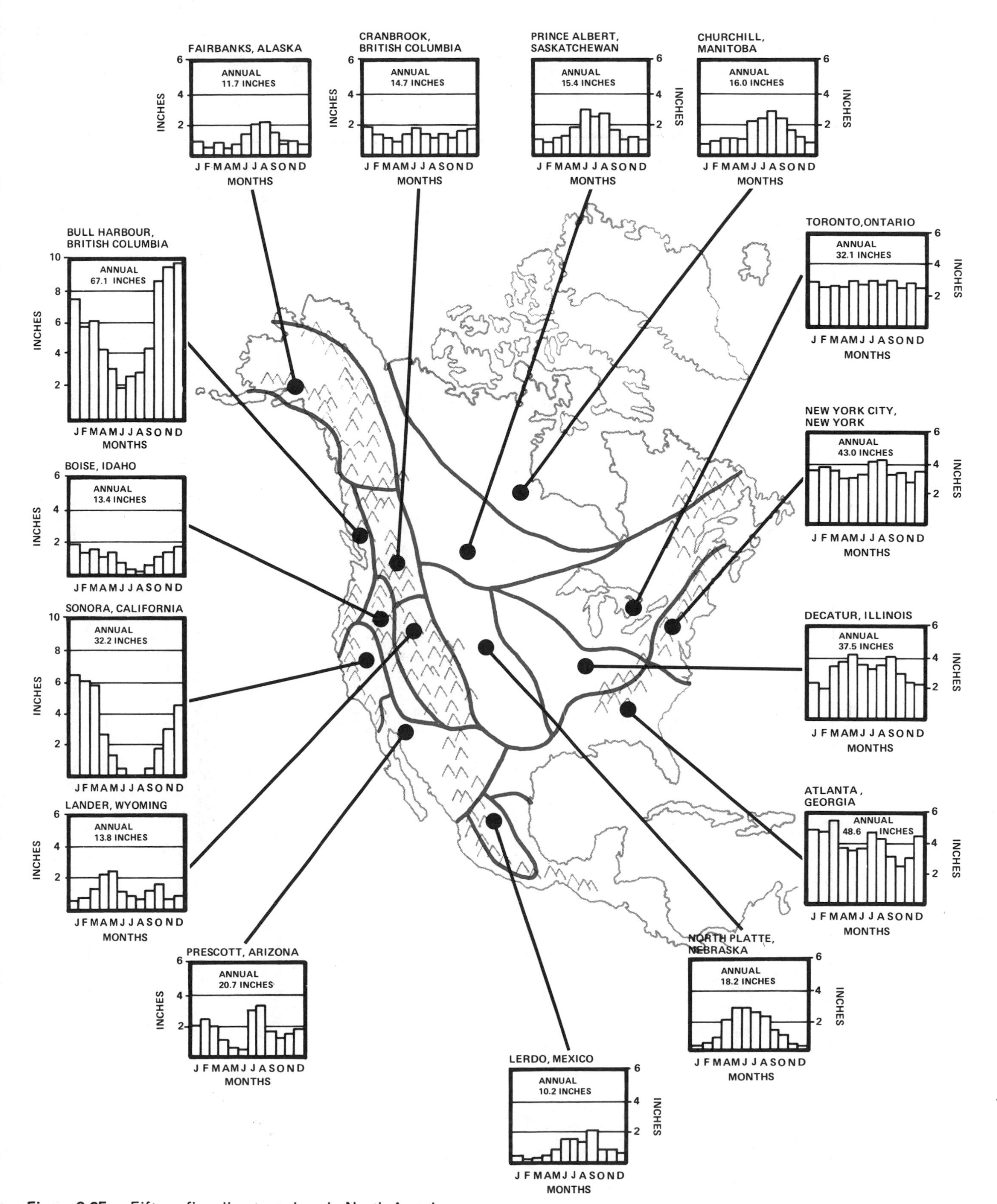

Figure 3.65. Fifteen fire climate regions in North America.

TABLE 3.20

Fire Climate Regions

Characteristics	*Vegetation*	*Fire Season*	*Annual Precipitation*	*Topography*	*Synoptic Weather Types*	*Special Conditions*	*Typical Fuel Models*
(1*) *Interior Alaska and the Yukon*	Spruce, aspen, and some tundra	May through August	10-15 inches; maximum in summer	High coastal mountains; Brooks Range in east; Cordillerian Highlands in south	Maritime polar	Dry thunderstorms	A, D, C, H, G
(2*) *North Pacific Coast*	Rain forest; heavy conifer stands	Late June through September	60-150 inches in British Columbia and in south; 15-20 inches in northern California	Coast Ranges and Cascade Mountains	Maritime polar often has influence; foehn-type winds; cold front with Pacific high bulge	Extremely heavy fuel volumes; strong, dry north-to-east winds may produce extreme fire danger in late summer and early fall	B, G, I, A
(3*) *South Pacific Coast*	Grasslands; brush in intermediate zone; conifer forest in mountains	Usually June through September; can be critical in any month	10-20 inches in lower elevations; up to 60 inches in mountains	Sierra Nevada, Cascade Mountains, and Coast Ranges; California Central Valley	High fire danger from Mono, Santa Ana, and east foehn winds highs and closed highs aloft over western U.S. cause high fire danger	Summers usually rainless; drought common in south; many dry thunderstorms in mountains in northern half	A, B, C, D, E, F, G, H, I
(4*) *Great Basin*	Sparse sagebrush and grass; pine and fir at higher elevations	June through September; may go into October	Less than 10 inches in parts of Nevada and Utah; 10-20 inches in east Washington and Oregon; 20-40 inches in higher mountains	Cordillerian Highlands, plateaus, and lesser mountain systems between Rocky Mountains and Sierra Cascades	Same as highs for South Pacific Coast; maritime polar fronts; closed highs from gulf of Mexico	Entirely in rain shadow of Coast Ranges; summer precipitation light; often much dry lightning	A, B, C, D, H, G, I
(5*) *Northern Rocky Mountains*	Heavy pine, fir, and spruce forests; some grass areas	June or July through September	10-20 inches in valleys; 40-60 inches in mountains; generally light in summer	Cordillerian Highlands; Rocky Mountains; many mountain ranges and rivers	Chinook winds on east slopes; similar to Great Basin; high-level thunderstorms	Frequent and severe dry lightning during fire season; July and August usually dry	A, B, C, G, E, I
(6*) *Southern Rocky Mountains*	Brush and pine in lower elevations; fir and spruce in higher elevations, interspersed grasslands	June or July through October; may be earlier or later	10-20 inches in valleys; 30-40 inches in mountains	Rocky Mountains and interior valleys	Maritime polar fronts; dry cold fronts. Humidities can be acutely low	Light summer thunderstorms; strong chinook winds on east slopes in spring and fall; some lightning fires	A, B, C, F, H, D, G
(7*) *Southwest (including adjacent Mexico)*	Mostly grass, sage, chaparral, and ponderosa pine	May and June most dangerous; can occur in any month	Dry; 5-10 inches in some areas	Plateaus and southern Cordillerian Highlands	Subtropical high aloft to the north causes dry thunderstorms	Dry thunderstorms in early summer	A, B, C, D, E

(8*) *Great Plains*	Grasses; grainfields; isolated pine forests; hardwoods and brush along streams. Many areas of heavy fuel loads.	April through October; can occur in any month	10-20 inches in northwest; 20-40 inches in southwest; west in rain shadow of Rocky Mountains	Plains sloping from Rocky Mountains (exceptions—Black Hills and plateaus)	Continental and maritime tropical; Pacific and Bermuda highs; Pacific arrives dry	Spring and fall chinooks in west	A, F, D, B, G, H
(9*) *Central and northwest Canada*	Spruce, pine, poplar, and aspen forest; south prairies are grass and grainfields	Varies from spring and fall to summer or a combination	8-10 inches in northwest; 20 inches in south; 30 inches in east	Broken; low foothills in west; prairies in south	Source region for continental polar air masses	Evidence of extensive past fire history; varying frequency of dry thunderstorms	A, B, G, D, H
(10*) *Subarctic and tundra*	Scattered spruce forest; scrubby open tundra in north	Midsummer	10-15 inches in northwest; 20-25 inches in east; mostly in summer	Low glaciated terrain	Source area for continental polar formations	Strong winds and low RH common; lightning important	A, C, D, H
(11*) *Great Lakes*	Aspen, fir, and spruce in north; mixed hardwoods in south	April through October; peaks in spring and fall	Generally over 30 inches; some areas higher in summer	Upland areas and heavy glaciation; Great Lakes influence weather	Hudson Bay and northwest Canadian highs often cause high fire danger	Lightning fires common	A, C, D, E, H, G
(12*) *Central states*	Hardwoods; mixed pine and hardwoods; interspersed farmlands	Spring and fall; longer in south	20-45 inches; summers occasionally dry	Mostly flat to sloping; hilly in Missouri and Arkansas Ozarks and in western Appalachia	Pacific, Hudson Bay, and northwest Canadian highs; sometimes Bermuda high	Lightning occurrence low	A, E, H, F
(13*) *North Atlantic*	Spruce forest in north to mostly hardwoods in south	April through October; peaks in spring and fall	40-50 inches; drought can be severe	Appalachians in west; fairly wide coastal plain	Pacific, Hudson Bay, northwest Canadian, and Bermuda highs	Both conifers and hardwoods subject to cumulative drying in fall	A, B, D, E, G, H
(14*) *Southern states*	Pines on coast; hardwoods in bays and bottoms; mixed hardwoods and pines in uplands; flash fuels	Spring and fall; can occur in any month	40-60 inches; 70 inches in Appalachians; 60 inches in Mississippi Delta	Low and flat along Atlantic and Gulf of Mexico; Piedmont in intermediate area; southern Appalachians and lower Mississippi Delta	Pacific high with dry cold front; Bermuda high causes drought	Cold dry fronts cause critical fire weather; lightning is minor	A, B, C, D, E, H
(15*) *Mexican central plateau*	Brush and grass with ponderosa pine at higher elevations	Mostly in summer	Low to moderate	High plateau between Sierra Madre Occidental and Sierra Madre Oriental to the east; generally above 6,000 ft.	Affected by moist air from both Atlantic and Pacific	Generally frequent showers in the summer	A, B, C, D, F, G

*Refers to map number (Figure 3.65).

4 SAFETY

INTRODUCTION

Fire fighting is always hazardous. In the United States, fire fighting has the highest accident rate of all professions. Wildfire control can be particularly hazardous unless the necessary safety procedures are practiced thoroughly and often. The safety and welfare of the entire fire-fighting organization are the responsibility of the fire boss. Each person in authority is likewise responsible for the safety of the men under his direction.

Chief R. E. McArdle originally issued the ten fire-fighting orders used in fire-fighting operations of the Forest Service of the U.S. Department of Agriculture. These orders were developed over a period of sixty years from analysis of the reasons for death and accidents of wildfire fighters. The ten orders were designed to provide for basic safety in wildfire control, and have stood the test of time. Where injury or death has occurred, one or more of these orders have not been practiced. They are repeated here as rules. Every person fighting a wildfire should know and always follow these rules.

THE TEN FIRE-FIGHTING RULES

(Fire behavior)

1. *Keep informed on fire weather conditions and forecasts.* Know and understand local variations. Know the best local source of weather information, and use it.

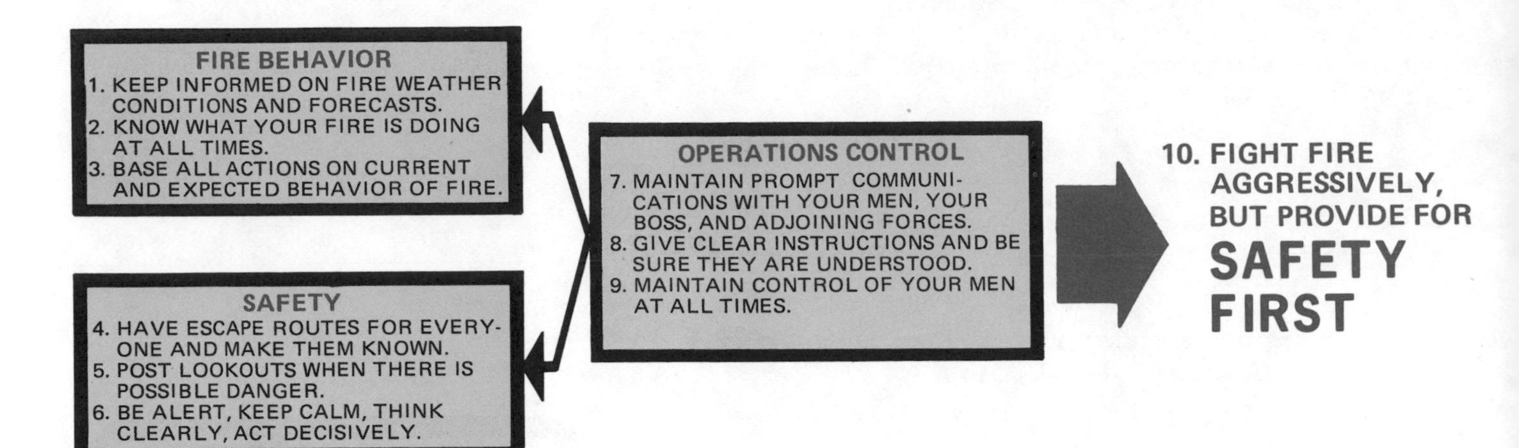

Figure 4.1. The ten fire-fighting rules.

2. *Know what your fire is doing at all times; observe it personally.* On small fires the fire boss should look at all sides of the fire. On larger fires the fire boss should be able to communicate with scouts who can adequately advise him of conditions in those areas he cannot see. He should have communication with all supervisory personnel on his fire. Patrolmen should be assigned to search and put out all spot fires. If possible, the fire boss should operate from the best observation point in the area. Scouts should carry a cutting tool in brush and maintain an adequate escape route.

3. *Base all actions on the current and expected behavior of the fire.* Change crew positions to safe locations if winds change or thunderheads appear or if other conditions make the original working location unsafe.

(Safety)

4. *Plan escape routes for everyone and make them known.* Plan the escape in advance of the need. A good escape route should be known to everyone, be within easy and quick reach, allow for fast emergency escape, be free from flame contact, and be free from hot gases of the fire. Do not try to outrun a fast-moving head. If possible, get into the burned area. The safest route may be uphill, downhill, or along the contour of a slope, depending on a fire fighter's location, the behavior of the fire, and the speed the fire fighter can sustain under the circumstances. If necessary, mark escape routes in advance so there will be no mistake.

5. *Post a lookout where there is possible danger.* Danger includes sudden increases in rate or direction of spread (as in blowups), falling snags, slurry drops, crowning out, whirlwinds, rolling rocks or logs on steep slopes, unexpected change in weather, spot fires, erratic and unexpected fire behavior, and fire sweeping uphill from below your position. Arrange for adequate communication between the lookout and the crew.

6. *Be alert, keep calm, think clearly, and act decisively.* This is difficult to do in the face of an oncoming fire, and firemen must train themselves not to panic. Those who leave the crew are the ones who are injured the most. All acts on the fire line require discipline.

(Operations control)

7. *Maintain prompt communications with your men, your boss, and adjoining forces.* The best means of communication is by radio. If there are

an insufficient number of radios, communication should be established by telephone, written messages, word of mouth, signal mirrors, and/or arm signals.

8. *Give clear instructions and be sure they are understood.* Have the listener repeat your message if necessary. Make sure your message reaches the intended receiver. Explain any unfamiliar names or terms. Make sure messages are delivered promptly.

9. *Maintain control of your men at all times. Stay together as a crew.* Make a head count at frequent intervals. Keep radios operating. In traveling, arrange for control at the rear end of a line of men. Never allow anyone to be completely out of contact with your group. Encourage and develop good esprit de corps.

(Objective)

10. *Fight fire aggressively, but provide for safety first.* This rule states the objective of the other nine rules.

DANGEROUS SITUATIONS

There are thirteen situations that shout, "Watch out!" Every fire fighter should know and observe them.

1. *You are building a line downhill toward a fire.* This situation can be very dangerous if the fire makes a fast run and overtakes the crew. Have a well-defined and usable escape route. Keep a sharp lookout on the fire below.
2. *You are fighting a fire on a hillside where rolling material can ignite a fire below you.* If fire is started below, you can be pocketed between it and the main fire. Patrol for spot fires below, and have a definite escape route. Maintain a lookout.
3. *The wind begins to blow, increase, or change direction.* Watch for spot fires, snags on fire, and possible mass transport of embers. Be ready for a blowup. Check escape routes.
4. *The weather turns hotter or drier.* Prepare for increased fire activity. Post lookouts and check escape routes.
5. *You are on a line in heavy fuel with unburned fuel between the fire fighter and the fire.* Be aware of flare-ups, spots across the line, and difficult, if not impossible, working conditions.
6. *You are in an area where the topography and/or cover makes travel difficult and slow.* Have an escape route planned in case a run occurs in your direction. Provide for observation of the fire's progress. Try to avoid this situation.
7. *You are in unfamiliar country.* Travel cautiously; watch out for cliffs, steep grades, and other travel hazards. Know the fire weather, and provide for communication.
8. *You are in an area where the fire fighters are not familiar with local factors influencing fire behavior.* Arrange for local information, and maintain communication.

9. *You are attempting a frontal assault on a fire with pumpers.* This method should never be attempted under extreme fire conditions. It should never be attempted in light or fast-spreading fuels unless the fire weather conditions are known to be low or low to moderate and you are sure the condition will remain low. If possible, work two pumpers together for mutual support. Preferably, attack the head from the burned area.

10. *Frequent spot fires are crossing the line.* Be sure an escape route is planned. Try to eliminate the source of spot fires.

11. *You cannot see the main fire, and you are out of communication with anyone who can see it.* Arrange for correction of these conditions, or at least maintain communications. Be certain a satisfactory escape route is available and known to the crew.

12. *You do not clearly understand your assignment or instructions.* Obtain clarification. Correlate your assignment with that of adjacent crews.

13. *You are drowsy and feel like taking a nap near the fire line.* Keep awake and keep moving. Do not lie down or go to sleep.

WELFARE OF THE CREW

Only men in good physical condition should work on wildfires. To be effective, the crew must be kept in good condition, and basic needs must be provided for.

Eight to twelve hours of work per day is all that should be expected. An overworked crew not only is much less productive but becomes accident prone. Both the fire line construction and the mop-up lines should be worked in shifts.

If the control job continues past daylight hours, provision for rest is necessary. The crew may be released to return home and replaced with a fresh crew, or sleeping facilities may be provided. On project fires, bedrolls and eating facilities are provided at camps set up to support the control action.

Food is essential to firemen. It is sometimes provided in the form of lunches or hot food delivered to the line. Other meals are provided at the camp or furnished at a local cafe. Good nourishing food and plenty of it are basic to good fire suppression.

Drinking water is also essential for fire fighters. This is usually provided by individual canteens for hand crews on the line. Resupply can be made by water boys, air drops, or any number of methods. The important point is that adequate, potable, and safe water must be provided to all firemen.

Logical rest periods maintain the energy of the crew. Line workers should not be overworked except in emergencies. If it is necessary to work under very hot, smoky conditions to control a hot spot or prevent a breakover, men should be alternated at that location and provided a "breather" period. Men should be rotated between tough and easy jobs.

Wildfire crewmen should wear suitable clothing. Heavy, laced leather shoes or boots with nonskid soles are a necessity. Rubber boots are unsuitable for wildfire fighters. The most flame-resistant clothing should be worn, such as denim, cotton, or wool. Loosely woven cotton burns quickly, and man-made fabrics melt into the flesh when they burn. Hard hats are a must in heavier fuels and should be worn on any fire. Nozzlemen working close to the fire should wear protective clothing and face shields or goggles. Wear extra jackets on cool nights and when resting after becoming overheated. Flame-resistant shirts should be required when embers and sparks may be prevalent. Gloves should always be worn.

Of course, immediate first aid must be supplied for any injury.

SAFETY BRIEFING

A safety briefing must be given to all fire fighters before they go on the line. The crew leader is responsible for the safety of the men under his command and should conduct the safety briefing. First, the crew leader should determine if any fire fighter is too old, too young, or physically unable to fight fire. Second, the men should gather in a tight, compact group. The leader should speak with emphasis and loud enough so that instructions can be heard. Third, safety supervision and instruction should be continued on the fire line where it can be stressed during shifts.

The following are suggested safety points to be given to crews of fire fighters.

- Your leaders are experienced. Stay with them, and do what you are told.
- The way to carry fire-fighting tools safely is down at your side (demonstrate). It is dangerous to carry tools on your shoulder.
- When you walk to and from the fire line, keep at least 6 feet apart in single file (demonstrate).
- When you are working with tools, keep a safe distance from other fire fighters; stay about 10 feet apart.
- Your feet are your worst hazard. Keep sure footing at all times to avoid injury.
- Stay with your crew. Men have burned to death by sneaking off for a nap.
- Watch out for tree branches which might injure your face or poke your eye. Don't smack someone else by letting a branch fly back in his face.
- Keep away from old dead trees, especially if they are burning. They may fall.
- Be alert for rolling rocks or rolling logs when you are walking or working on slopes.
- Avoid stepping in burned-out stump holes that will be full of hot coals.
- When you are hot and thirsty, drink water slowly, and do not drink too much at one time.

- Sit down when you are traveling in a truck equipped with seats. Otherwise, sit on the floor.
- If you cut yourself, blister your heels or hands, or incur any other injury, report it to your leader immediately, and receive first aid.

Safety is a matter of common sense. Use it, and you will keep yourself and others out of trouble.

HAND TOOL SAFETY

The following precautions should be observed by all fire fighters to insure safe use of hand tools.

Figure 4.2. Hand tools must be maintained, carried, and used safely. None of the accidents shown here should ever occur if the proper precautions are taken.

- Carry hand tools at the balance point of the handle, and in the downhill hand alongside the body. With one exception, never carry hand tools on the shoulder; crosscut (ribbon) saws should be carried on the shoulder with the teeth pointing away from the body and preferably with a guard over the teeth.
- Sharp tools should have guards over the blades when they are not in use (e.g., axes, Pulaskis, saws, etc.).
- Keep tools sharp; dull tools are dangerous. See the chapter on presuppression for maintenance of hand tools.
- When they are not in use, place tools so that the blades will not injure passersby. Avoid placing a tool where feet may be cut. Lean tools against a tree, rock, or stump in plain sight.
- Keep handles tight in the heads and free of splinters. Do not use tools with damaged or broken handles until they are repaired.
- Walk at least 6 feet from others when you are carrying tools.
- Work at least 10 feet from others when you are using tools.
- Use tools only for the purpose intended. Use the right tool for the job.

- When you are using cutting tools such as axes, brush hooks, Pulaskis, hatchets, machetes, etc., (a) have a firm grip and a firm footing; (b) always chop away from your body and be ready to check a glancing blow (if it is necessary to cut toward your body, have complete control, use lighter blows, and be ready to avoid a glancing blow); (c) remove all underbrush and overhanging limbs that might interfere when you are swinging a tool; (d) use a natural stance with plenty of room to swing the tool, and never chop cross-handed; (e) guard against chips that may hit the eyes; (f) be especially careful on hillsides; (g) do not use chopping tools as wedges; and (h) when you are using files to sharpen tools, use a handle and a knuckle guard.

PUMPER AND TANKER SAFETY

The following precautions should be observed by all fire fighters to insure safety when pumpers and tankers are used.

- Pumpers and tankers must be maintained in top condition at all times. They must be kept clean and free of accumulated grease and oil. Particular care should be given to maintaining gas and hydraulic lines without leakage. Tight-fitting gas caps should be used.
- Crews should be fully trained in the use of the apparatus.
- Pumpers should be manned by at least one operator, one hose puller, and one nozzleman.
- Tankers should be kept at a reasonable distance from the heat of the fire. In a pump-and-roll stance, this may require that the nozzleman carry the hose while he is walking alongside the pumper.
- Pumpers and tankers should be positioned on the side of the road away from the oncoming fire to reduce heat on the equipment and to allow passage of other equipment.
- Fire apparatus parked on a highway at a fire should be marked by flags, flares, or red lights both front and rear.
- Adequate supervision of and communications with the pumpers, including use of hand signals, should be maintained.
- Pumper crews should wear protective clothing. Goggles or transparent face shields should be worn by nozzlemen.

If you are caught in front of a head fire with a pumper, don't panic. Remember fire-fighting rule 6. Be alert, keep calm, think clearly, and act decisively. Hopefully, each fireman and each officer will operate so that no pumper will ever be caught in front of an oncoming fire. However, in spite of the best efforts, pumpers have been caught in front of a running fire, and in some instances the vehicle engine has quit running. Usually, the engine quits because of a vapor lock or because the vehicle becomes stuck. It helps to install an electric fuel pump on pumpers and other fire vehicles.

It is strongly recommended that any wildfire pumper have a separate engine for the pump; never should the only pump be driven by the truck engine. A separate engine doubles chances that the pumper will continue to pump water.

If you are caught and water can be pumped, wet down the area around the truck and the truck itself. Use backpacks or gravity flow if the pumps are inoperable. Then, as the head fire approaches, use fog streams to protect you from the heat and flame and to punch a hole in the oncoming front. Even if only a small area around the truck has been wet down, firemen have been known to survive in the cab of the truck in grass fuel fires. Do not try to outrun the fire, unless you are positive that you are very near one flank; then, you might escape away from the flank. Stay together to assist one another.

Protective coats have several good breaths of air between them and the body. As a last resort, get your nostrils, eyes, and mouth under the coat, and make a run through the flame front. On reaching the burned area, help each other put out the flames in clothing. Running is a last resort; usually, it is best to stay with the truck, if possible.

The best air is closest to the ground. High heat and flame are the killers; it takes a while for protective clothing to catch fire. It is important to think your way out by using what you have to best advantage—wet down or burn out the area you are in. The burned area is the safest escape route if it can be reached.

TRACTOR SAFETY

The following precautions should be observed by all fire fighters to insure safety when tractors are used.

- Guides, spotters, or helpers for dozers should be selected for their physical fitness as well as their other abilities. At least one helper should be assigned to each tractor.
- Anyone working around tractors should be specifically instructed on their job requirements.
- At night, men on the ground assigned to tractors should wear two head lamps, one shining in front and one shining in back, so that the tractor operator can see them at all times. Tractors must be furnished with lights for night work.
- All men in the vicinity of a tractor should protect themselves from the tractor operations instead of depending on the tractor operator to keep away from them.
- In dozer operations in advance of the fire, a safety strip should be built for a retreat in case the fire makes a run. This strip is especially necessary when you are working along a ridge top above the fire in the valley below.
- Machines must not work directly above each other or at close intervals when lines are being constructed upslope or downslope.
- Men should never sit or bed down near a tractor.
- When the tractor is idling or stopped, the blades should be on the ground.
- No one should ever get immediately in front of or in back of a tractor in operation.

- No one but the operator should be allowed to ride the tractor, except a spotter or a tractor boss when they are necessary in heavy brush.
- Men must not work directly above or below a tractor on a slope.
- Tractor hand signals should be learned and used for direction and safety.
- All tractors should have approved spark arrestors.
- Long sustained grades on the fire line should be broken in order to avoid excessive erosion.
- Tractors should be equipped with safety canopies in wooded areas.
- No one should ever get on or off moving equipment.

AIRCRAFT SAFETY

Some of the more important limitations on aircraft use are the following.

Flying time: Light aircraft, particularly, have limited fuel capacity. They should always return to base with an adequate reserve of fuel in case of an unforseen circumstance.

Pilot's flying time: Constant alertness and the inability to change positions in the seat build up tension and fatigue. Flight time should not exceed five hours per day normally and should not be over seven in emergencies, with not less than a one-hour rest period during any daily schedule.

Mountain flying: Low-level flight in mountain topography is a science all its own. Updrafts, downdrafts, and crosscurrents are only a few of the hazards. Canyons and ridges are potentially dangerous areas to aircraft. Experience and training are requisite to mountain flying. Although they are useful when they are operated correctly, light planes can get into trouble very quickly.

Altitude: Each type of plane has definite limitations on how close to the surface it can fly under various conditions. Each has its own rate of climb, flight altitude, and safe air speed. If these are not strictly observed and correctly judged, disaster occurs quickly. Mountain pilots have a saying that a crash is inevitable when you run out of altitude, air speed, and ideas all at the same time.

Landing strips: Light planes land at a speed of 40 to 60 miles per hour and require a relatively smooth surface that is free of holes, rocks, stumps, etc. for a distance of 600 feet. Larger planes require longer and wider runways.

Observation: Land surfaces have a flat appearance from the air. Obstructions and ground conditions are not observable in detail from fixed wing craft. Helicopters can get much lower and fly more slowly, so the detail can be fairly well observed.

Night flying: Except into and out of standard airports, night flying is generally not practical. The pilot has the final decision on whether his craft will be used. Ground crews should not make requests that tax the limitations of the aircraft.

The following precautions should be observed to insure safety at heliports and helispots.

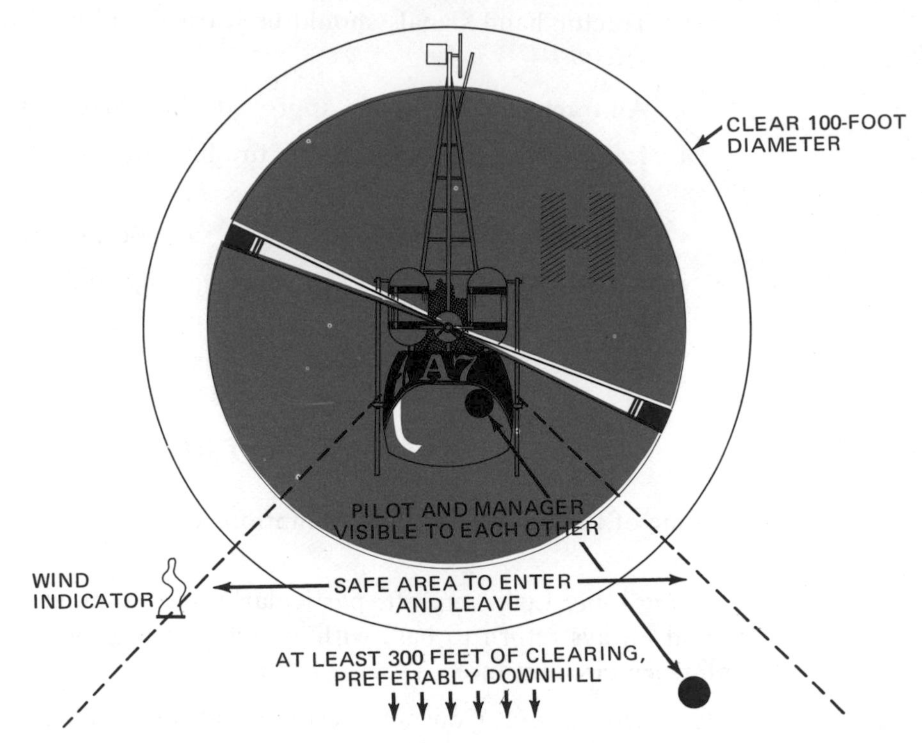

Figure 4.3. Safety precautions at a helispot.

Each location must have a manager to supervise loading and unloading, enforce safety rules, and schedule flights.

Personnel working around helicopters should wear goggles.

Doors should be opened only on signal from the pilot.

Loading and unloading of personnel should be done only on signal from the pilot.

Passengers should wear their hard hats only if the chin strap is secured. Otherwise, the hard hats should be carried when the passengers are approaching or leaving the helicopter.

People, equipment, and loose objects should be kept out of the area of a heliport.

A wind indicator should be used for landing helicopters.

No smoking should be allowed within 50 feet of the aircraft, fueling operations, or fuel storage.

Passengers should stay clear of the main and tail rotors at all times. They should approach and leave the helicopter from the front toward the sides, walking in a crouched position and in plain sight of the pilot. The rear of the helicopter should never be approached. Passengers should cross from one side to the other in front of the aircraft and should always walk downslope from a helicopter, never on ground that is higher than the spot on which the helicopter has landed. On leaving the helicopter, passengers should be clear of the main rotor for 50 feet before resuming the upright position.

FEDERAL AVIATION ADMINISTRATION CLEARANCE

As with ground transportation, the air over a fire area can become crowded with sightseers. Air collisions have a finality about them that makes them particularly undesirable. The Federal Aviation Administration (FAA) can enforce air space restriction to all craft except those employed on the fire. Requests for air space restriction should be made through your local FAA office. There is one problem, however: the system is not foolproof. Some aircraft may not get the message and may come in anyhow. This sometimes happens with military aircraft already in flight. While you are in a helicopter, observing one or more jets below you cruising along at 600 mph just above the surface can spoil your day! The best defense is always continued vigilance and evasive action if it is necessary.

IMPACT OF SLURRY DROPS BY LOW-FLYING AIRCRAFT

Injury can result if a person is caught in a slurry drop. Men should stay clear of the area where a drop is made. If you are caught in a slurry drop, observe the following precautions:

- Lie face down with your head toward incoming aircraft and with your hard hat in place.
- Discard hand tools to the side, preferably behind and/or downhill from your body.
- When you are lying on the ground, grasp something solid to avoid being rolled about by the drop.
- Do not run unless escape is assured.
- In timber stay clear of dead snags, tops, and limbs in the drop area. Watch out for rocks or other material that may be dislodged.
- Slurry is slick. Watch your footing after a drop has been made.

Figure 4.4. Slurry drops can be dangerous. If you are caught under a drop, follow the safety rules.

PARACARGO DROPS

The danger zone for paracargo drops is a strip 200 feet on each side of the flight path, 300 feet in the direction of approach, and approximately 1,300 feet in the direction in which the plane leaves the target. All animals, vehicles, and people should be cleared from this zone before arrival of the aircraft. A camp should be at least 600 feet from the drop zone. The zone should be kept clear until the plane has definitely departed.

VEHICULAR TRAFFIC

Fire fighters traveling to a fire in an automobile vehicle should observe the following precautions.

- Know where you are going. Know the best route, and make sure you know the location of the fire or assigned sector. The first crew into the fire should mark the route for succeeding crews.

- Drive defensively and follow traffic regulations. A few extra minutes in travel are preferable to not showing up.

- Crews should be seated in vehicles. Ride the tailboard with care. Do not put on clothing during the ride.

- Notify the fire boss or dispatcher when you arrive at the fire.

- Try to have what you need in equipment and crew members before leaving for the fire, but do not spend valuable time waiting for either.

- The driver of a vehicle should walk around it before driving off in order to make sure the area is clear. Use chock blocks when the vehicle is parked.

- In cross-country driving, watch for hazards such as bogs, tail water areas, stumps, rocks, cliffs, drop-offs, pits, gullies, and other obstructions to travel.

- Keep tankers and pumpers free of accumulated grease. Keep the motor and undercarriage clean. Check gas and hydraulic lines frequently for leakage, and correct any problem immediately. Use tight gas caps.

- Each pumper should be equipped with a 20-pound ABC extinguisher for use on the truck as well as on structures, a pair of fence pliers (fences should be cut rather than broken down with the truck), and an electric fuel pump. This pump will not prevent all the vapor locks, but it will help considerably. At least one backpack should be carried on each tanker or pumper for emergencies.

- Bridges in your area should have been checked for capacity during presuppression planning or prefire planning surveys, and those that are unusable for your apparatus should be shown on the map of the area. If there is any doubt about a bridge being capable of holding your apparatus, do not use it. When a vehicle breaks through a bridge, not only is it lost for use, but serious injury can result to the men riding the truck.

- If a vehicle must be left unattended for any length of time, leave the keys in it and park it in the safest place possible with respect to the fire, that is, out of the flow of traffic, preferably facing downhill away from the fire. It is a good idea to leave a note indicating where the driver can be reached.

FOOT TRAVEL

The following safety precautions should be observed by all fire fighters traveling on foot.

- Travel at a sensible pace. It is useless to arrive at the fire worn out from hiking. Avoid steep up-and-down travel as much as possible.
- Stay at least 6 feet apart, and carry tools properly at your side.
- The crew must keep together.
- Communications should be maintained.
- Lights should be provided for night travel.
- Pass burning fire-weakened trees, or "leaners," on the side uphill or opposite to the lean. Then watch it closely.
- Stay away from snag-felling areas. Only a qualified snag-felling crew should be allowed in these areas.
- Watch for rolling rocks and logs.
- Watch for rock slides, cliffs, outcrops, and other hazards to travel, especially at night.
- Fast travel through dense unburned brush or reproduction is practically impossible.
- In fast-burning fuels, watch out for fast spread of the fire in any direction. Communicate with a lookout, and have an escape route planned at all times.
- Travel far enough apart so that swinging branches will not slap the face of the person behind.

SAFETY ON THE LINE

The following precautions should be observed regardless of fire size or the number of fire fighters.

- Overhead (squad boss or fire boss) should instruct crews on area hazards and safe working practices before they start work. The men in charge must have experience and/or training in these practices. Each person in charge should be identified to all crew members. The crew must understand his authority to issue instructions and must follow instructions at all times, particularly in emergencies.
- Escape routes should be chosen to avoid traps. Each crew should be thoroughly instructed in escape route use. Lookouts should be posted where necessary. If it is obstructed, the escape route should be cleared in advance to make it usable.
- A safe place should be provided for resting, lunching, or bedding down. If necessary, a lookout should be posted.
- Night crews should arrive on the line in daylight to familiarize themselves with the area. They should be advised of unsafe working conditions by the crew that they are relieving.
- Every fire fighter should be alert to the action of the fire, since it can overtake men day or night.

- Individuals on the line should work at least 10 feet apart.
- Reasonable rest periods should be provided. Shifts should not exceed twelve hours. Some reserve energy should be saved for emergencies.
- Immediate first aid must be administered to the injured. First aid equipment and knowledgeable first aid people should be on the fire. If it is a large fire, an emergency medical technician or a physician should be immediately available.
- Fire weather information should be used.
- Men or machines should not work directly above one another.
- Extra fusees should be packed in the hand, not in clothing. Use a stick in the fusee ferrule to get the flame close to the fuel. Keep falling slag off clothing.
- If possible, power lines in the fire area should be cut off by the power company. If they cannot be cut off, men should not work underneath a power line when fire is producing any appreciable amount of heat or smoke under the line. Back off 100 feet or so until the main heat and smoke have passed. Nozzlemen must not direct a straight stream toward the electric wires. Never approach a broken line, since it can whip over a wide area. Do not consider either end of a broken line dead until the power company has

Figure 4.5. Safety precautions must be observed around electric power lines. A fallen line may carry an electric charge to a fireman through a fence, through smoke, or water.

grounded both ends and has advised that it is safe. Fence wires can become charged by broken power lines coming in contact with them. Be aware of this possibility and keep everyone away from the fences until the situation is corrected.

- Do not place drip torches or flame throwers where they will get hot, and never open them while hot. Use only the recommended mix of fuel.
- Avoid stepping on hot ashes; there may be a hole burned underneath, and it may be full of hot coals. Check around stumps for concealed burning roots. As a general rule, do not walk on ashes unless the spot is checked with a stick or a tool.
- Be alert for sudden flare-ups that may scorch hands or face.
- Only experienced crews should be used in snag felling; the area must be cleared of other workers.
- Unauthorized persons should not be allowed on the fire line, especially underaged boys. They can be used on mop-up or in the camp but not on the hot fire line.
- Observe the ten fire-fighting rules and avoid the thirteen situations that shout, "Watch out!"

SAFETY AND ENVIRONMENTAL HAZARDS

AUTOMOBILE FIRES

Follow these rules in approaching automobile fires.

- Saving of life is the first priority.
- Avoid approaching the car in the area of the gasoline tank.
- Disconnect the battery as soon as possible. Turn off the ignition key.
- Chock the wheels.
- Use a dry chemical extinguisher, preferably ABC if the fire is small, or use a fog pattern with a horizontal sweeping motion. Be aware of fire lingering in cushions and upholstery.
- Crimp broken gas and hydraulic lines.
- Remember that passengers can be seriously injured in the process of removing them from the burning vehicle. Use care and caution. If you suspect neck, back, or internal injuries, seek expert help.

RAILROAD AND TRUCK FIRES

Railroad and truck fires may cause wildfire or may result from the spread of wildfire. The limited capability of fire-fighting apparatus in this type of fire dictates that the primary consideration is the safety of persons in or near the area.

If the train or truck contains hazardous material, the individual car or trailer will be placarded on four sides. On trains, the placard will be a diamond-shaped yellow card about 16 inches square with red printing. The type of hazardous material will be handwritten on the card. If you can read the writing, you are too close! Trucks display a streamer about 6 inches wide and 3 feet long on all four sides. These can be read at a greater distance than the placards on trains, but you may still be too close if you can read the streamer on the truck. If a train or truck accident occurs and placarded units are involved, observe these precautions:

- Evacuate the area for at least 2,000 feet in all directions.
- Unless your fire service has specifically drilled for this type of fire, do not attempt control.
- Do not breathe smoke from these fires, as it may contain toxic fumes.

There are three available sources of information concerning shipment by public carriers. To obtain information on the hazardous material, have the name and location of the carrier and the name of the manufacturer of the material. The information is usually available from the manifest held by the conductor of the train or the driver of the truck. For hazardous materials, call collect 800-424-9300 (Chemtrec, Washington, D.C.). For chemicals, call 513-961-4300 (National Safety Insecticides Corporation, Cincinnati, Ohio). For explosives, call 202-293-4048 (Bureau of Explosives, New York, New York).

HAZARDOUS MATERIALS

Hazardous materials include insecticides, herbicides, pesticides, fertilizers, plastics, and most man-made materials. There are a great many in use today. Antimony, formaldehyde, halogens, and some other chemicals used as fire retardants may produce toxic fumes if they burn. Hazardous materials may be contained within a structure or a storage area that is impinged upon by wildfire. If possible, "Read the label and live." All hazardous materials are required to have a label on each package detailing the hazards to use. An aerosol can can explode viciously if it is exposed to enough heat. Aerosol cans may be found in structures, dumps, or lying on the ground in wildfire fuels. Fields recently treated with chemicals can give off lethal fumes when they burn—stay upwind, and do not breath the smoke if there is doubt. Temperature inversions can spread vapor clouds several times as far on a sunny day as on a cloudy one.

FLAMMABLE LIQUIDS

Storage tanks containing flammable liquids such as gasoline, butane, propane, and other petroleum products may be located within or adjacent to a wildfire area. When it is at all possible, keep the wildfire away from such installations. Be aware of the following precautions.

- Never approach a flammable liquid storage tank for the purpose of extinguishment unless that particular hazard has been preplanned for using the apparatus on hand and the crew has had sufficient training and practice.
- Evacuate all people and animals for a distance of at least 2,000 feet in all directions.

- Always approach horizontal flammable liquid storage tanks from the sides.
- Never approach the storage tank unless the safety valve is operating. There will be no doubt of whether it is operating because it creates a vibration when working properly.
- Tanks with steel supports will fail in less than five minutes. Stay away.
- Abandoned and buried tanks can be hazardous because they may still contain an explosive mixture.
- Except for very small flammable liquid fires that can be extinguished with a dry-chemical hand extinguisher, the principal means of control is by closing a valve in the line supplying fuel to the fire or by crimping a small line. Unless you are trained in this type of fire, do not take chances–move out and let the material be consumed.

DUMPS

Dumps usually do contain hazardous materials. When you are controlling a dump fire, observe the following:

- Approach the fire with caution and endeavor to knock it down from as great a distance as possible with the equipment on hand.
- Be prepared to take evasive action if explosions occur.
- Approach from the upwind side. Do not breath the smoke.

LIGHTNING

Although lightning strikes about 100 times every second, it is not likely to strike you if you follow a few simple safety rules.

- If there is any choice of shelter, choose in the following order: large metal or metal-framed buildings, buildings with lightning protection installed, large unprotected buildings, and small unprotected buildings.
- Stay away from open doors, windows, fireplaces, stoves, chimneys, and electrical equipment such as radios, televisions, lamps, and other plug-in devices.
- When you are outdoors, choose the protection of a cave, get under an overhanging cliff, lie down in a depression in the ground, get into a deep valley or canyon, or stand in the interior of a dense forest of trees of different heights. Do not stand under the tallest or dead trees, but instead stand near short trees that are surrounded by taller ones. If you are near isolated trees or trees growing in small groves, it is preferable to crouch down in the open maintaining a distance greater than the height of the trees.
- Stay away from lone trees, wire fences, poles, hilltops, metal pipelines, wide open spaces, and lakes.
- Remain in a truck or automobile; they offer good protection during a lightning storm.

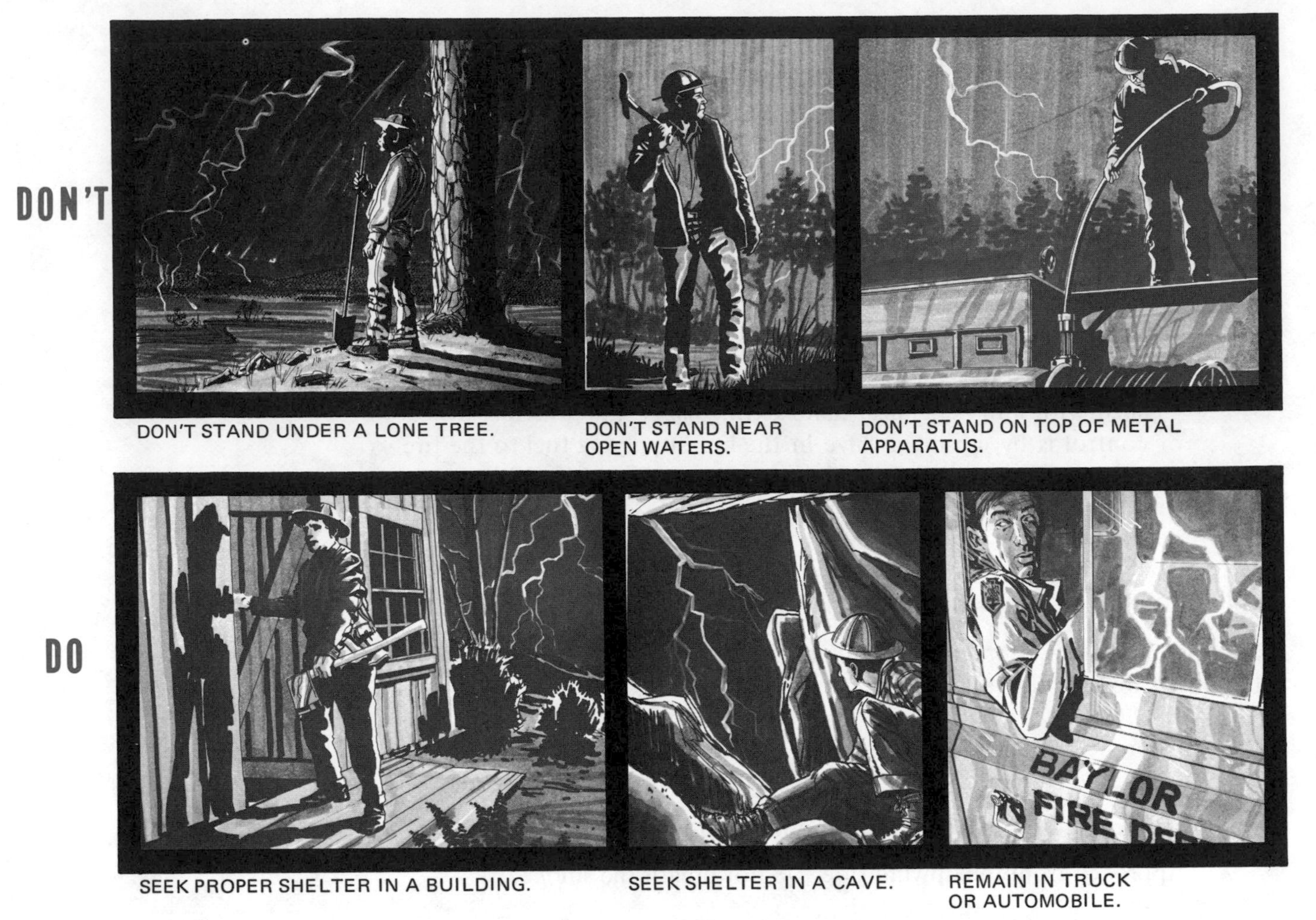

Figure 4.6. Lightning is not likely to strike you if you follow a few simple safety rules.

- Do not ride a tractor when lightning is evident, especially if the attached equipment is in the ground.
- If the storm is within 1 mile, never use the radio except from within a rubber-tired vehicle, and never extend the aerial on a pack set.
- Do not hold metal tools during a lightning storm. Stack them and seek shelter.

EMERGENCY FIRST AID

All regular fire fighters, both paid and volunteer, should have completed the standard Red Cross first aid course. Supervisory overhead should have completed that course or a similar one in order to qualify for wildfire service. Pocket first aid kits, and snakebite kits where they are needed, should be carried by all squad bosses and other overhead. A good first aid kit should be carried on each wildfire pumper.

The following is a first aid reminder list only and not a substitute for completing a good first aid course.

I. Examine the victim thoroughly. Treat injuries immediately in the following order.
   A. Serious bleeding.
      1. Remove clothing from the wound.

2. Apply pressure.
3. Elevate injured part unless it is broken.
4. Bandage.
5. Treat victim for shock.

B. Stoppage of breath.
1. Give artificial respiration.
2. Treat victim for shock.

C. Burns.
1. Remove clothing from burn if it is not stuck to the skin.
2. Cover burn with a sterile dressing.
3. Treat victim for shock.
4. Give victim frequent small drinks of water.

D. Broken bones.
1. Immobilize victim (use splints or padding) before moving him.
2. Treat victim for shock.

E. Shock
1. Keep victim lying flat.
2. Keep victim warm enough to prevent shivering.
3. Give victim a cup of warm water, tea, or coffee, unless he is nauseated, unconscious, or internally injured.
4. Reassure victim.

F. Heat exhaustion.
1. Place victim in shade, and loosen his clothing. Keep him warm and lying down.
2. Stimulate circulation by rubbing victim's body briskly.
3. Use ammonia inhalant.

II. Get help immediately. Report all injuries to the fire boss.

III. Arrange transport to a first aid station or to an ambulance.

IV. No less than fifteen men should be assigned to carry a litter through rough timbered country.

ADVISING CIVILIANS

OBJECTIVES

The first priority on any fire is to save life, and the second is to save as much property as possible, particularly the more valuable and important property. The third priority is to prevent panic. The fourth is to permit unhampered fire-fighting operations.

Property owners can aid in the protection of their homes and property. Water supply, earth-moving and water-dispensing equipment, and hand tools should be made ready. Fire fighters should advise property owners how their actions can aid fire-fighting operations.

ACTION BY PROPERTY OWNERS

Property owners can aid operations by taking the following actions.

- Close all doors and windows on the outside of the structure as well as all doors inside to slow room-to-room fire travel.

Figure 4.7. The four priorities in fire fighting.

- Connect garden hoses and leave them loosely coiled in plain sight. Conserve water and fill any large containers, such as tanks, swimming pools, etc.
- Place available ladders against buildings.
- Roll up automobile windows, and back car into garage. Cars left in the open should be placed in cleared areas and located so that they will not obstruct fire apparatus.
- Leave the lights on in structures. In case of general evacuation, leave the front door unlocked.

- Consider tearing down and removing combustible objects (wooden fences, wood piles, light patio furniture, bamboo screens, etc.) that present an exposure hazard to adjacent structures. Seal up attic and ground vents.
- Chop down highly combustible shrubbery and place it where it will not become hazardous to adjacent structures; examples are cypress hedges and dead trees.
- Remove leaves from roofs, rain gutters, and combustible fences. Remove dry grass from around structures, butane tanks, etc.
- Lower and close venetian blinds. Remove flammable window curtains and other readily combustible items (newspapers, furniture, and throw rugs) from rooms where heat and drafts might logically break windows.

PERSONAL SAFETY

To insure personal safety, civilians should observe the following precautions.

- Keep the family together in a safe place such as a large clearing or burned area, or stay in the house.
- Stay away from hillsides above brush fires. Remember that smoke and heat drafts may be fatal.
- If it becomes necessary to drive through a brush or grass fire or through a smokey area, roll up car windows, close vents, and drive slowly with the lights on.
- Be aware of congestion on narrow roads. Do not drive down steep and unfamiliar roads.
- Carry a shovel, an ax, and a water bucket in each car.
- If retardant drops have been made across roads, use extreme care because the retardants are slippery.
- If you are caught in an emergency, stay in your car with windows rolled up and vents closed.
- Usually, it is safer to stay in the average residence than to flee uphill from a sweeping fire. A house may eventually be destroyed, but, before it becomes untenable, a great amount of mass heat will have swept by so that later a person can survive outside even though the structure is eventually lost. Wetting down the roof and eaves enhances the probability of survival.
- Be alert, keep calm, think clearly, and act decisively (fire-fighting rule 6).

5

TACTICS

CONTROL

Tactics are defined as skillful means used to gain an end. Tactics are the methods used to confine, and thereby control, wildfire. Strategy is the employment of these methods. The basic concept of all tactics used in wildfire control is to establish a "control line" around the area of fire, confine the fire within that area, and rob the fire of additional fuel by preventing any spot fires from occurring outside the line or by controlling those that do occur. Primarily, wildfire control is executed by working on the fuel side of the fire triangle.

CONTROL LINE

The control line is the name given to all segments of action on the perimeter that were used in confining the fire to a definite area. It includes natural barriers and hand-built, machine-built, pumper-built, and air-supported lines that totally encircle the fire area.

FIRE LINE

The term fire line ordinarily refers to the lines that are built by men, machines, apparatus, and air facilities and does not include natural barriers. It is quite often and erroneously used as a synonym for control line.

Figure 5.1. Fire control essentials include (1) establishing a control line, (2) containing the fire within the control line, and (3) preventing or controlling spot fires.

The fire line, or the line as it is often called, is built by removing the vegetation and burnable material on top of the ground so that the mineral soil is exposed. The line may also be made by using a water spray to wet the fuel in a strip of necessary width in those fuels and in those locations where this method is applicable, such as grass, crops, short brush, leaves, needles, and weeds. Spraying water is most effective in light flashy fuels where the topography is such that pumpers can be driven along the burning edge. It is important to mop up the fire edge before the water dries on the fuel. The control line may also include natural barriers and airborne retardants applied in strips. In all cases, the fire must be kept inside the fire line until control is certain; this containment requires men to patrol and hold the line. The width of the clearing of vegetation as well as the width of the line in the

Figure 5.2. The width of the fire line is determined by the intensity of the flames and heat that may reach across the line and ignite unburned fuels by radiant heat and falling debris. Sparks blown across the line by wind or air currents are not ordinarily considered in determining the width of the line. The fire line should only be dug down to the mineral soil, unless a trench is needed to catch rolling material.

ground depends on the kind of vegetation, the topography, the burning conditions, and the location in relation to the spread of the fire; i.e., along the flanks or in front of the fire. The line may vary in width from that of an ordinary cow trail in light grass to that of several bulldozer blades in heavy tall timber. Ordinarily, the initial line in the ground is about 12 to 18 inches wide. The fire line in the soil is usually made with hand tools unless bulldozers, blades, plows, or other suitable earth-moving machinery is available. Control may also be accomplished by burning off strips or areas in advance of the fire. This process is called fighting fire with fire (backfiring). This method requires a control line around the intentionally burned area so that the set fire can be contained.

When the fire is confined to a definite area by a control line, it must be kept within that area. Control is not established until (1) the fire is mopped up to the point where it cannot cross the control line and (2) the area outside the control line is known to be free of any spot fires. Until these two conditions are known to exist, control of the fire is not established.

PARTS OF A WILDFIRE

Wildfires form definite patterns depending on the fuels, weather, and topography. With a knowledge of fire behavior and by appraising the conditions at hand, the course and pattern of a wildfire can be forecast fairly accurately. Generally, there will be a well-defined portion or portions that are spreading faster than other parts of the edge. Much will depend on the velocity and character of the wind – whether it is turbulent, shifting, or steady. Or the pattern may be affected by the steepness of the slope, drainage patterns, changes in fuel types, and the length of time the fire has been burning. The longer a fire burns, the more it is affected by the diurnal change and the change in weather conditions. The parts of a wildfire are described as the head, rear, flanks, fingers, perimeter, islands, pockets, and edge (see Figure 5.3).

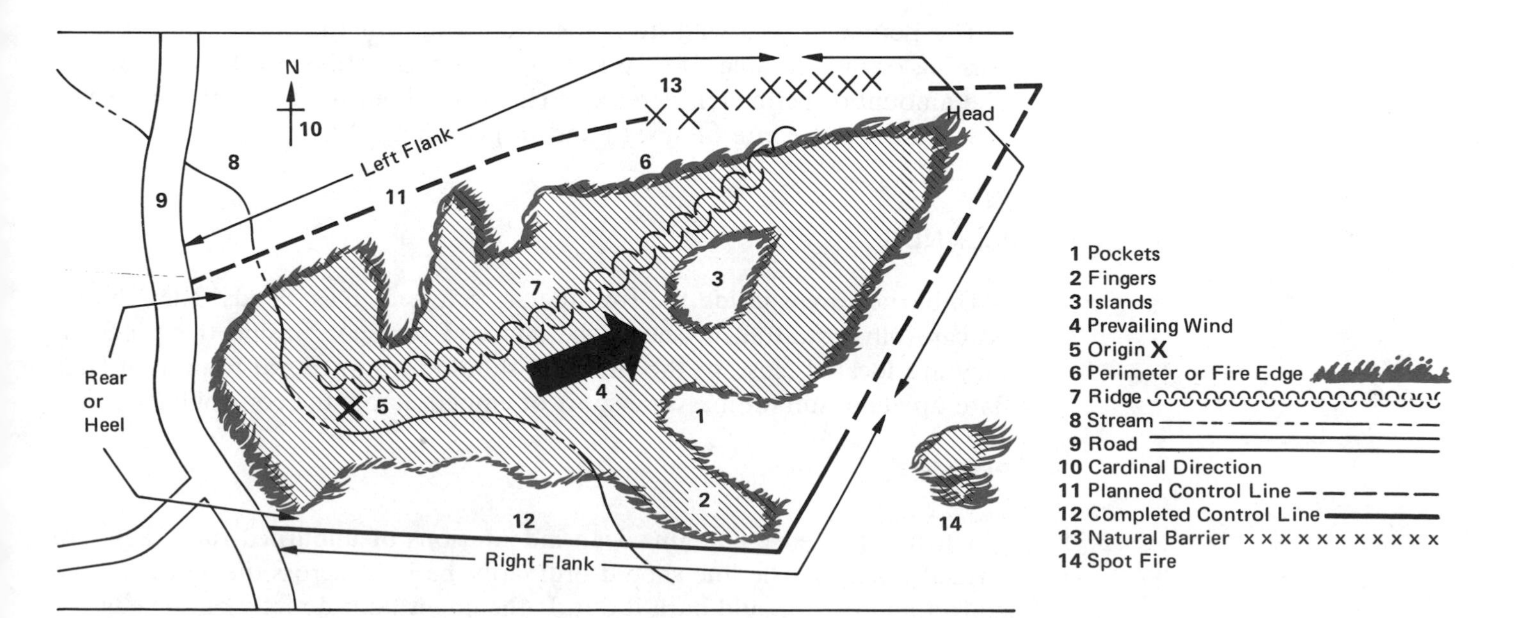

Figure 5.3. Parts of a wildfire shown by fire map symbols.

HEAD

With any wind the wildfire moves in the direction in which the wind is blowing, and this advancing section is termed the head of the fire. There may be one or more heads. The head is the running edge where the fastest rate of spread occurs, and it is the most difficult to control. The most intense burning will be found at the head and along the forward portions of the flanks.

REAR

The edge nearest the source of the wind is termed the rear, and it is opposite the head. The rear is usually the nearest to the point of origin if the wind has remained in one direction. The rear edge may burn slowly into the wind or slowly downhill. It is the easiest edge to control.

FLANKS

The sides between the head and rear are designated the flanks. They are stated as right or left when you are facing the head from the rear. If the wind should shift markedly, one of the flanks may become the head. If that flank is long in distance, a fire of extreme proportions will probably develop. The flanks burn less intensely than the head. The intensity of burning usually increases gradually from the rear to the head along the flanks.

FINGERS

Quite often, shifting winds and changes in topography and fuel cause points, or fingers, to develop behind the head and along the flanks. The head may consist of a number of fingers. These fingers at any location may develop into a new or a secondary head unless they are stopped. Fingers often develop from hot spots along the perimeter.

PERIMETER

The perimeter of a wildfire is the total length of the outside edge of the burning or burned area. It is constantly changing until control is established. The amount of perimeter increase in chains per hour is a useful measurement in calculating the rate of spread and in planning control with the resources available.

ISLANDS

Unburned areas inside the fire line are referred to as islands. They should be carefully cold-trailed, should be checked for spots inside them, and, if they are located close to the fire line, should be burned out to prevent later flare-up and resultant mass transport of embers across the control line.

POCKETS

When a fire edge contains deep indentations of unburned fuel, these are termed pockets. The line should ordinarily be built across the mouth of the pocket and then should burn it out. If the line is built directly on the edge of the fire, much more line is required, and the unburned area in the pocket is subject to double jeopardy from fire pressure on both sides.

EDGE

The fire edge is the burning or burned-out outside edge of the burned area at any given moment. The fire edge and the fire line are not always the same. When they are not the same, action is required to burn out between the line and the fire edge, build a scratch line on the fire edge, or cold-trail the edge.

FIRE MAP SYMBOLS

Some of the most frequently used symbols on planning maps for fire control are shown in Figure 5.3.

SIZE-UP

The first action upon arrival at the fire is to size it up. Actually, size-up begins on the way to the fire as soon as the smoke is seen and the location is determined.

Size-up is the estimation of a condition to arrive at an opinion. It is a continuous action throughout an ever-changing situation. It is a constant process from the time the alarm is received to the time the fire is completely controlled. Wildfire control is largely a process of problem solving and decision making.

First, the problems must be analyzed taking into consideration all of the facts and conditions that can be seen or determined. Second, on the basis of the analysis and expected fire behavior, a course of action must be formed to control the fire. Third, instructions must be issued to those who will do the work of control. Follow-up is necessary to make sure correct action is being taken. If prefire plans have been made for the particular area, they will be of enormous help in the size-up.

Reconnaissance (scouting of the fire and its immediate area) is necessary to appraise the situation. If the fire is relatively small, appraisal can be done in a short time or may be possible on the way in to the fire or while the crews are unloading. If the scouting takes a longer time, it is best to commit your forces to the most logical location against the fire as you see it at that point and time. Later, as more information is gained, it may be necessary to change the plan. The important point is to work where it will eventually assist in the final control and to begin the work immediately. If the fire is larger than the initial attack crew can handle, work on the locations where the most loss can be avoided or where the most spread will be checked. If reinforcements are necessary, make the request as soon as possible so that they will arrive in time to be most effective.

Size-up is of great importance because it provides emphasis of essential information and develops a definite plan of action for effective control. Without a reasonable size-up the attack may be completely ineffective.

After the size-up has been made, the dispatcher should be kept advised of those who need assistance. He should be informed immediately of the arrival of any fire fighter, and he should be given a thumb-nail sketch of the situation.

If the fire location is any distance from an access road that can be easily described and located by incoming crews, the route of travel into the fire area should be adequately marked by the first crew. Much time can be lost searching for the way into a fire, especially at night. It may even be necessary to send someone back to the takeoff point on the access road to mark the way in; the time thus spent can save much time and confusion later on.

MAKING THE SIZE-UP

Knowledge of fire behavior is basic to making the size-up. Refer to prefire plans if they have been made for the area. Analyze the fuel, weather, and topography and how they will affect the fire's behavior. Think of these six items:

1. Fire itself
2. Safety (life and personnel hazard)
3. Exposures (property)
4. Resources
5. Calculation of probabilities
6. Plan and execution of control

Figure 5.4. In making a size-up of a fire, think of (1) the fire itself, (2) safety, (3) the danger to exposures, (4) resources available, (5) calculation of probabilities, and (6) plan and execution of control.

FIRE ITSELF

Ask the following questions about the fire's behavior. What is the direction of spread? Is the wind steady or gusty and changeable? What is the pattern of the fire area, its size, and its length? How intense is the burning and the rate of spread? Are there fingers or danger spots that need immediate attention? The lean of the smoke indicates direction and wind speed. What is the fire weather forecast? Are changes expected? White- or gray-colored smoke indicates that the fire is starting or slowing down. If the smoke is dark or black, the fire is burning into new fuel and is advancing.

What kind of fuel is adjacent to the burning area and ahead of it? Are sparks causing spot fires? Can anything be done to reduce spotting? What is the principal fuel type classification, and how does it burn?

What is the topography? How will it affect the spread of the fire? What is the access to the fire edge? How many natural barriers can be used? What do you estimate the perimeter to be at time of control?

SAFETY

Life hazard is the first priority in any fire (see the previous chapter on safety). What is the life hazard? If buildings are in the path of the fire, have they been evacuated? Are there any other areas that need to be checked for life hazard? What are the hazards to firemen (steep slopes, blind areas, rolling rocks, falling snags, power lines, access, etc.)? What additional information is needed? Where should scouting be done?

EXPOSURES

Property threatened by the fire is an exposure. Ask the following questions when exposures are involved in a wildfire. Where are the flash fuels? Is it reasonable to expect to keep the fire from spreading to those fuels? What is the spotting potential? What property is involved or threatened? How valuable is this property (after the life hazard is determined, buildings, flammable liquid storage, fences, special areas, livestock, etc. have the highest priority)?

Building fire that causes a wildfire. A building fire that causes a wildfire represents a classic dilemma in fire fighting: whether to fight the original fire (in this case, the building), or to devote the effort to the exposure (the wildfire).

If the wildfire fuels are plentiful and uniform enough to carry the fire, and especially if slopes are a factor and the fire in the building is well established and spreading, concentrate on the wildfire by selecting the key point of attack. Probabilities are that the building cannot be saved. Therefore, avoid the danger of a major wildfire, and do not waste effort on property already lost.

If the fire is just beginning to start spot fires, first extinguish them and then concentrate on the building. Except in the very early stages, wildfire pumpers are usually not very effective on a structure fire unless they can deliver larger volumes of water than are ordinarily used in wildfire control. Keep a sharp watch for spot fires. This action saves as much of the building as possible and prevents the fire from spreading to adjacent fuels and/or structures. Call for help from the local fire department pumpers if they are available.

Much will depend on the resources available. If a pumper is present, it would be best used on the structure as long as manpower is available to catch the spots. The fireman's first responsibility is to the community; therefore, he must control a fire that could spread to other property and prevent ignition of the exposure.

Property in front of a wildfire. If the wildfire is burning in uniform fuels toward property and if it is probable that it cannot be controlled before reaching the property, concentrate on saving the property. Buildings are set afire by sparks landing on the roof or catching under the eaves. Wood shingle roofs, especially shake roofs, are particularly vulnerable. Put a line around the building towards the fire. The distance will depend on the type of fuel and the effects of heat radiation. If possible, wet down the roof and walls of the building just before the main rush of heat will affect them. Also, wet down the area between the line and the building, or hold the water to knock down the fire as it reaches the line and to catch spot fires. This decision will depend on how much water is available and the amount and kind of wildfire fuels. Garden hose is effective in keeping a roof wet down. Consider burning back from the line toward the fire if circumstances are favorable and you can control spots.

RESOURCES

The resources available to control the wildfire are an important factor in making the size-up. How many fire fighters are available for assignment? What kind and amount of equipment can be assigned? How many laborers and what kind of pick-up labor can be expected, and where can they be safely assigned? How many and what kind of reserves are available, and when can they be expected?

Can any help be expected from weather changes? What is the time of day and expected diurnal changes in relation to the size of the job? What are the natural barriers and sources of water that can be used? What communications are available? Are maps or aerial photos available on which to plot the fire and strategy of control? What are the environmental considerations?

CALCULATION OF PROBABILITIES

There are a variety of methods that can be employed to control the wildfire. To calculate which will be most effective in a specific situation, the rate of spread must be determined, the type of fuel must be classified, the size of the fire must be estimated, and the line control forces needed must be determined. Weather, time of day, and time of year are also factors in planning control.

Water, of course, is the best and most effective control method if it is available and can be applied with reasonable efficiency. Don't forget to follow up and mop up behind tankers. Hand tools are the most usable method of building the fire line in the majority of locations. Earth-moving equipment is productive and usually efficient if it is available and can be applied. However, it does need to be followed up with hand labor and it may cause more damage than is justified.

If possible, natural barriers should be used so that manpower and equipment can be applied to those sections where only they can be effective. Use all factors available to a given situation for fast and efficient control.

Estimating guides for your particular area are available from your state forester. They have been developed for the particular fuel types and conditions for the various regions of the continent.

Rate of spread. Rate of spread, calculated in units called chains, is the per hour increase in the perimeter of a fire that may be expected under moderate conditions. Very general rates of spread for fuel classifications are shown in Table 5.1. It is a guide only and would need to be adjusted for steep slopes and higher fire danger classes.

TABLE 5.1
Rate of Spread for Moderate Fire Danger

Fuel type class	*Increase in perimeter, chains (per hour)*
L—Low	1
M—Medium	5
H—High	25
E—Extreme	125

Fuel type classification. Fuel type is a factor in the fire's rate of spread. Check with your local state forester for classifications used in your area. The ratings according to fuel type shown in Table 5.2 are for all exposures and slopes. Generally, you should drop one rate of spread if the fire occurs on flat or gentle slopes. The ratings are generalized and need to be adjusted for local conditions.

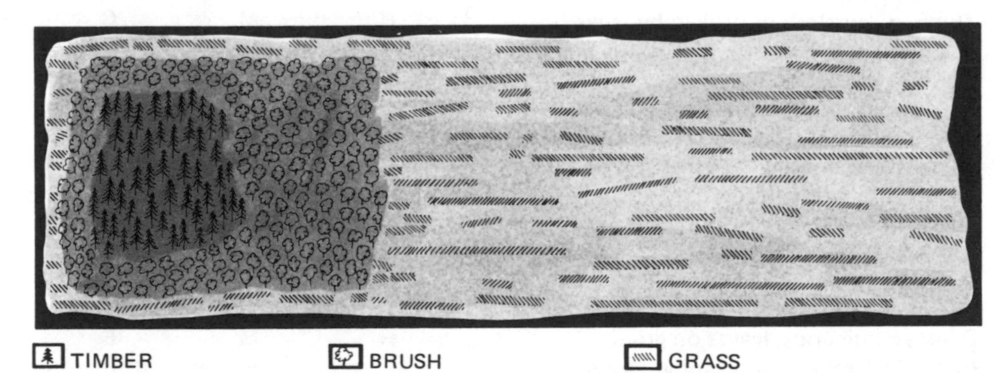

Figure 5.5. Burning rates vary with different types of fuel. This illustration shows the relative area burned in different fuel types during the same time interval.

Size of the fire. If the fire is small enough, its perimeter can be paced; otherwise, it should be estimated. The average pace (two steps) is 5½ feet, or twelve paces per chain (66 feet). This average will vary with individuals and must be corrected if pacing is done on a slope. One should practice his pacing to get his personal number of paces per chain.

If the perimeter is paced or estimated, convert the distance to chains. From Table 5.1, determine the increase in chains per hour, and multiply by the estimated number of hours it will take to control the fire. Add the two results (in chains) together to determine the total perimeter at the estimated time of control. (On the one hand, the tendency is always to overestimate acreage of a fire, but, on the other hand, the estimate should be sufficient so that control will be obtained.)

TABLE 5.2
Fuel Type Classifications

Fuel	Rate of spread*	Resistance to control*	Comparable fuel models (NFDR)
Conifers – mature			
Open, over-mature, parklike, continuous grass	E	L	A, C
Open, sparse ground fuels	L	L	F
Dense, moss, low limbs	E	H	D, G
Dense, duff, moss, snags, logs	H	M	G
Moderate density, heavy ground fuels	M	M	C, B
Conifers – uneven aged			
Heavy understory, heavy fuels	E	M	B
Openings	H	M	D
Conifers – slash			
Continuous	E	H	I
Not continuous	H	M	C, B
Light	M	L	D
Treated, with roller or bunched and burned	M	M	G
Bug killed and reproduction	H	M	C,B,G
Conifers – reproduction			
Dense, unthinned, continuous	E	M	D,E
Thinned, heavy slash	E	H	B
Thinned, less slash or broken by lanes	H	M	C
Hardwoods – broadleafed			
Pure stands	L	L	H, E
Mixed with conifers	L	M	E
Mixed with brush	H	L	C
Slash	H	M	G
Brush			
Dense, continuous, leaves on ground	H	M	E
Dense, tall, rough topography	E	E	B
Patchy, little grass in openings	L	L	F
Sagebrush, dense, continuous	M	H	A,D
Sagebrush, with grass openings	E	M	C
Sagebrush, patchy, low, heavy grazed	L	L	F
Grass			
Green	L	L	A
Cured, dense	E	L	A
Cured, less dense, some exposed soil	H	L	A
Cured, sparse with much exposed soil	M	L	A
Cured, wet meadow type	L	L	A

*E = Extreme, L = Low, H = High, M = Moderate

Use the perimeter in chains at estimated final control to calculate the line control forces needed. Table 5.3 shows conversions of perimeter to area and vice versa if you wish to estimate total acreage involved.

TABLE 5.3
Area – Perimeter Table*

Area (Sq. chains)	Minimum perimeter 1C[1]	Probable perimeter 1.5C[2]	Maximum perimeter 2C[3]	Area (Sq. chains)	Minimum perimeter 1C[1]	Probable perimeter 1.5C[2]	Maximum perimeter 2C[3]
1	3.5	5.25	7.00	210	51.4	77.10	102.80
2	5.0	7.50	10.00	220	52.5	78.75	105.00
3	6.1	9.15	12.20	230	53.7	80.55	107.40
4	7.1	10.65	14.20	240	54.8	82.20	109.60
5	8.0	12.00	16.00	250	56.0	84.00	112.00
6	8.7	13.05	17.40	260	57.1	85.65	114.20
7	9.4	14.10	18.80	270	58.3	87.45	116.60
8	10.0	15.00	20.00	280	59.4	89.10	118.80
9	10.6	15.90	21.20	290	60.4	90.60	120.80
10	11.2	16.80	22.40	300	61.5	92.23	123.00
11	11.7	17.55	23.40	320	63.4	95.10	126.80
12	12.3	18.45	24.60	340	65.4	98.10	130.80
13	12.8	19.20	25.60	360	67.2	100.80	134.40
14	13.2	19.80	26.40	380	69.1	103.65	138.20
15	13.7	20.55	27.40	400	70.9	106.35	141.80
16	14.2	21.30	28.40	425	73.1	109.65	146.20
17	14.6	21.90	29.20	450	75.2	112.80	150.40
18	15.1	22.65	30.20	475	77.2	115.80	154.40
19	15.5	23.25	31.00	500	79.3	118.95	158.60
20.0	15.9	23.85	31.80	550	83.2	124.80	166.40
22.5	16.8	25.20	33.60	600	86.8	130.20	173.60
25.0	17.7	26.55	35.40	650	90.4	135.60	180.80
27.5	18.6	27.90	37.20	700	93.7	140.55	187.40
30.0	19.4	29.10	38.80	750	97.0	145.50	194.00
32.5	20.3	30.45	40.60	800	100.2	150.30	200.40
35.0	21.0	31.50	42.00	850	103.4	155.10	206.80
37.5	21.7	32.55	43.40	900	106.3	159.45	212.60
40.0	22.4	33.60	44.80	950	109.3	163.95	218.60
42.5	23.2	34.80	46.40	1000	112.1	168.15	224.20
45.0	23.7	35.55	47.40	1050	114.8	172.20	229.60
47.5	24.5	36.75	49.00	1100	117.5	176.25	235.00
50.0	25.0	37.50	50.00	1150	120.2	180.30	240.40
52.5	25.8	38.70	51.60	1200	122.8	184.20	245.60
55.0	26.3	39.45	52.60	1250	125.4	188.10	250.80
57.5	26.8	40.20	53.60	1300	127.8	191.70	255.60
60.0	27.5	41.25	55.00	1350	130.3	195.45	260.60
65.0	28.6	42.90	57.20	1400	132.6	198.90	265.20
70.0	29.7	44.55	59.40	1450	134.9	202.35	269.80
75.0	30.7	46.05	61.40	1500	137.3	205.95	274.60
80.0	31.7	47.55	63.40	1550	139.6	209.40	279.20
85.0	32.6	48.90	65.20	1600	141.8	212.70	283.60
90.0	33.6	50.40	67.20	1650	144.0	216.00	288.00
95.0	34.6	51.90	69.20	1700	146.1	219.15	292.20
100.0	35.5	53.25	71.00	1750	148.3	222.45	296.60
110.0	37.2	55.80	74.40	1800	150.4	225.60	300.80
120.0	38.7	58.05	77.40	1850	152.5	228.75	305.00
130.0	40.4	60.60	80.80	1900	154.6	231.90	309.20
140.0	41.9	62.85	83.80	1950	156.5	234.75	313.00
150.0	43.3	64.95	86.60				
160.0	44.8	67.20	89.60	2000	158.6	237.90	317.20
170.0	46.2	69.30	92.40	2050	160.5	240.75	321.00
180.0	47.5	71.25	95.00	2100	162.5	243.75	325.00
190.0	48.8	73.20	97.60	2150	164.4	246.60	328.80
200.0	50.2	75.30	100.40	2200	166.3	249.45	332.60

*Perimeter of fire corresponds with area enclosed by it. Perimeter is shown in linear units of the same kind as the square units used for area. If area is in square chains, perimeter is in chains. 10 square chains = 1 acre.

[1] Perimeter is that of a circle corresponding with the area.
[2] Perimeter is 1.5 times that of a circle corresponding with the area.
[3] Perimeter is 2.0 times that of a circle corresponding with the area.

Figure 5.6. Calculation of needed line control forces.

Calculation of needed line control forces. The forces needed for line control can be calculated by the following formulas:

$$C/P = M \qquad \text{and} \qquad M/W = Z$$

where C is the number of chains of perimeter, P is the number of man-hours of production, M is the number of man-hours needed, W is the work-hour period, and Z is the number of men or units needed. Table 5.4 is an example of a calculation for a line control force. Adjustments for specific conditions would be required wherever it is used. Factors must be added for exposures and travel time, and the mop-up crew must also be planned for.

TABLE 5.4
Calculation of Line Control Forces

Factors	*Example*
Estimated number chains of perimeter at control or first burning period	120 chains
Estimated length of held line to be constructed per man-hour or machine unit hour	2 chains
Number of man-hours needed to control	120/2 = 60
Estimated available work period (not over 8 hours)	6 hours
Number of units or men needed provided they all start work at same time	60/6 = 10
$\frac{\text{Chains perimeter}}{\text{Man-hour production}}$ = Man-hours needed	$\frac{\text{Man-hours}}{\text{Hour work period}} = \frac{\text{Number of men}}{\text{or units}}$
Add for exposures and travel time Plan for mop-up crew	

Line construction output in chains per man-hour or machine unit hour is shown in Table 5.5. These calculations are for line construction only and do not include holding, cleanup, or mop-up operations.

TABLE 5.5
Line Construction Output*

	Resistance to control†			
Control unit	*Low*	*Medium*	*High*	*Extreme*
Full-time trained wildfire fighters	3.3	2.0	.8	.25
Volunteers – trained	2.5	1.5	.6	.2
– untrained	1.25	.8	.3	.1
– little or no training	.8	.25	.08	.02
Dozer unit	50.0	20.0	8.0	4.0
Plow unit	240.0	150.0	40.0	10.0
Pumper unit[1]	240.0	80.0	10.0	5.0

*This table concerns construction only; it does not include holding or mop-up.
†Measurements are in chains per man-hour or in chains per machine unit hour.
[1] It is not always possible to traverse the fire line with a pumper unit. If resistance to control is low or medium and pumper travel is hampered, reduce pumper one-half to one-tenth. If resistance to control is high or extreme, hose lines should be laid. Safety may prohibit their use during initial attack.

Priority of control action. Factors to be considered in deciding priority of action are the following:

- Evaluate life hazard.
- Estimate property values.
- Estimate relative value of cover and/or resultant damage.
- Cut off fire from most dangerous fuel.
- Confine fire to one side of drainage or keep it within drainage if possible.
- Keep fire to smallest area consistent with cost and values.
- Make all work contribute to final control by becoming part of final control line or by delaying spread until the final line location can be built.
- Use equipment in areas that are too hot for manpower or where apparatus can be used effectively.
- Provide for a line of retreat.
- Estimate relative cost of control and evaluate alternate action.

PLAN AND EXECUTION

When you have considered all the factors involved in making the size-up, including the calculated control probabilities, make a plan of attack and make it work.

If you have an aerial photograph or a map of the general area, plot the perimeter of the fire on it as nearly as possible. Divide the perimeter into

Figure 5.7. The last step in size-up is to plan and execute control. Make a control plan such as this one and make it work.

logical segments according to the size of the fire. Assign definite segments, including natural barriers, to crews, tankers, dozers, etc. Give clear instructions to the leader of each crew or unit and arrange for communication.

METHODS OF ATTACK

There are two basic methods of attack:

1. Direct—fighting the fire itself directly on the edge by using water spray, throwing dirt, using beaters, or building a line to mineral soil by throwing the burning edge inside the fire and then widening the line as necessary.
2. Indirect—building a line some distance from the edge of the fire when the fire is too hot to fight directly.

There are many variations to the indirect attack because of fuel, topography, and access considerations. A principal variation is fighting fire with fire. This is also called backfiring. The term backfire is misleading and indicates a simple form of control, which it is not. Fighting fire with fire is hazardous and complex. It should be used only by experienced fire fighters and usually is a last resort.

DIRECT ATTACK

The direct attack is used mostly on ground or surface fuels like grass, grainfields, leaves, needles, some brush, or duff; on underground fires; on the flanks or rear of larger fires; in the later stages of large fires; and on any fire where the burning intensity, heat, and smoke are not too much for the fire fighters to work on the fire edge.

Figure 5.8. Direct attack.

Advantages. The advantages of direct attack are the following:

- There is minimum area of burn.
- Crew can usually escape into the burn area for an escape route.
- Full advantage is taken of burned-out areas.
- No additional area is intentionally set on fire.

Disadvantages. The disadvantages of direct attack are the following:

- Fire fighters can be hampered by heat, smoke, and flames.
- The control line can be very long from following an irregular edge.
- Natural barriers may not be taken advantage of.
- Hot spots may cause breakover and spot fires.
- More mop-up and patrol may be necessary.

The direct attack is commonly used on the head of smaller fires and on the flanks or rear of larger fires where the heat intensity is such that the fire edge can be worked directly. It is also used on most grass fires of any size where pumpers can be applied directly.

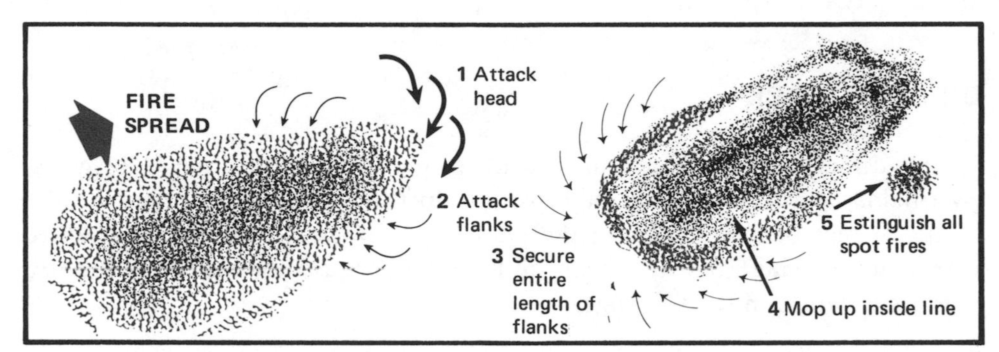

Figure 5.9. Procedure to be used in direct attack on a small, hot fire.

If the fire is small and if the head can be attacked with safety, the control action is applied at the head first. The flanks are then worked back to the rear. The entire length of the line is made secure. The area inside the line is mopped up a sufficient distance towards the center so that burning sparks and embers cannot be blown across the line. This mop-up zone will vary with the type of fuels, slope, and weather conditions. The entire area outside the line must be thoroughly worked to extinguish completely any and all spot fires.

When the head is spreading fast and it is unsafe to get in front of it, the best method is to flank the fire on one or both sides. This method uses a pincer movement to eventually cut off the head. The line is built from an anchor point near the rear and proceeds along the flank or flanks toward the head. The pincer action can be done from one flank, but it is quicker if both flanks can be worked at the same time. If enough resources are available, one crew or tanker can be started at a point along the flank where the fire is continuous to the head. Another crew must immediately work the line from that point to the rear, making the line secure. They then follow the tanker to secure the line it is working. Until this rear portion of the flank is secure, the other crew could become pocketed between two heads if the fire jumps the

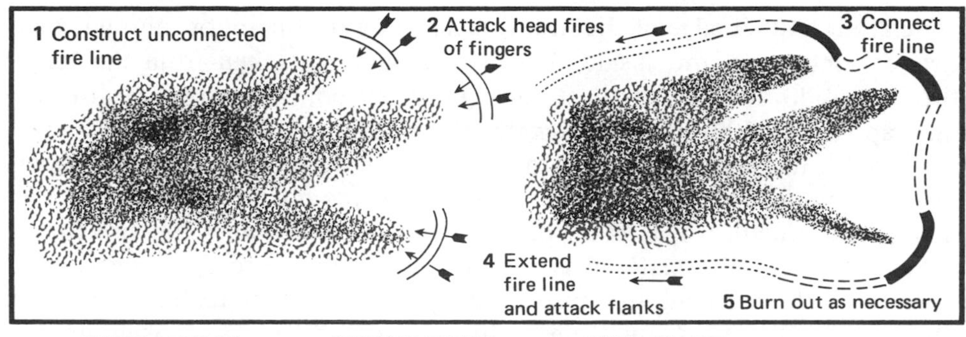

Figure 5.10. Procedure to be used in hot spotting, or the point and cutoff technique.

rear of the flank. When the control line is established, mop-up, spot fire control, and patrol complete the operation.

Hot spotting or point and cutoff technique. Hot spotting, or the point and cutoff technique, is a combination method that may be used in direct attack but is more often used in indirect attack. It first attacks the fingers or hot spots along the fire edge by constructing unconnected lines across those edges that are burning fastest. These short sections of line are then connected, and the flanks are tied into the rear with effective fire lines. To a more or less extent, experienced fire fighters use this technique in all attacks. Its purpose is to contain the fastest running edges first and then to complete the line for full control. Burning out (an indirect method of attack) is quite often necessary in the second phase of completing the line between the hot spots. Instead of building a longer line by working the fire edge, it is usually best to build the shortest line from point to point and fire out the intervening spaces. Escape routes are particularly important with this technique.

Cold trailing. Cold trailing is the process of checking out the edge of apparently burned-out sections of the fire. This may mean building or examining a scratch line. A scratch line is a simple line usually about 6 inches

Figure 5.11. Procedure to be used in building a scratch line (cold trailing).

Figure 5.12. If there is doubt about any remaining fire, feel the area with your hand.

wide along a quiet edge of the fire. It is made by beating out or throwing any fire or embers to the inside. Usually, all material taken from the line is thrown inside. It is worked fast and lightly but continuously. The fire edge may appear dead but its entire length should be carefully checked. If there is any doubt, feel with your hand to make sure. Using common sense, you will not burn your hand because your reflexes are too quick, but you will know if any fire remains. This is a positive method of knowing the fire edge is dead. When burning material is found, build a line across its edge and, if possible, extinguish all fire inside. Otherwise, allow it to burn up under observation. Potential fuels near these edges should be broken up and thrown far inside the burn. These fuels include logs, limbs, cow chips, and other accumulations of fuel.

Rule of Thumb: The most effective place to stop a fire is at the advancing flame edge. However, this may not be practical or possible in many situations.

INDIRECT ATTACK

With the indirect method, the line is located some distance from the fire edge. How far it is located from the fire is of prime importance. All of the factors of fire behavior must be used in making the decision. Since the intervening material must be burned out, the line must be located where it will be effective when the fire reaches it. Yet, it must keep the intervening area small enough so that no more area is burned than necessary; otherwise, the set fire can build up enough that it may jump the line. The location involves experience and good judgment. The line must be wide enough so that the radiant heat developed by the kind and amount of fuel inside will not ignite fuels on the outside of the line. The ability of the line workers, the time of day, the intensity of burning, the speed and stability of the wind, and the slope are primary factors of indirect line locations (see a previous section in this chapter on calculation of probabilities).

Since the indirect method is used where the fire edge is too hot to approach directly, it is the method that is most used on larger fires and at the head of hot, fast-running fires. It is also the method that is most used during the higher classes of fire danger. The line is shortened by running it through pockets or indentations in the fire edge and between natural barriers. Because of limited resources, only the shortest line is possible in the time allowable. The indirect method is often combined with the direct method in the total line construction. The indirect method may be used during the time of day when the fire danger conditions are highest, but, when conditions ease, the attack may return to the direct method.

Advantages. The advantages of indirect attack are the following:

- Indirect work is easier on the line workers.
- There is more time to construct the line and to develop the best teamwork and coordination.
- Indirect attack makes maximum use of natural barriers.
- It develops the shortest line.
- It takes advantage of the easiest location for line construction.
- In burning out, the crew is available when hot spots develop.

- There may be less danger of slop-overs in indirect attack, depending on how forces are deployed and the effectiveness of their work.

Disadvantages. The disadvantages of indirect attack are the following:

- More acreage is usually burned.
- The burning-out fires may break over the line and add more area.
- Careful watch must be maintained along the entire line.
- Crew can be outflanked.
- No advantage is taken of burned-out edges.
- Changing weather conditions, especially wind, could change the direction of spread, making useless much of the work previously done.

Figure 5.13. Procedure to be used in indirect attack (parallel method).

With the indirect method the line is built some distance away from the fire edge. The fire lines must tie to a secure anchor point, i.e., road, stream, rockslide, etc. Variations may be the 2-foot method, the parallel method, or the backfire method. The 2-foot method is building the line at that distance from the fire edge and burning out as the line is built. It is a cross between the direct and indirect methods and is applied when it is too hot for direct work. With the parallel version, the line is built from 2 to 50 feet from the fire edge and as straight as possible to avoid indentations in the fire edge. This takes advantage of natural barriers and topography and gains time to build the line. Ordinarily, the indirect method is used where it is too hot to construct the line directly on the edge. The distance away from the fire edge depends on the fuel, the intensity of the fire, and the topography. If the parallel lines can be constructed fast enough, they can eventually connect and contain the fire; or one line can be built along one flank and obliquely across the front to a point on the other flank and then back along it to encircle the fire. In any indirect line construction, the intervening space between the line and the fire edge should ordinarily be burned out to secure the line. This is termed burning out, in contrast to backfiring. Burning out on the flanks isn't nearly as hazardous as burning out from a line in front of the head. Wide variations are possible with the indirect method, owing to the

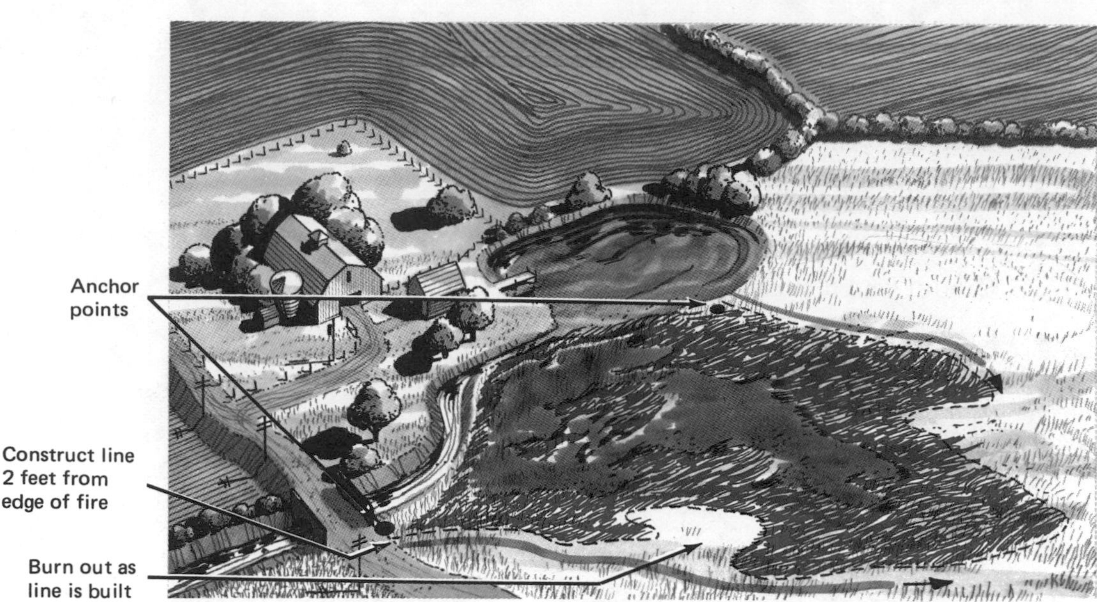

Figure 5.14. Procedure to be used in indirect attack (2-foot method).

many combinations of conditions. With this method, the fire fighter chooses the location where the fire will be fought; with the direct method, the location is dictated by the fire.

Another form of the indirect attack is backfiring. The principal difference between burning out an indirect line and backfiring is the extent that the new fire is started and the location where the started fire is used.

In summary, fire can be used in three principal ways to help control a wildfire:

1. Burning out, firing out, or clean burning (the terms are synonymous) the area between the fire line and the edge of the fire is not a tactic but a technique of the indirect method.
2. Strip burning is the process of burning an area in controlled strips between the fire line and the fire edge. By burning in strips, the capacity of the set fire to cause spots or jump the line is materially reduced if the fire has been well planned and enough manpower and equipment are available to control spotting. Strip burning is used primarily on the flanks as a method of burning out, but it is also used as a method of controlling backfires.
3. Backfiring is setting an area on fire in advance of the head fire for the purpose of stopping its spread, changing its direction, or slowing its progress.

Burning out or clean burning. Burning out, or clean burning, is not considered a separate tactic but rather is considered a part of line construction. It consists of starting a fire along the inside edge of the control line so that the fuel in the area between the fire line and the fire edge will be burned. Pockets and islands may be burned out after the line is built so that they do not pose a threat at a later time. This burning forms a wider barrier to the spreading main fire and burns the fuel while the crew is there to keep all fire inside the line. The burnout is started with a drip torch or a fusee or by pulling burning material along with a rake. It is important to conduct the

Figure 5.15. (*Left*) In burnout or clean burning from a road or natural barrier, use water to control radiated heat and mass transport by burnout fire and to control any spots that do start across the road. (*Center*) In burnout from a hand line, have enough manpower to keep burnout fire inside the line. Adjust firing to holding. (*Right*) Burn out pockets and islands that may cause spots across the line if they burn later when the main crew has moved on.

burning during times when a good burn will be made. If the burnout is patchy and not complete, it may be more hazardous at a later time when burning conditions increase and a complete crew is not there to hold it inside the line. The burnout may not be complete and yet may dry out overhead fuels, thereby setting the stage for a crown fire when conditions change. Therefore, the burnout should be done when burning conditions favor a complete burn of surface fuels. If possible, start the burnout so that the convection column of the main fire pulls the convection of the burnout fire in toward the main fire. On hillsides, the burnout should start from the top and work downward; otherwise, an excessive head may be built up which will overrun the line. Avoid crown fires unless desired.

The use of the burning-out procedure must be determined by the type of fuel, particularly in relation to several storied fuels, and your ability to obtain a clean burn. The hazards to clean burning are snags, piles of heavy ground fuels, and live trees with branches extending to the ground or with moss. Remove, break up, or build a line around such trouble spots when they are close to the main fire line.

Strip burning. Strip burning may be used on the flanks and also in front of the head. In this practice the area to be burned out is burned in strips. It is particularly adaptable in light fuels where a water line can be laid down inside the fire line with a pumper so that the burning is confined to a strip. The strip may be only 5 to 10 feet wide to start with. It should be only as wide as can be controlled with the resources available. It should be done in

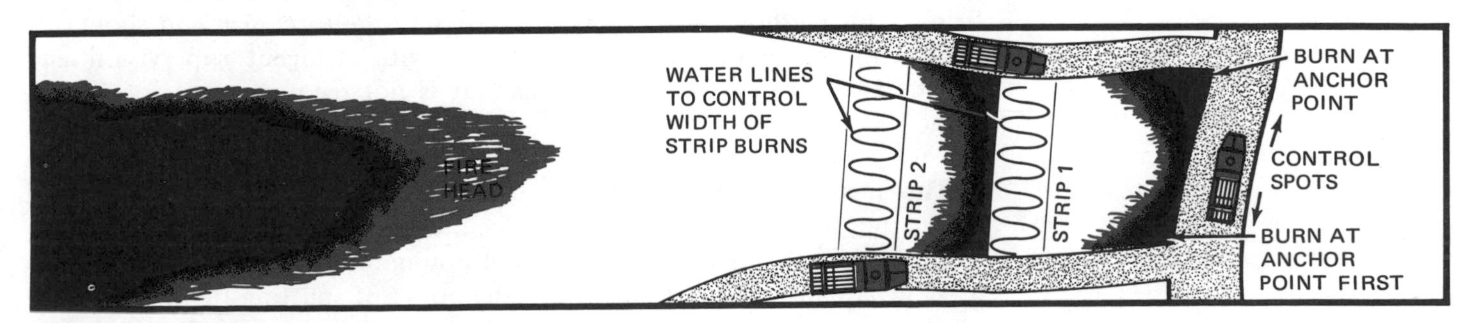

Figure 5.16. Strip burning in front of the head of a fire.

such a way that the radiant heat will not start fire across the line and that any spots that are started can be controlled with the men and equipment available. The second strip is usually twice as wide as the first one. A third strip can be several times the width of the second one, depending on the spotting potential. This method provides a means of controlling the fire that you start, which must be done if fire is to be used effectively against the wildfire.

In strip burning parallel to the fire line, one of the strips should be started at the line, and progress should be made only as fast as the holding crew can control it. If too much flame and spotting develops along the line, use dirt and water, if it is available, to cool down the edge next to the line. Also, burn in shorter sections so that the flame and heat will be such that the holding crew can keep the fire inside the line. If the fire started at the line can be readily held, it may be advisable to carry a second string of fire inside and ahead of the first one in order to pull the flame of the one at the line

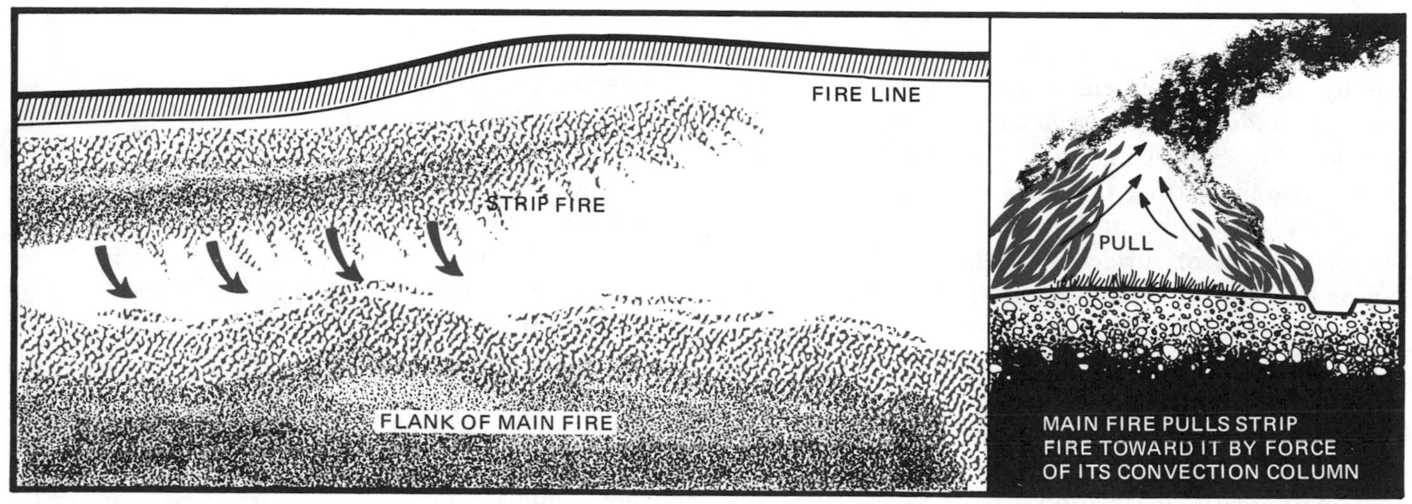

Figure 5.17. Strip burning parallel to the head fire.

away from the line and into the second string of fire. This operation is rather hazardous for the second torchman, and thus he should always be ahead of the torchman on the line and not too far inside (usually 10 to 20 feet) so that he has a ready escape route available. The type of fuels and the terrain will largely dictate how the strip fire can be executed. Never start the strip fire so far inside the line that a sufficient front of flame and heat can build up by the time the fire reaches the line. If that happens, the strip fire may overrun the line.

Another method of strip burning is called "center" firing, or "blowhole" firing. This method can be very dangerous to the torchman and should be used only by experienced fire fighters with stringent supervision and constant observation of the torchman. It is not recommended for general usage.

This method, when it is used in advance of the head fire, is a backfiring technique; when it is used on the flanks, it is a form of strip burning. The set fire is strung at right angles from the control line toward the head fire. If it is used in advance of the head, it has the effect of splitting the head fire by placing a pie-shaped burn in front of it. It is particularly useful on rapidly

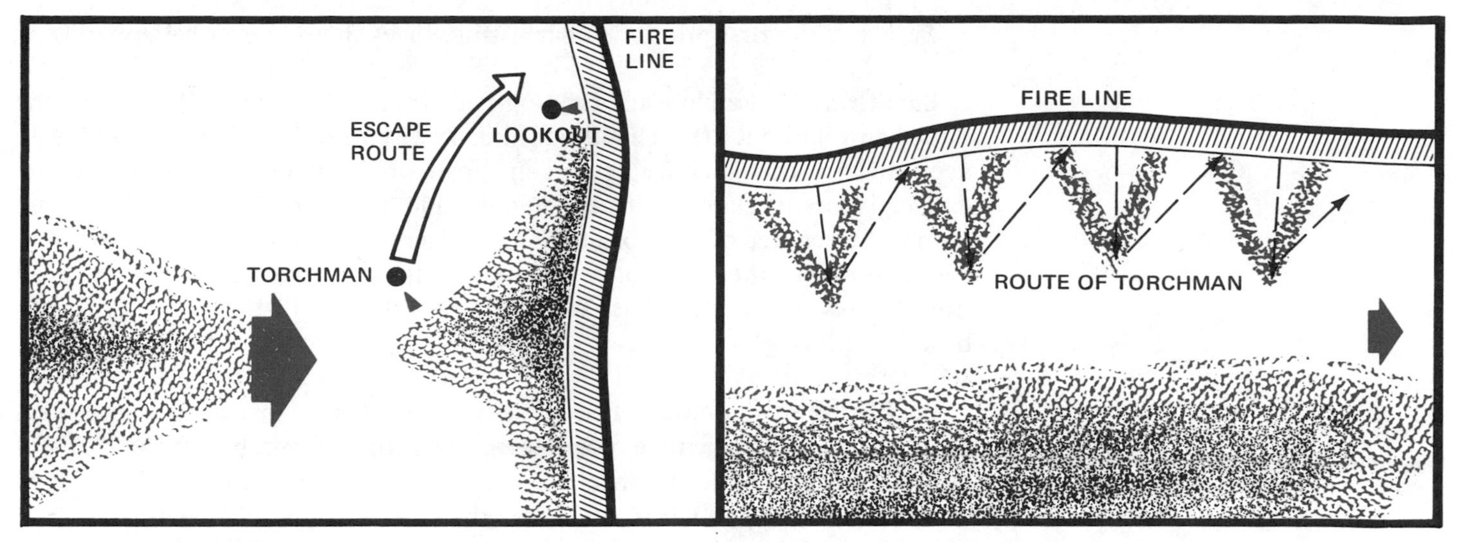

Figure 5.18. (*Left*) Center firing in front of a narrow fire head. (*Right*) Center firing along flank.

moving narrow heads. The set fire must be strung from the line to the head. To set it in the opposite direction would simply speed up the head fire. This type of firing places the torchman in danger. A turned ankle could be fatal. Center firing should only be practiced where the area can be traversed fairly easily on foot and when someone is assigned at the line to watch the torchman constantly in case he should need help. Sufficient distance should be left between the center fires for return of the torchmen.

When using the center firing method on the flanks, the same precautions should be observed. This technique puts shorter sections of flame along the fire line, making it easier for crews to hold; but it can be hazardous. The strips into the main fire should be far enough apart to allow ample time for return of the torchman. The line may then be fired between the strips with some pulling action from the strips. Center firing is more useful on the flanks than on the head.

Backfiring. Backfiring is a form of the indirect attack. It is the process of intentionally starting a fire along the inside edge of a fire line or fire barrier in advance of a head fire or along the forward flanks of a rapidly spreading fire. The area between the control line and the head fire is burned to eliminate fuel in advance of the fire and thus widen the control line, to change the direction of the fire, or to slow the progress of the fire and gain time for line construction. The backfire is usually set considerably in advance of the head fire. Some area is deliberately sacrificed to prevent a larger fire.

The word backfire connotes a simple operation of starting one fire to bring another under control. However, the term is used too loosely. It is one method of using fire to fight fire and can be effective if it is used properly. Unless the fire that is started can be controlled and the spots from it can be extinguished quickly, there is every possibility that simply more fire will be created and the problem will be compounded. Backfiring should only be used as a last resort or to prevent a situation that would later be untenable.

- Backfiring should only be executed by experienced fire fighters.
- Preparation, organization, and coordination are mandatory.
- A calculated risk is almost always involved.

Safety is the first priority in backfiring. Lives should never be jeopardized.

Backfiring is used on hot running fires that cannot be controlled by the direct and indirect attack methods already discussed in this text. There is no need to use the backfire tactic on small or ordinary fires when the fire danger is in any class except the highest one; true backfiring is used only on a small percentage of extraordinary fires. The fire boss may have to decide to backfire before the situation warrants if it is estimated that, when the head fire reaches the control line, sufficient conditions will then exist to justify backfiring.

Backfiring works best in light, flashy fuels. Heavy surface fuels and snags should be avoided where possible. Crown fires from backfiring are not desired except in rare instances. In some instances there will be no choice in the type of fuel; and in such cases, the fuel type should be treated, and preparations should be made to prevent a crowning backfire.

Organization is of basic importance. One qualified person must be responsible for controlling and directing the backfiring operation and all other personnel must understand this. On small fires, the backfiring will be done with direct instructions from the fire boss. In large situations, he would delegate the operation to a qualified crew boss or sector boss for the section involved. Constant communication between the man in charge and the crews is necessary. Preplanned escape routes and safety zones are absolutely necessary. It is best to avoid the use of large crews in firing because their control is often difficult. However, the holding crew and crew to hunt spot fires should be large enough to accomplish their functions. The firing crew should be supported with tractors, pumpers, hose lays, and holding crews as needed according to the conditions (see the sections in this chapter on the use of water, bulldozers, and plows).

The timing on setting the backfire depends on a favorable combination of (1) fuel, (2) weather, (3) control factors, and (4) the need to accomplish a satisfactory burn. A careful estimate of all factors involved is required. If conditions are too severe to start the backfire, it can be anticipated that the fire line will not stop the fire. Waiting too long to set the fire may result in an unsatisfactory burn. Backfiring is best accomplished 1 or 2 hours after the peak burning period. Burning early in the morning and at night is usually unsuccessful.

Basic backfiring procedure. In hilly or mountainous terrain the best location for considering the use of backfiring is just over the top of a hill from the slope where the main fire is located. The location of the line or barrier should be just over the crest of the ridge on the slope opposite the fire where the grade is less than 20 percent. This location also takes advantage of any upslope wind present on the opposite slope. It should not be so far down the opposite slope that the area involved will be too large to hold the set fire. A fire line of sufficient width to hold the set fire must be built in advance of firing, or a natural barrier such as a ridge top road or a preplanned and prebuilt firebreak may be in place. Anchor points must be in place prior to firing. Anchor points are the locations where lines or barriers to both flanks of the main fire join the line across the front of the fire; examples include a road, a rock slide, a cliff, a stream, a cold burn, or a man-made line. Lines along the flanks should be built so that the total area

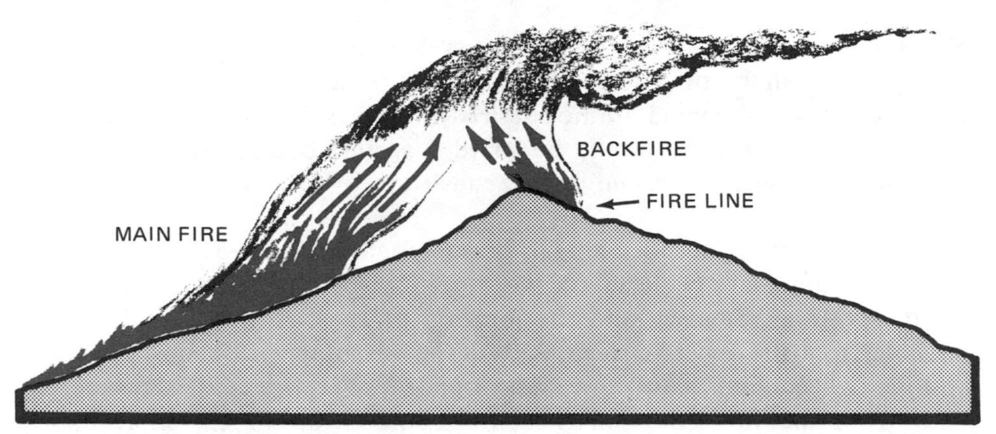

Figure 5.19. In backfiring in hilly terrain, start the fire just over the top of a hill and burn toward the approaching main fire.

of fire can be contained when the main fire reaches the backfire. The backfire must develop satisfactorily prior to the arrival of the main fire toward the area of the control line. If the backfire is too late, the impact of the main fire on the control line may be worsened. *The objective is to have the backfire drawn into the main fire at a safe distance from the control line.*

Concentrations of fuel inside and adjacent to the control line should be scattered or removed so that there will not be too much radiant heat and/or flame across the line for the resources to hold. Snags should be felled, especially inside the line. Live trees with branches extending to the ground or with moss should be pruned to a sufficient height so that the surface fire will not jump to the crowns. Fuel immediately adjacent to the outside edge of the line should be treated by felling standing snags and neutralizing fuel concentrations. Stumps, fallen logs, and punky material in the same area should be covered with dirt on the side toward the line.

The corners between the anchor points and the backfire line should be burned out first. These are the points that are most difficult to hold. Then the edge of the backfire line is established by burning from the anchor points to the center of the control line. Next, burn from the anchor points downhill along the flanks. Where there is a choice, always start the backfire at the top

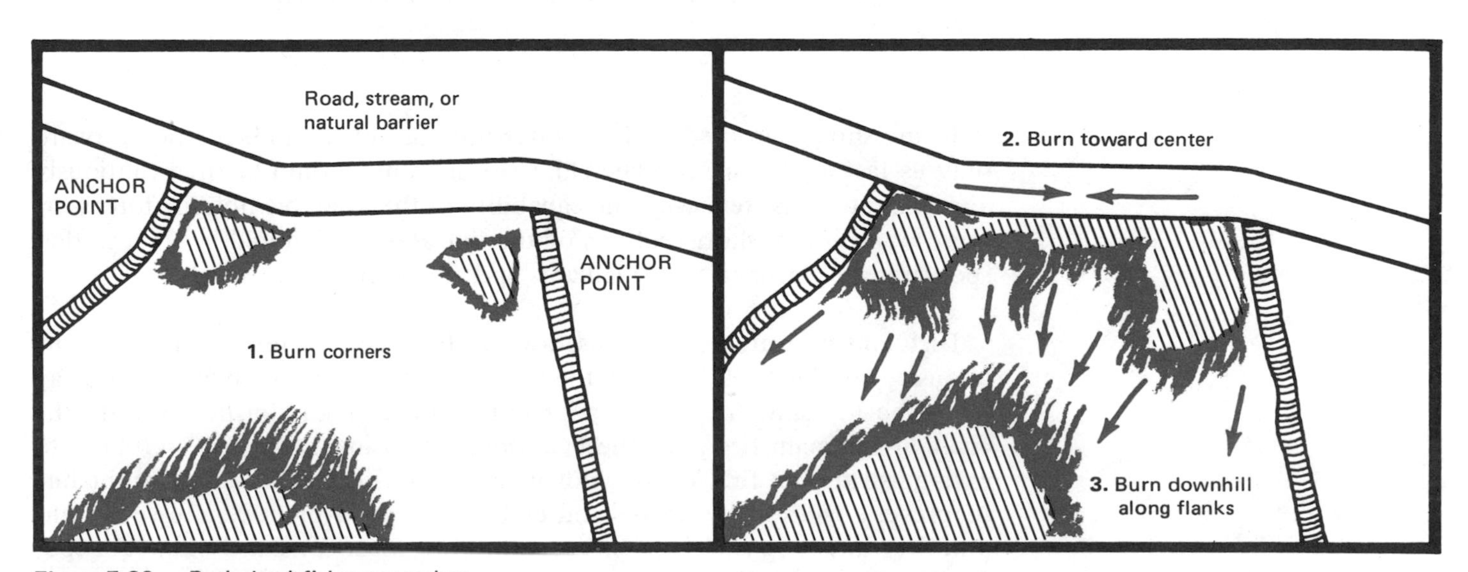

Figure 5.20. Basic backfiring procedure.

of a slope and ignite toward the bottom or downhill to prevent more buildup of fire than can be held along the line. Try to get a reasonably fast burn. Very slow rates of spread are not desired with backfires. On the other hand, hot runs of the backfire should be guarded against because they may cause much spotting, slop-overs, and intense radiation.

If the head fire is approaching in fingers, the backfire should be started at locations immediately ahead of the fingers. Much depends on the wind speed, spotting potential, and the location of the backfire line. If you are working from the top of a ridge, there is the advantage of possibly having an upslope wind coming up the slope opposite the fire and opposing the wind from the fire side. This keeps the column at the top more or less perpendicular or somewhat bent back towards the main fire thus reducing the spotting potential into the unburned slope. So far as is possible, burn only when you are sure the wind direction and speed will remain steady.

Figure 5.21. The objective of backfiring is to have the backfire drawn into the main fire at a safe distance from the control line. A wind blowing uphill on the backfire slope will help in drawing the convection column of the backfire into the convection column of the main fire.

Firing should proceed without interruption, yet it can be carried forward only as fast as the line can be held. Critical points should be fired cautiously and at a rate consistent with the capability of the available holding forces. At these points, only short sections of the line at a time should be fired so that spot fires are confined to a small area and are easily suppressed.

Dirt and/or water can be used along the backfire line to decrease the intensity of the backfire near the line until the backfire burns well away from the line and the intensity of radiated heat is diminished or until the draft of the main fire pulls the backfire in. In using water, be careful not to extinguish the set fire. Wetting down an area outside and adjacent to the line to prevent ignition by sparks and embers is a good practice. However, this practice cannot be solely relied on, and a crew must constantly hunt for spot fires (see Figure 5.22).

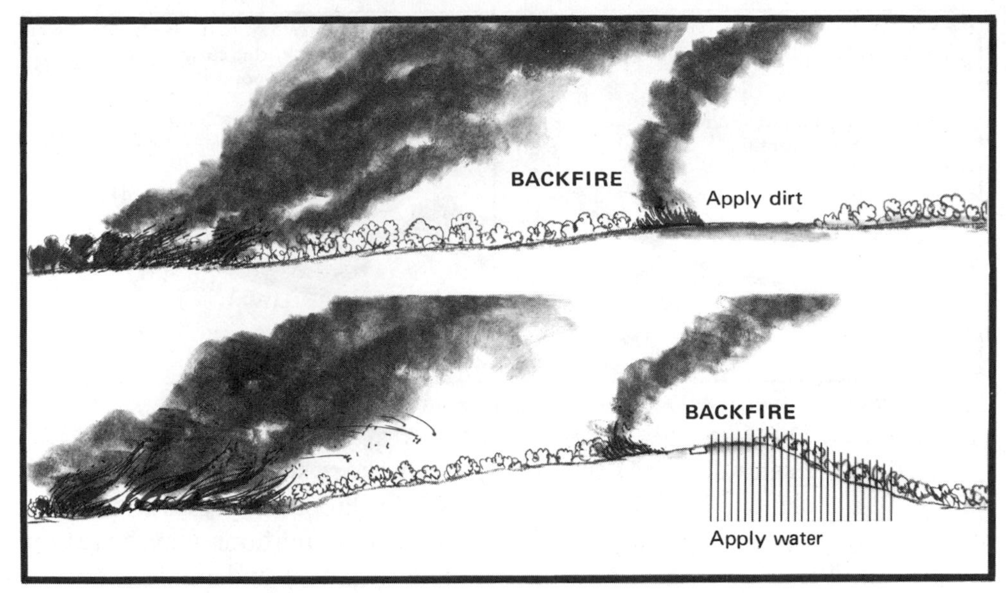

Figure 5.22. Dirt and/or water can be applied to the backfire near the control line to keep it under control and to reduce spot fires.

To avoid crowning, the backfire should progress slowly downhill against the wind. Reproduction and lower limbs on trees and fuel accumulations that would cause crowning should be removed and broken up. If a crown fire is wanted, fire from the bottom uphill and with the wind; thus the intensity of the backfire is speeded up. To intensify burning, add fuel by cutting and stacking or rearranging and add diesel fuel.

On rolling or flat terrain and in fuels where equipment such as pumpers and/or plows can be used, the same general principles and procedures as those given for hilly terrain are used. However, in these situations there is somewhat more flexibility in operations. Rates of spread and patterns are not affected as much by slopes. Wind speed and direction may have less variation. Also, there may be more choice of line location.

Exceptions to basic backfiring procedure. Following are exceptions to basic backfiring principles:

1. Where a fire is burning briskly up side ridges and is creeping through canyons and valleys while a midafternoon wind is blowing across the ridge, timing and aggressiveness in backfiring are particularly necessary. Such conditions may exist in brush, reproduction, and grass fuels. Build the control line just over the top of the main ridge and set the backfire as promptly as possible. This action takes advantage of favorable midafternoon wind and humidity conditions. The same prompt action should be taken for any fire on any slope when the normal outlook is that with nightfall the main fire will die down, leaving ragged, irregular edges that are difficult or unsafe to hold.

2. A saddle or pass between two high points on a ridge should be viewed cautiously, since it is a natural chimney. Burn into saddles rather than away from them; i.e., burn downhill with the backfire. When the backfire reaches point A (see Figure 5.23), continue slowly with firing from point A toward point B. Send part of the crew to point C at the same time and continue firing from it toward point B so that both crews will reach

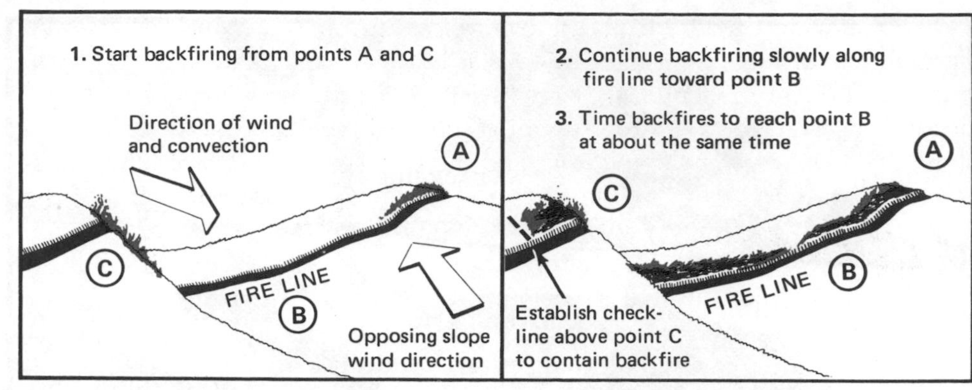

Figure 5.23. Procedure for backfiring a saddle between two ridge peaks.

point B at the same time. At the same time, backfire slowly to the left from point C or hold the backfire with a check line. Proceed very cautiously in crossing the saddle, since temporary adverse wind conditions may be set up by the backfire. This situation assumes that the wind and convection are moving upslope as they ordinarily would when burning conditions are favorable for backfiring during the day. At night when the downslope winds have started, burning should be done in the opposite direction. However, in night firing, the fire danger would need to be high enough to sustain the backfire. This same method would be used to cross deep canyons or ravines with a backfire.

Figure 5.24. Backfiring in V-shaped canyons.

3. If the main fire is progressing down one slope of a V-shaped canyon, construct the fire line at the bottom a short distance up the opposite slope so that the bottom of the drainage will catch the rolling material. Set the backfire a short distance up the slope on which the fire is burning so that it will spread both up into the main fire and down to the bottom. This action insures a rapid and clean burn by the backfire and reduces the threat of spot fires by allowing the backfire to burn downslope as well as upslope.

4. Hooks in the line, such as sharp bends and square corners, are particularly vulnerable to spotting, especially in the vicinity of ridges. Strip firing may draw sparks and flame away from the line. Simultaneous firing from each side into the hook (similar to saddle firing) may sometimes be helpful. These sections should be neutralized before firing the rest of the line.

5. Carrying the line from the top to the bottom of a slope under severe burning conditions is particularly hazardous. Assume the following conditions: uniform moderate to heavy fuels; wind up a moderately steep slope, possibly quartering at the top; fire edge farthest advanced at the top of the slope; favorable burning conditions; and the backfire line in place. Proceed as follows. Start at the top of the ridge and carry the backfire downhill to the canyon bottom, letting one stretch of backfire burn well in toward the main fire before setting the next section of backfire. Fire in the order shown in Figure 5.25, letting the first firing burn into the main fire before setting the second. The main fire is most advanced at the top and stands the best chance of crossing the control line near the top of the slope. By firing in this order, the backfire will have the least chance of gathering headway and making a big front, which may threaten to sweep over the line. Let each stretch of backfire burn well in toward the main fire before setting the next stretch.

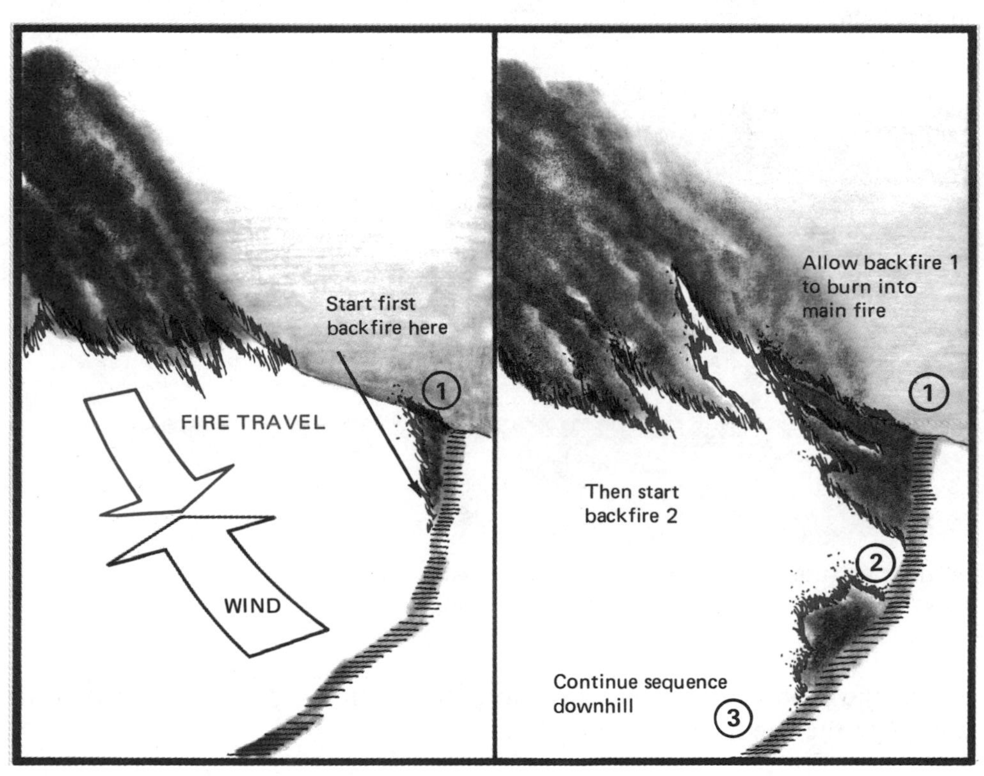

Figure 5.25. Procedure for carrying the line from the top to the bottom of a slope under severe burning conditions.

6. In cases where the main fire has died down and burning conditions are poor (the wind is gentle and there is a downslope draft on moderately steep slopes), the backfire is set some distance down the slope so that it will gather headway as it runs uphill. This is then repeated by sections along the line. In this way, enough heat and draft are obtained to neutralize unfavorable burning conditions and to obtain a good burn. The purpose is to establish a secure line where it would be impractical to complete control by direct attack before the next burning period. This is a tricky practice requiring careful handling and should be used only when other practices appear wholly unsatisfactory.

7. When the line is being constructed uphill with the plan of backfiring downhill as soon as the line is completed but the spread of the fire does not give enough time to complete the line, run a temporary line in, across, and

Figure 5.26. Procedure for backfiring under unusually poor burning conditions: (1) start backfire; (2) repeat by sections moving downhill; (3) allow backfires to spread uphill into burnout area.

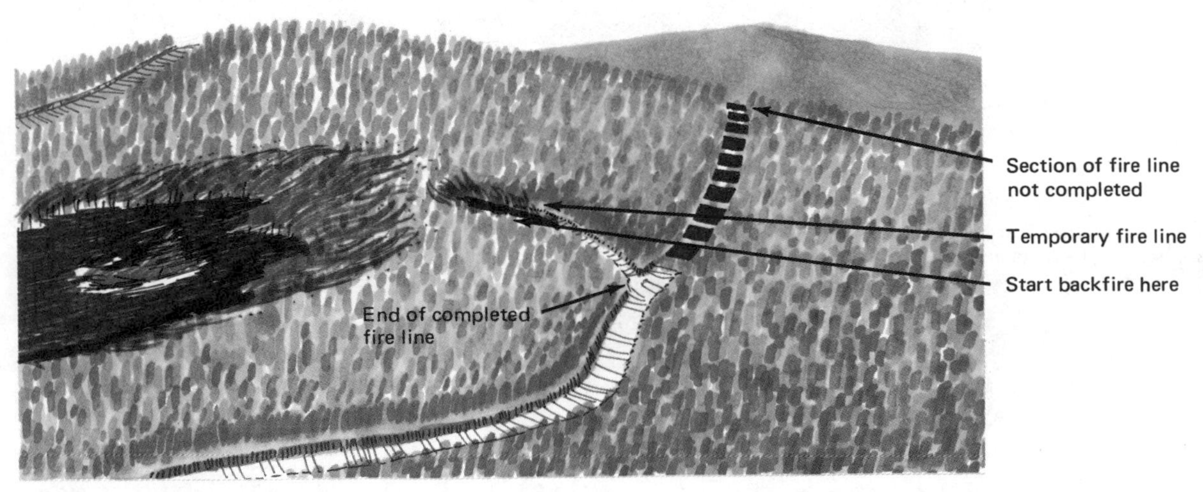

Figure 5.27. Emergency backfiring procedure for use when incomplete line is threatened by the main fire. This procedure slows progress of the main fire and permits completion of the line.

above the head fire and back fire along this temporary line to gain time to complete the line to the anchor point.

8. In backfiring to control the rear of a fire on a steep slope, start the construction of fire lines at the top of the slope and bring them down nearly perpendicular to the slope so that there will be no need for undercut lines. Also, the firelines can be flared out so that rolling material cannot roll across them. Fire out both lines on the flanks promptly as the line progresses downhill. Install a line at the base of the slope to control any rolling material either by placing it a short distance up the opposite slope or far enough away from the toe of the slope so that material will not roll across. If the fire has a tendency to spread laterally at the bottom, it will be necessary to work from the bottom up. Under erratic and adverse conditions of weather and in heavy brush, safety will dictate the latter course.

Backfiring tools and devices. Many devices are usable for backfiring, including fusees, drip torches, powered torches, and flame throwers; but fire fighters should be prepared to improvise if these are not immediately available. Waiting for their delivery may make it too late to fire at the right

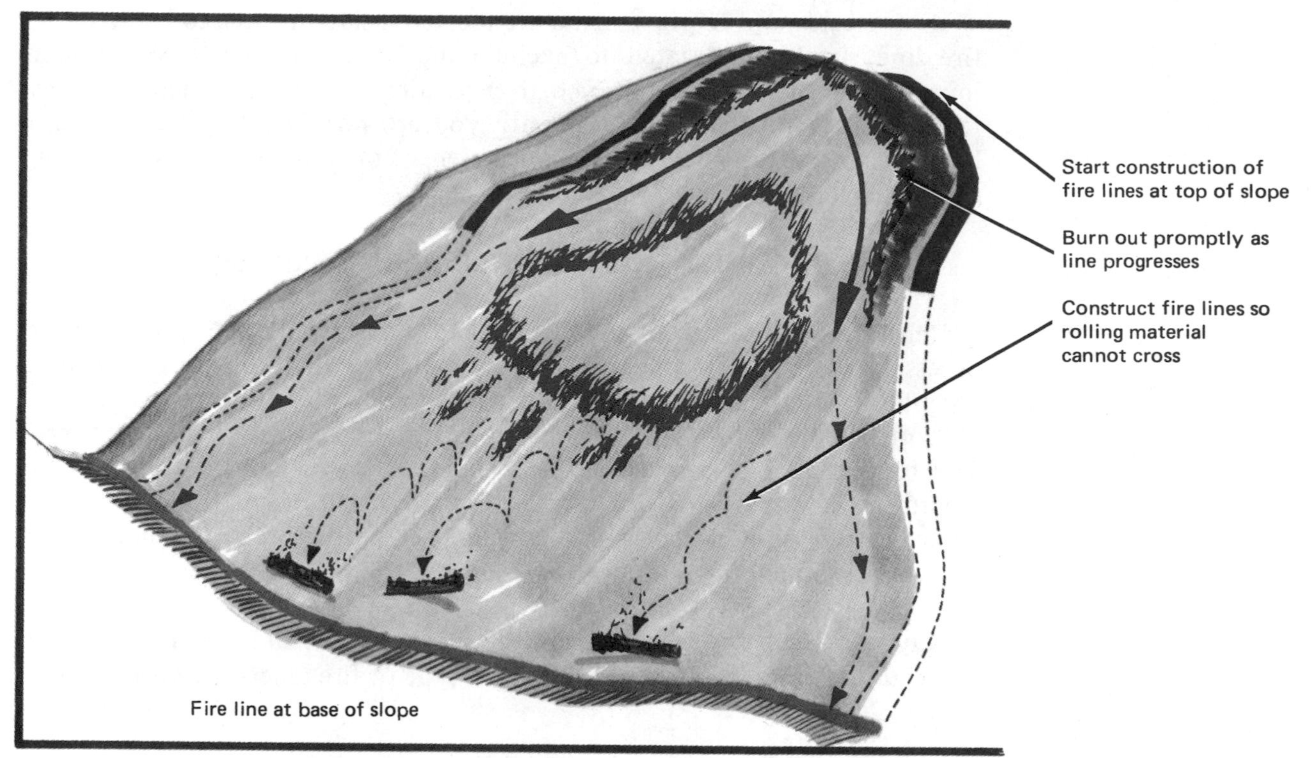

Figure 5.28. Backfire procedure to control the rear of a fire on a steep slope. In heavy brush, safety may require burning from the bottom to the top.

time. Hot coals can be moved with shovels and rakes or McLeods. Also ignited yucca stalks or pieces of cedar bark can be dragged over the area to be fired. Tire casings charged with fuel oil and ignited can be dragged by a pickup, horse, or tractor. A rag soaked with fuel oil attached to a wire or stick can be used to ignite an area. These devices are limited only to the ingenuity of the fire fighter.

A word of caution. Before backfiring check local ordinances and laws. Some areas have laws preventing the use of backfires; in other areas, the law may require the fireman to pay damages if the backfire escapes. There are areas where the law protects the fireman as long as due diligence is used. If the backfire is to be started on private property, the owner's permission must be obtained. Over the long run, try to obtain local ordinances to give trained firemen blanket authority to use backfires in emergencies. In some localities there is a tendency for local people to start backfires without coordinating with the fire service. Most often, the results are disastrous. If this is the case in your community, take steps to prevent such actions and maintain control of your fire-fighting operations.

The use of fire to fight fire is often the only practical solution in wildfire control, especially with extraordinary fires. At other times, backfiring amounts to taking the easy way out, with more acreage being burned or burned hotter and more deeply than was necessary.

The general public has every right to expect the fire fighter to be the least likely source of this type of fire. Conservation and civic groups, as well as the affected property owners, will look with a critical eye on these types of fires regardless of the values involved.

If structures and improvements are located inside the area to be fired, the fire line should be located to exclude the improvements. Firing around improvements first can minimize danger to homes and other structures. *Never fire* in the vicinity of structures until you are positive that the occupants have been evacuated. Often the presence of numerous and valuable structures forces a change of tactics.

ORGANIZATION FOR WILDFIRE SUPPRESSION

Regardless of the size of the fire, certain basic management principles are necessary to obtain prompt and efficient control of the fire with the smallest acceptable fire damage and with reasonable cost. Positive and known lines of authority and a dependable response to instructions must exist at all times. The fundamental principles of fire management are the same on all fires; regardless of size. As a fire increases in size from a small one to a very large one, the only change is in the complexity of the organization, which is built on the same principles.

Any fire-fighting agency must be able to expand its management organization to meet the demands of the type of fire emergency it may face.

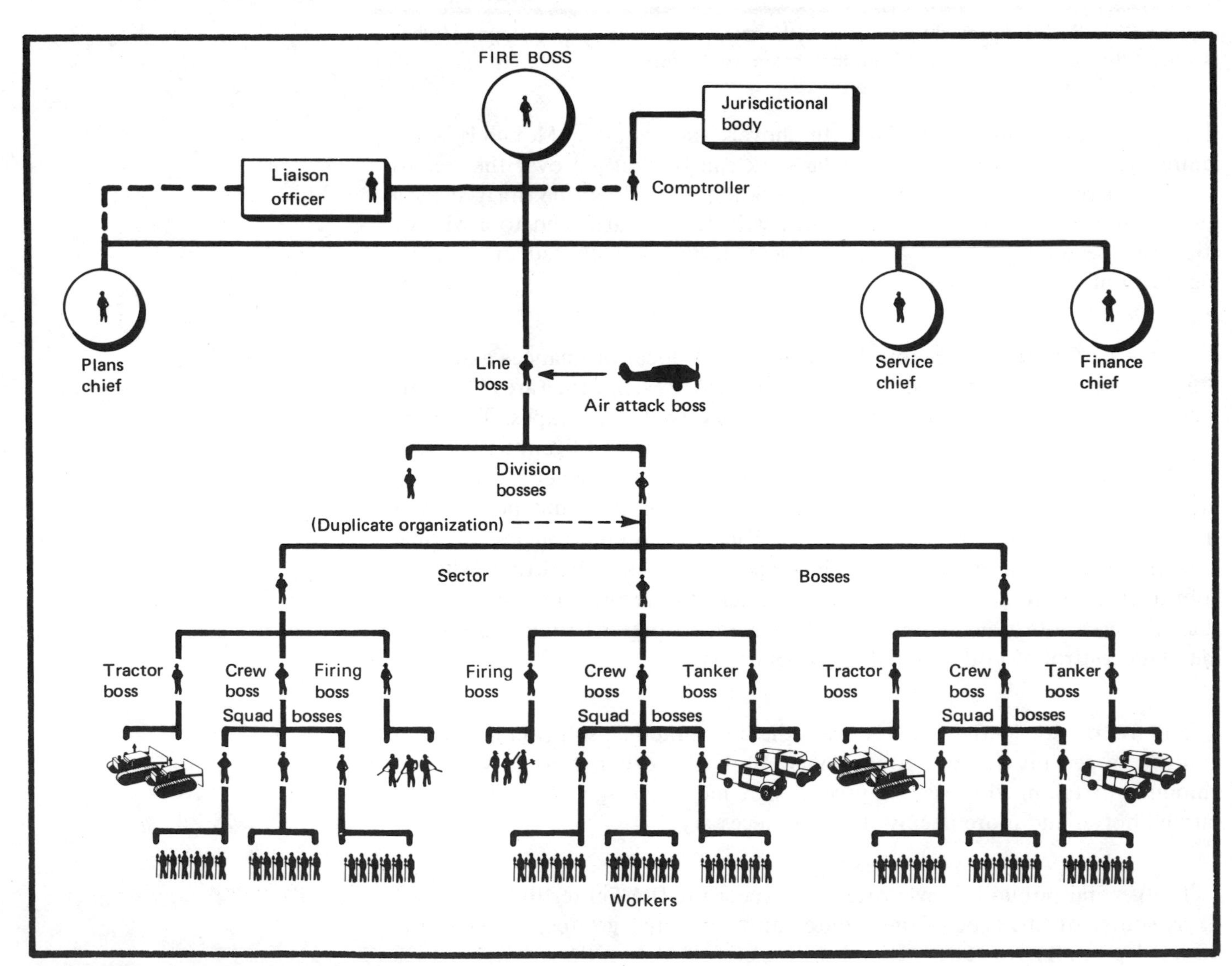

Figure 5.29. Organizational chart of line control forces needed on a large fire.

It must do so with a predetermined and fundamental yet simple plan; otherwise, the result can only be mass confusion.

Usually, the problems involved in controlling a wildfire become larger and more complex as the fire increases in size and the rate of spread increases. Therefore, the manpower, equipment, and apparatus must increase, as must the specialists needed to handle the increased work loads of the several functions that would be administered by the fire boss, sector boss, or crew boss on smaller fires.

The descriptions given here are the standard units of organization used in controlling wildfire; they may vary somewhat or have different names in the multitude of fire services working on wildfires, but these descriptions are the ones most likely to be used on campaign fires involving several different agencies or departments.

Regardless of the size of the fire or whether only one man, a squad, a crew, or a larger organization is the initial attack force, the following three jobs must be performed.

1. The resources must respond to the fire area with the right tools to do the job. This is the service function.
2. The size-up must be made and a plan of action developed and executed. This is the planning function.
3. A control line must be put around the fire, spot-fires must be controlled, and the necessary mop-up must be accomplished. This is the doing or line function.

For smaller fires all three functions are handled by the squad boss, crew boss, sector boss, or officer-in-charge who is acting as fire boss. For larger fires the three functions would be accomplished by a service chief, a plans chief, and a line boss with a suitable assignment of personnel for satisfactory performance of each function. The fire boss would exercise command by administration through the chiefs of the three functions. A finance chief with suitable personnel would also be assigned.

As the size of the fire job increases, multiples of squad units are formed into crews, then into sector teams, and, further, into divisions. For exceptionally large fires, zone organizations may be formed. The needed distribution of pumper crews, tractor crews, plow crews, firing crews, and felling crews are assigned to each unit of organization. Air support may be added to this ground organization.

SQUAD BOSS

A squad boss is sometimes called the foreman, straw boss, or crew boss according to local usage of terms. In the fire service organization of a four- to five-man pumper company, the officer in charge of the first-due company would be the squad boss.

A squad is the basic wildfire control unit. It is generally made up of three to ten men and is directed by a squad boss. The squad boss reports to the crew boss if more than one squad is fighting the fire; otherwise, the squad boss is the fire boss.

The squad boss is responsible for keeping the squad fully employed on their assigned jobs. His duties are to:

- Carry out the orders of the crew boss (who may also be assigned as fire boss)
- See that his men have proper tools and safety equipment and know how to use and care for them
- Help the crew boss in off-line and camp duties
- Keep a list of names of the squad members and keep time if the crew boss makes that request
- Look after the safety of the squad on the line
- Observe and enforce smoking rules
- Take necessary action with lazy, incompetent crewmen or agitators
- Make sure that the squad is transported safely
- See that the squad has food and water

If the squad boss is also acting as fire boss, he would carry out all of the fire boss's responsibilities.

If a plow or dozer is used on the initial attack, the squad boss might organize with an operator, a burnout man, and two or more men assigned to holding while he does the locating of the line. A pumper attack might be organized with a driver-pump operator, a nozzleman, a hose puller, and one or more followup men. If used in a pump-and-roll stance, the hose puller might also be assigned to the followup.

CREW BOSS

A crew boss supervises three squad bosses and their crews. He reports to his sector boss or to the fire boss, or he may be the fire boss. His area of responsibility includes the performance of his crew, their safety, and their general welfare. He usually retains the same crew for the duration of the fire; therefore, he has responsibility for them both on the line and off the line.

His on-the-line duties are the following:

- At the beginning of each shift, explaining the nature of the work to be accomplished, the expected duration of the shift, and the chain of command
- Organizing the crew to accomplish specific tasks efficiently
- Locating and assigning individual tasks to squad bosses or crew members within the assigned section of the line
- Explaining and demonstrating safe and efficient techniques when the crew members are inexperienced or unskilled
- Looking ahead and locating escape routes and safety areas for the crew
- Keeping the men working at proper intervals along the line
- Frequently inspecting the assigned area to ensure required standards of performance

His off-the-line duties are the following:

- Keeping the crew together as a unit and mobilized to answer any request at any time
- Inspecting the men's physical condition, clothing, and equipment before and after each shift and taking necessary action to keep the crew in usable condition
- Preparing and maintaining time reports
- Supervising transportation and travel of the crew
- Learning from the bulletin board or registrar the layout of the camp and the camp routines for feeding, sleeping, commissary, sanitation, etc. and instructing and controlling the crew to the established routines

The crew boss should also arrange for tools, equipment, water, lunches, first aid kits, and travel and should ensure safe vehicle travel, careful driving, and safe foot travel and tool carrying.

SECTOR BOSS

A sector boss supervises three crews (nine squads). He has responsibility for all attack actions on his sector, which may include pumpers, tractors and plows, firing crews, felling crews, and line crews. If the fire warrants, air support operations may also be the responsibility of the sector boss. He might also perform as a fire boss when only nine squads are assigned to the fire and it is anticipated that these resources will be sufficient to control the fire.

DIVISION BOSS

On extremely large fires, the line work is divided into divisions. Each division consists of two or three sector teams and is under the command of a division boss (Figure 5.30) supervised by the line boss.

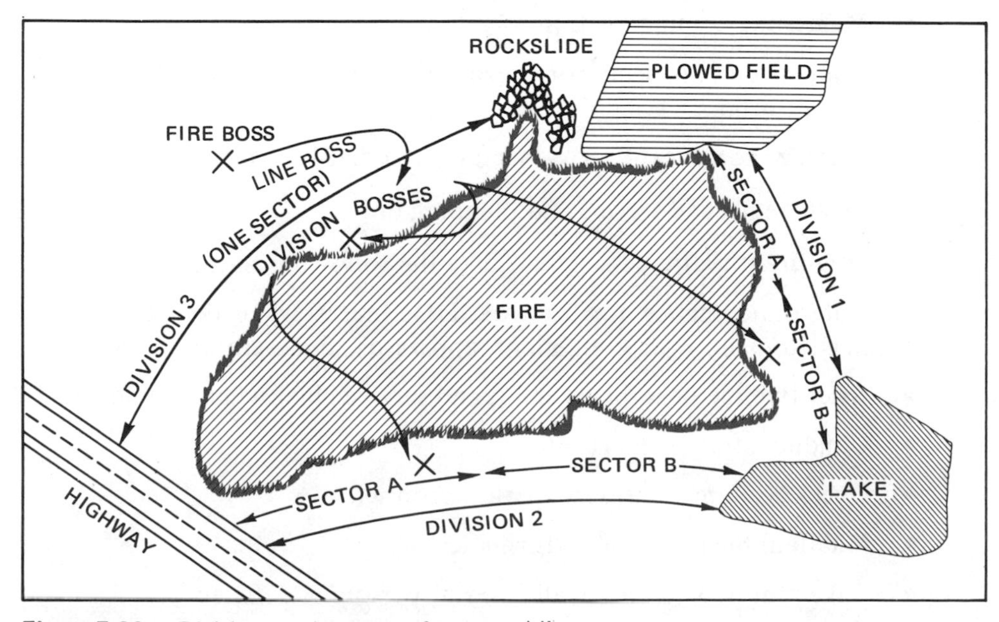

Figure 5.30. Divisions and sectors of a control line.

FIRE BOSS

On every fire, one and only one individual must be recognized as the fire boss at all times. The fire boss has the primary responsibility for organizing and directing the fire organization for efficient and complete control of the fire. Before arrival at the fire, he should obtain the best information on the location of the fire, the forces and equipment being employed and the standby forces and equipment available, and the current and forecasted weather conditions. He should travel by a route that will provide a vantage point for sizing up the fire and assign scouts as necessary to keep informed about all sides of the fire.

Regardless of the size of the fire, the fire boss should exercise good principles of management. He should gear his thinking to the size of the fire and its development. He should be able to adjust his tactics to the expected as well as the present situation, develop alternate plans of action to fit changes in conditions, and be prepared to make adjustments and sacrifices. Finally, he should know the location of the resources at all times.

On arrival at the fire, the fire boss should take the following actions:

- Determine probable spread during the first burning period.
- Notify the dispatcher of the adequacy of resources assigned to the fire, and give an estimate of the fire potential and the additional resources needed.
- Inventory and analyze possible environmental damages, both as a result of the fire and possible control strategies.
- Prepare a plan of control.
- Organize resources according to plan.
- Assign men and equipment on arrival.
- Direct and coordinate fire suppression.
- Keep the dispatcher informed.
- Maintain a communication schedule.
- Prior to crew or sector organization, assume responsibilities of crew or sector boss.
- Organize forces into crews, sectors, and divisions as needed.
- Organize needed service and planning units.

On a going fire, the fire boss should take the following actions:

- Determine tactics for control. Issue necessary orders to get the needed manpower, equipment, apparatus, and facilitating gear.
- Brief the line boss, plans chief, and service chief where employed.
- Conduct strategy meetings.
- Check welfare and safety of all personnel.
- Maintain high level of performance.
- Take required action on all cases of personnel deficiency.
- Anticipate need for and assign work to functional assistants.

- Determine need for enforcement of safety requirements.
- Make sure that functional assistants understand and complete their work.
- Remain in communication except in short periods. Designate acting fire boss when necessary.
- Make sure the acting fire boss and other subordinates obtain needed rest.
- Visit critical and problem areas personally.
- Make sure that fire control actions are carried out with a minimum of damage to the environment.
- Determine demobilization time. Use organization channels.
- Make sure that fire camps and firelines are restored and that felled material is removed from streams.
- Write performance ratings for personnel as requested or required.

The individual who is assigned as fire boss may change as the complexity of the fire increases because of the necessity of assigning a more experienced and trained manager.

INTERAGENCY LIAISON OFFICER

Where an interagency liaison officer is assigned, he acts as a communication link between the agency in charge of the fire and the cooperating agency or department that he is representing. He is responsible to either the fire boss or the plans chief. He must be familiar with the policies and procedures of the agency in charge of the fire and well versed in all aspects of fire suppression activities.

Maintaining close contact with the cooperating agency, the liaison officer will:

- Keep posted on the plan and control progress of the fire.
- See that radio and telephone communications are installed.
- Keep currently informed of needs and requirements.
- Relay all requests for additional manpower and equipment to the fire headquarters.
- Keep the fire headquarters informed of all surplus manpower and equipment available for transfer.
- Acquaint himself with the quantities of men and equipment provided by the cooperator, reporting such information daily to the plans chief.
- Inform the cooperating agency of the time and place of joint strategy meetings.
- Maintain contacts with home units to ensure flow of information on locations, assignments, and positions served by crews or individuals.

- Coordinate and assist the public information officer in news releases to home units.
- In general, serve as a link between the agencies involved in control of the fire.

LINE CONSTRUCTION

The fire line is constructed by a variety of methods, depending on the fuels involved, the available equipment and manpower, and the terrain as it dictates the use of resources, i.e., manpower and equipment.

Fire lines are built in four basic ways:

1. By using hand tools only.
2. By using earth-moving equipment, which is termed equipment or machines in this text, i.e., bulldozers, blades, disks, trenchers, plows, etc. All of these must be followed up with hand tools or pumpers where they can be used.
3. By applying water from ground tankers directly from the vehicle (pump and roll) or by laying hose lines. These fire lines also require follow-up with hand tools even if only with a backpack pump or shovel to assure mop-up in light grass fuels.
4. By aerial application of retardants using slurry planes, helicopters, or spray planes, followed up by ground forces using either hand tools, pumpers, equipment, or any combination of these.

TECHNIQUES OF LINE CONSTRUCTION

Undercut lines. Where the fire line must be built horizontally across a slope and below the fire, it should be built as a trench or ditch to catch the roll from above. Cones, burning chunks, rocks, and even logs often roll downhill as the fuels burn around them, thus scattering burning material downhill. In some areas, this scattering may jeopardize the entire control operation.

On the other hand, undercut lines are less than desirable if they can be avoided. However, considerable area may have to be sacrificed below such a line before a better location is reached, or the area below may be completely impractical for line building.

An undercut line should be built as a deep trench, that is, well banked with earth on the berm and along its entire length. Logs, rocks, and any available material can be used to form the berm, but the surface should be mineral soil. It should be deep enough to catch and hold any material that may roll into it. Move logs above to lie up and down the slope, or block them so they cannot roll. Thus, the amount of trenching can be reduced.

Fire on both sides of a ravine. In this situation, control the smaller head first; then go after the larger one. Construct the fire line on the side of the ravine opposite the main fire. The same action should be taken where separate fires are burning on each side of the ravine, where a single fire is threatening to cross a ravine, or where fire is burning on both sides of a road or barrier.

Figure 5.31. Construction of an undercut fire line.

Figure 5.32. Location of fire line in a ravine fire.

Flammable fuels outside the fire line. Rotting stumps, logs, and other materials outside the line, which are not burning but are subject to radiant heat and sparks from inside the line, can be treated with dirt. Cover stumps or logs with enough dirt to insulate them or wet them down if possible. This fuel is a dangerous type for spot fires.

Use of dirt in mop-up. Do not cover burning material with dirt except to cool it down so that it may then be extinguished by other methods. Dirt heaped on burning chunks inside the line will later fall off as the material continues to burn, exposing the glowing fuel. The wind can then whip sparks out of it. Complete burying where the burning material is below ground level may be satisfactory in some instances, but complete extinguishment is more effective.

Temporary checklines. Often the edge of a fire occurs in stringers and spots too ragged and too numerous to control by direct attack. When the edge advances, time is needed to complete the control line. Temporary checklines may be built around the points of the fingers and around the hot spots and spot fires to hold them or slow them down until the final line can be completed. This is a form of the point and cutoff method where the checklines may be used as part of the final line; in this instance, a separate final line should be built some distance away and the intervening area should be burned out.

Figure 5.33. To prevent ignition of stumps and logs on the other side of the fire line, cover them with dirt.

Figure 5.34. Temporary checklines slow down spot fires or fingers until a final fire line can be completed.

Checklines are used in pine needles, duff, or grass where the forces available are insufficient to complete a final line at once.

Ringing a snag. A snag is a standing dead tree or part of a dead tree from which leaves and small branches have fallen. If snags are inside and close to the line but are not burning and there is insufficient time to fell them, ring the snag with a line and remove all burnable material inside the ring. The ring should be at least 10 feet in diameter, depending on the type and amount of ground fuels adjacent to it. This action prevents the snag from catching on fire and throwing sparks across the line.

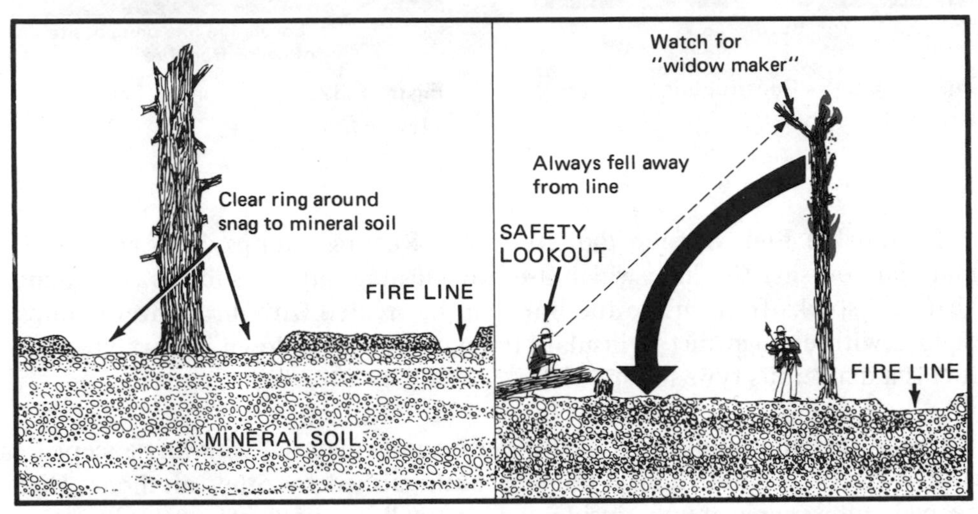

Figure 5.35. Ringing a snag.

Figure 5.36. Felling a snag.

If fire in a snag is burning briskly, it may be too hot or too dangerous to cut the snag down. The ring should be large enough to catch falling limbs or chunks. Be alert to falling material and use a lookout. Drop back 1¼ times the height of the snag, build a line, and burn out inside the line. The snag will thus be contained when it falls. Fuel may be piled at the base of the snag to speed its burning if conditions are such that this can be done without mass transport of embers. Keep a constant lookout for spot fires.

If it is determined that the snag may be safely felled, clean an area inside the main fireline large enough to catch the snag and fell it into this area. Chop off burning material and extinguish it with water or dirt. Watch for spots.

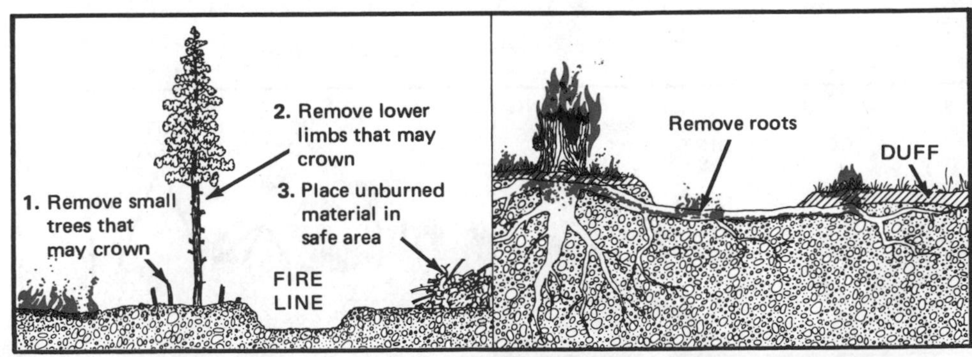

Figure 5.37. Procedure to prevent crowning.

Figure 5.38. Roots that spread across a fire line must be searched for and removed.

Always post a lookout to warn fellers of the falling top or limbs. Have a clear escape route and make a dry run before starting to cut. Only experienced fire fighters should be used on felling crews.

Crowning. To prevent crowning, throw dirt or use water at the base of trees or brush that are threatened. Scatter fuels that cause heat. Remove small trees that may crown. If it is not on fire, throw cut material well outside the line; otherwise, scatter it in burned-out areas. Remove lower limbs on large trees that could crown. To the extent possible, prevent any crowning; if this is not possible, use every means to prevent the spread of the crowning.

Roots across the line. Cut a section out of all roots extending from the fire edge across the line. This means digging for them. Fire can burn unnoticed along roots underground and can break out many feet away across the line and as much as two weeks later, causing a rekindle.

Rotary system. On a particularly bad hot spot along the line where only dirt is available, a rotary system may be used employing several shovel men. One shovel man picks up a shovel full of dirt, carries it to the fire, and spreads it on the fire. The second man immediately follows him and is followed by others as available and needed. They walk in a circle, following each other. It may be necessary for one or two men to stockpile dirt if it is difficult to dig up the dirt. This can provide an almost constant fan of dirt against the hot spot.

Figure 5.39. Spot fires occurring outside the fire line must be extinguished immediately.

Extinguishing spot fires. Spot fires often occur some distance away from the fire line. Immediate action is needed on them. Construct a line around them, and, if at all possible, extinguish them completely. Backpack pumps are very useful. A trail should be marked back to the fire line, and a sign should be posted, indicating the direction and distance to the spot fire.

LINE CONSTRUCTION WITH HAND TOOLS

Organization. In the majority of wildfire situations, hand tools are the only means of attacking the fire. In some situations, the use of hand tools must be combined with other methods of attack. The best organization for using hand tools has developed from experience on fires of all kinds throughout the continent. The organization will be different in the distribution and kind of tools used in different sections, but the basic methods are the same.

Hand tool crews initiate the line by cutting, fuel dispersal, scraping, burning out, and holding. Later, their function is mop-up and patrol. In some fuels, all functions are performed by the same crew as it works along the line.

One method of organizing the line work is to assign a specified length of line to each worker. He performs the cutting and digging on it until it is complete; then he travels past the rest of the crew to another location. This method of line organization has several disadvantages.

There are several methods of line organization now practiced throughout the different areas. All of them use the principle of working the crew in a line or "indian file" that moves ahead as a unit performing the several tasks enroute. This is usually described as the progressive method and is an outgrowth of the "one lick" method developed on the Rogue River National Forest in Oregon about 1940.

Line location. The lead man or line locator may be the crew boss. In most instances, the crew boss can best perform his responsibilities by moving back and forth along the length of the line crew, adjusting work loads, correlating functions with adjacent crews, and using another man as line locator.

The line locator marks the location for the clearing and the line. Much of the success or failure of the attack depends on his ability to observe conditions, understand fire behavior, appraise crew capability, and locate the line to accomplish the job with a minimum of time and energy compared to the acreage burned.

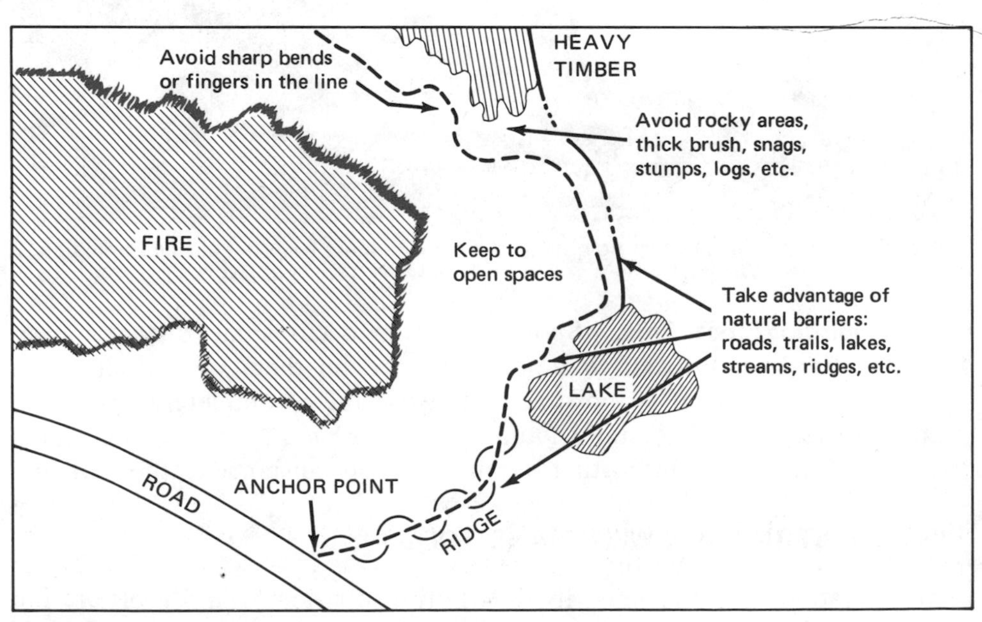

Figure 5.40. Locating the line for line construction with hand tools.

Good practices of line location for safety and efficiency include:

- Locate the line to speed the work of the crew with a minimum of effort. Choose best route for easy and rapid construction.
- Locate the line as close to fire edge as possible under the burning conditions and as planned in the strategy.
- Avoid heavy fuel areas and keep them outside the fire area if possible. Property values will often dictate the location of the line.
- Pick the easiest way through without giving up too much acreage. Take advantage of natural barriers and keep to open spaces as much as possible.
- Avoid undercut lines as much as possible. Locate where burning material cannot roll across.
- If possible, avoid sharp bends and fingers in the lines.
- Generally, try to avoid steep slopes and locate the line just over the top of a ridge rather than across the slope except under low fire weather conditions or other suitable reasons.
- When the indirect method is used, locate the line a sufficient distance from fire edge so that the line can be constructed and burned out before the main fire reaches it.
- Locate far enough from snags to catch most of the sparks from them and to include the length of snags when they fall or are felled.
- Always anchor lines to the best natural barriers or control points immediately available.
- Take advantage of diurnal changes and predicted fire weather.
- Endeavor to contain the final perimeter to a single watershed or one side of a watershed.

Immediately following the line locator is the initial cutting crew. Their job is to open a lane in standing fuels, clearing away the smaller standing or downed trees and brush so the line can be built in the soil.

Material that is on fire is thrown as far inside the line as possible and is scattered. Avoid building up fuel concentrations inside the line. If there is any doubt about fire existing in any material, throw it inside. Otherwise, the fuel removed from the line is thrown outside the line and away so it will not be bunched. Keep the thrown material far enough outside the line so that radiated heat, when the fire burns to the line, will not ignite it. In some cases, the removed fuels may be needed inside the line to assist in burning out. In this case, throw inside, but avoid bunching.

Rule of Thumb: The width of clearing should be at least half the height of the fuel. Many times it will need to be wider.

Each man on the cutting crew cuts material in the 10-foot section of line he is working in at the moment, and then he moves up on signal to the section just occupied by the cutter ahead of him. The number of cutters and the work load for each must be adjusted for the fuel type. When the cutting crew has advanced along the line, the necessary width of clearing should be accomplished. The holding crew that follows may widen the clearing where necessary.

The next unit of line workers is the digging or scraping operation. Depending on the fuel type, they will be using shovels; shovels and Pulaskis; shovels, Pulaskis, and grub hoses; McLeods and shovels; rakes; or most any kind of combination that is suitable for digging the line down to bare mineral soil (through the humus). The soil type, amount of rock, and the terrain will also affect the kind and amount of tools used.

If necessary, the first men on the digging unit use mattocks, adze hoes, or Pulaskis to loosen the top surface of the ground so that the shovel men who follow can pick up the material. The shovel men scoop up and throw out the duff, litter, leaf mold, and humus so that the trail is dug down to bare mineral soil. They work at a uniform rate of speed to keep the same relative position to one another as they move along the line. They should leave a finished line behind them. When it is adjusted correctly, this continuous progressive movement eliminates crowding, interference, passing, and duplication of effort. A narrow line is first made to mineral soil and is widened by each succeeding shovel man until it is the width desired for control. Many times a "scratch line" about 6 inches wide is built the first time through, and it is widened and improved later by another pass with the crew. Another method is to drop a man off here and there to widen the line and to allow the main crew to proceed with construction of the initial line. The work load of the individuals and the speed of progress are adjusted by the squad boss or crew boss.

If burning out is part of the operation, the torchman follows the digging crew using whatever means are available to start fire. He is usually assisted by shovels or backpack pumps to hold the started fire inside the line. Burning out is usually a critical operation that requires careful timing. The torchman must keep up with the digging unit and, at the same time, must not crowd the men who are holding behind him with too much hot line. Where possible, the line should be fired as soon as it is made. If the line is being built up a slope, it should be fired downhill (during the daytime) in sections to avoid too much buildup of heat against the line. Full advantage should be taken of drafts moving toward the main fire.

One to several men are required to follow up the torch crew to complete the holding operation, check for spot fires, and secure the line. Actually, mop-up begins in this operation. The mop-up may be done by an assigned crew following the line-building crew, or the line-building crew may be assigned to this important phase after constructing the line.

A crew organization for hand tools in grass and light sage or other light ground fuels is shown in Figure 5.41. Note that the first part of the line is narrow and is later widened to a width required for the height and amount of fuel.

As the fuel type changes, it will be necessary to change the kind and amount of tools. Sample organizations for trees, heavy brush, and scrub oak and for heavy brush are shown in Figures 5.42 and 5.43, respectively.

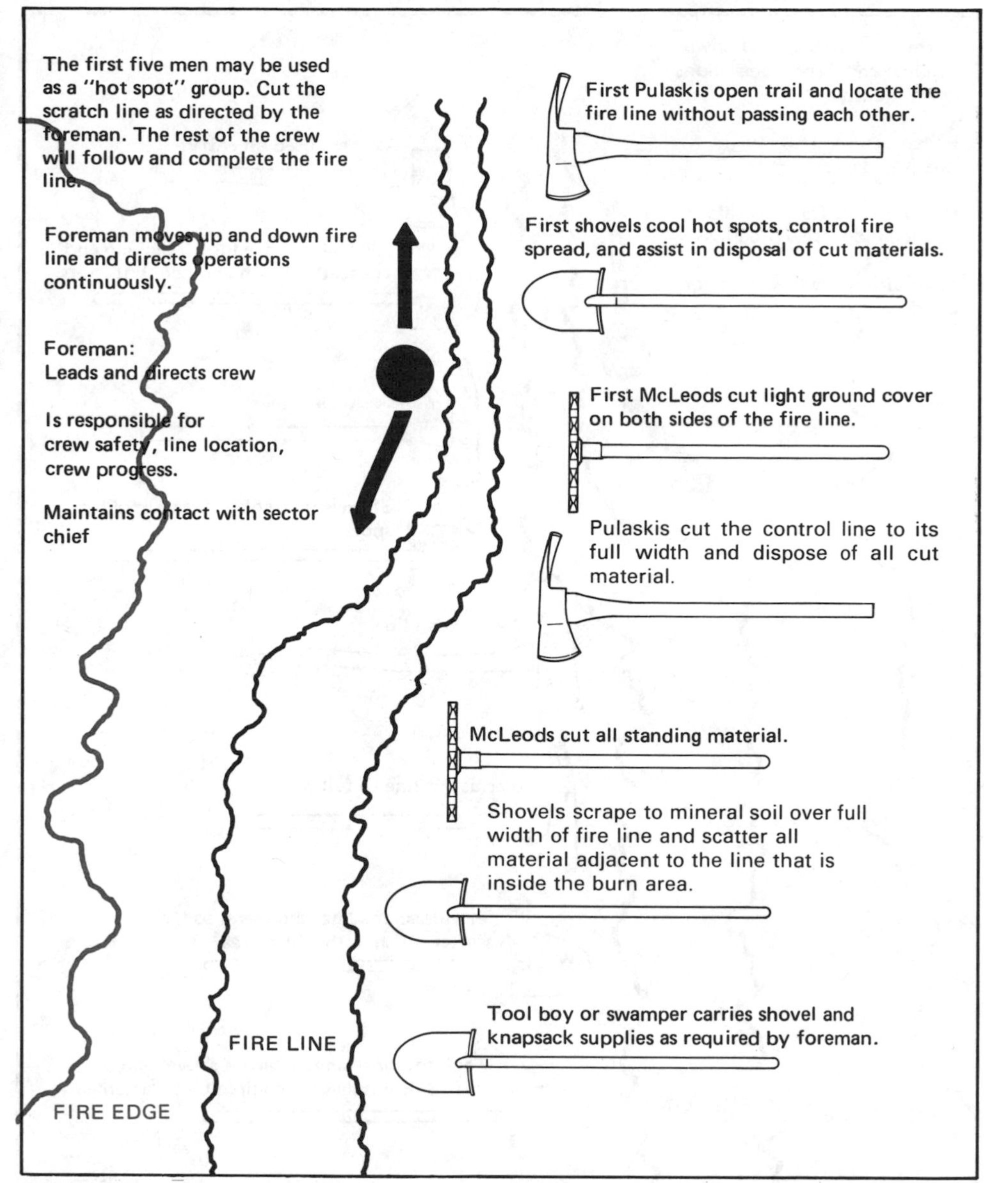

Figure 5.41. Crew organization for line control in grass or other light ground fuels.

Either hand or power saws may be necessary where any amount of large-sized fuels is encountered. Adze hoes are needed in heavy duff. The kind and amount of hand tools change with the fuel type and location.

USES OF WATER IN LINE CONSTRUCTION AND REINFORCEMENT

Water is the most widely used extinguishing agent for fires because it has a high capacity to absorb heat. It is usually readily available. If it can be reasonably and efficiently applied, it is the least expensive agent for putting out fire. In wildfire control the problem is how to get it to the fire in sufficient quantity and how to make the most efficient use of it. Because of the location and size of many wildfires, water has limited application, and often its use is impractical. In fires in light fuels on terrain that can be traversed with pumpers, it is the fastest and most practical means of control if suitable pumpers are available to apply it. In the heavier fuels away from roads, water must be applied with hose lays or by aerial delivery, thereby

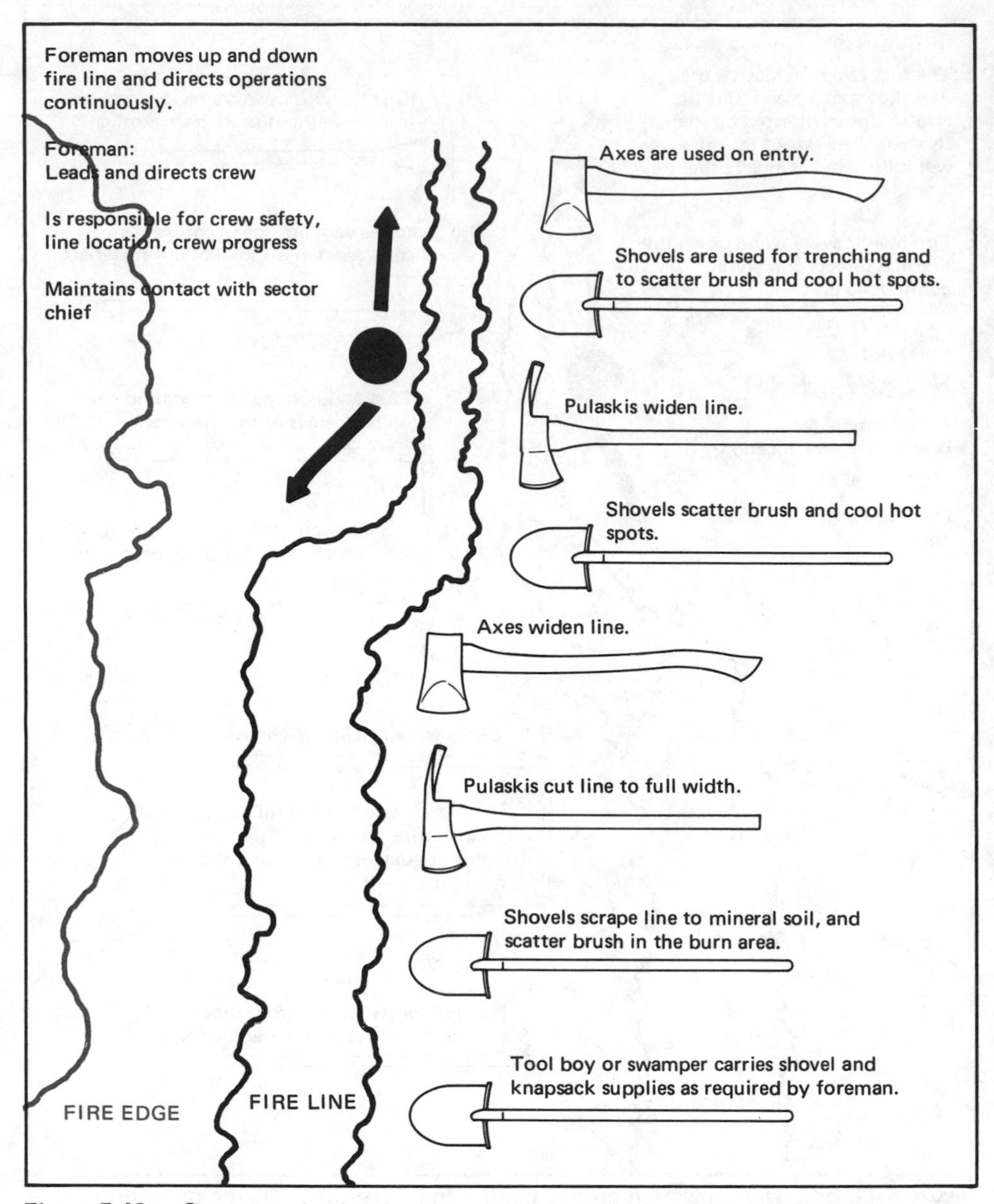

Figure 5.42. Crew organization for line control in trees, heavy brush, and scrub oak.

making its use more difficult and costly. With the exception of light fuels, water is rarely the only means used to control the fire. Water must always be followed up with manpower.

Rule of Thumb: A knocked-down fire is never out until it is thoroughly mopped up.

When it is applied to the burning edge of a wildfire, water works on all three sides of the fire triangle to extinguish the fire. (1) Water absorbs heat and thereby reduces the heat of the fire below the ignition point, and the fire goes out. (2) Water applied directly on the edge of a fire area contains the fire to a definite area and robs it of additional fuel, thus breaking the fuel side of the triangle. (3) Water has a smothering effect when it is spread over the fuel, and part of it turns to steam and thus replaces some of the air next to the fuel.

Therefore, to varying degrees, water also works to break the oxygen side

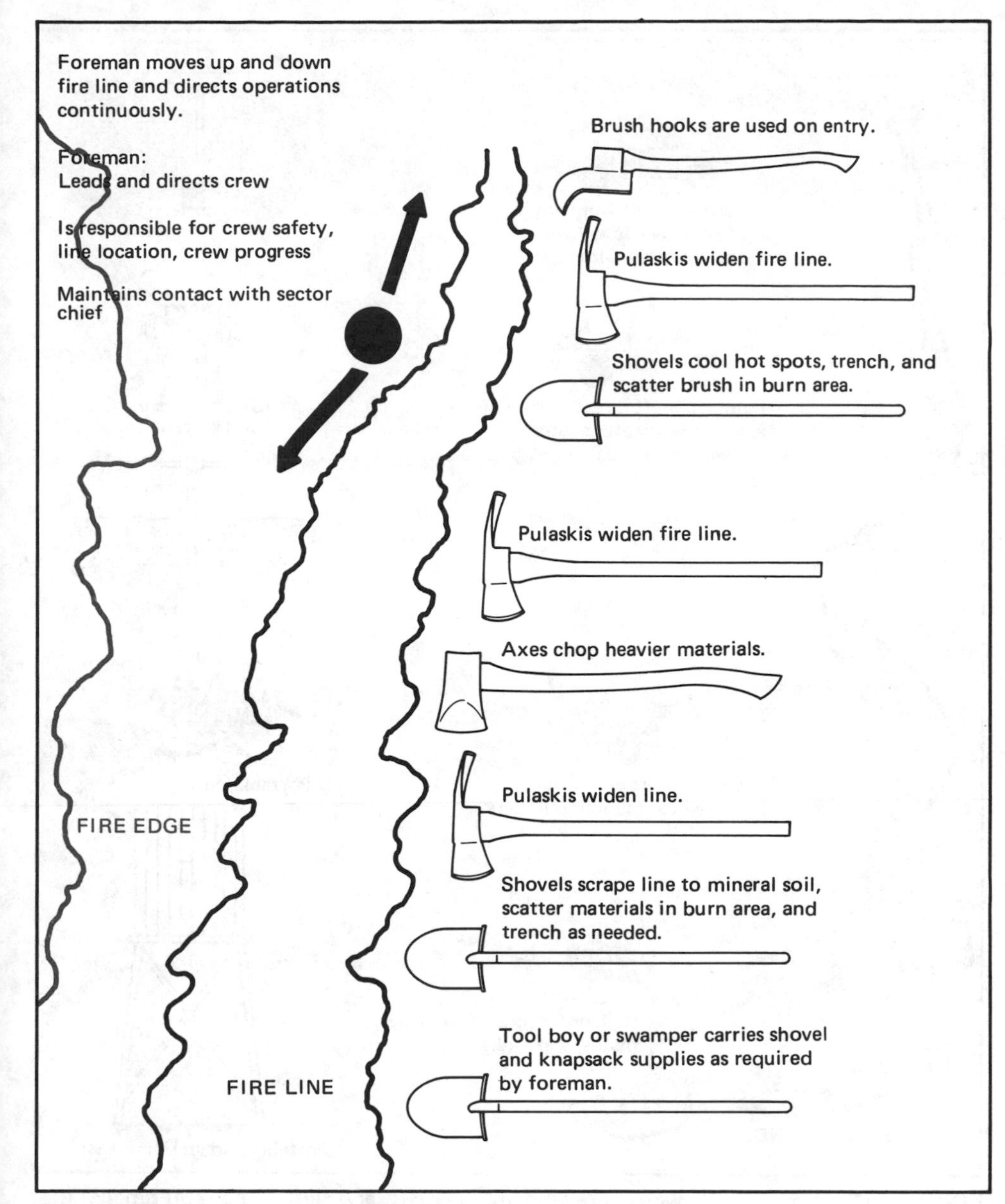

Figure 5.43. Crew organization for line control in heavy brush.

of the fire triangle. The most important effects are on the fuel and heat sides and, secondarily, on the oxygen side of the triangle.

Principles of water use. The principles of water use are the following:

1. One volume of water will extinguish 300 volumes of burning fuel if it is properly applied; the trick is applying it properly. A bucketful of water will put out a match, but a drop will also put out the match if the drop is properly applied. The ratio of 300 to 1 is true under controlled conditions of the laboratory; in actual field use, the ratio is probably closer to 100 to 1.

One gallon of water in cube form is 5 inches by 5 inches by 9 inches with an exposed surface of 230 square inches. Break that gallon of water into $^1/_8$-inch cubes, and the exposed surface area is 10,800 square inches, or 47 times the exposed surface area of the cube. Break that gallon into steam, and it has 1,700 times the exposed surface area of the gallon of water in cube form. It is the exposed surface area of water that picks up the heat.

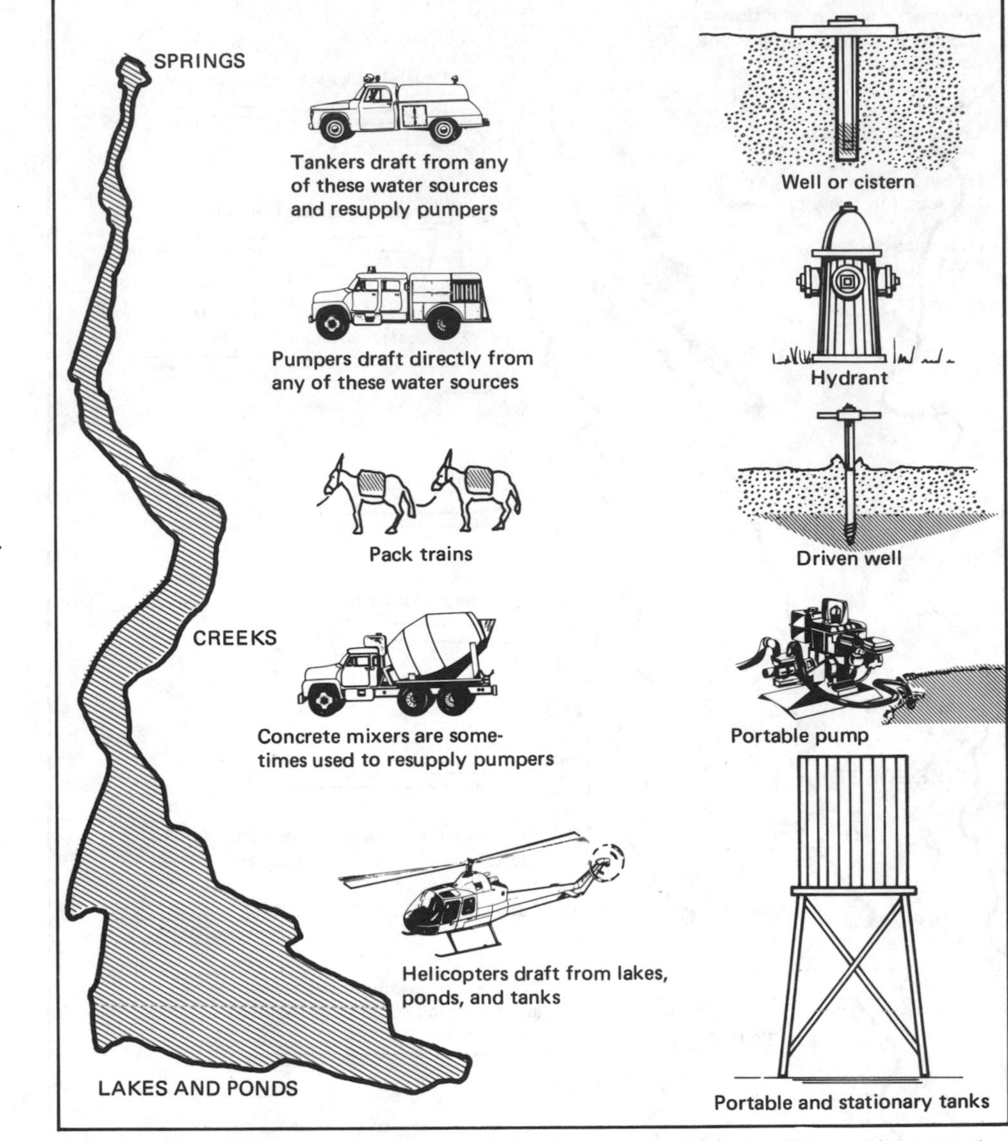

Figure 5.44. Sources of water are shown at right and left. The equipment that transports water from these sources is shown in the center.

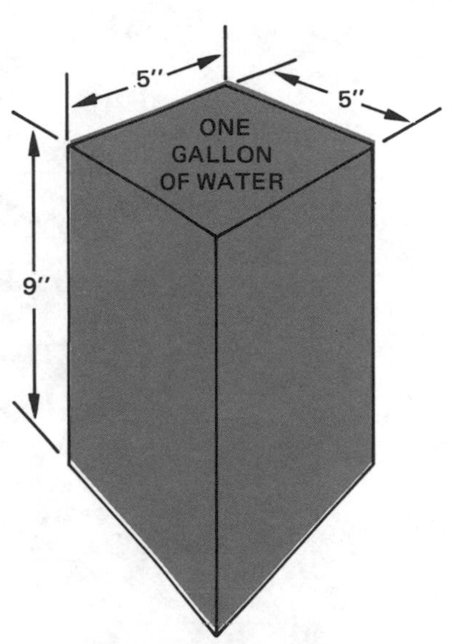

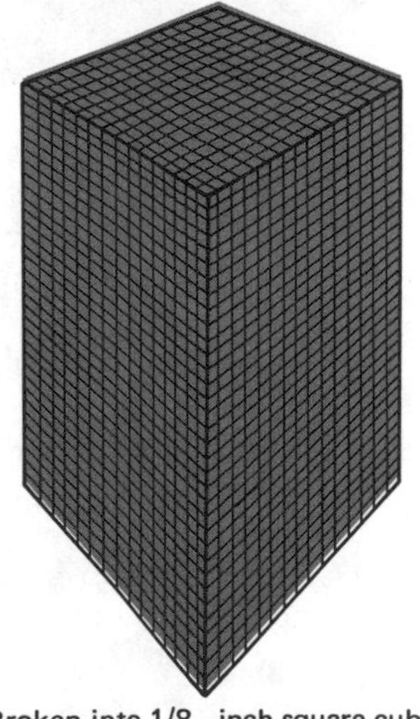

Figure 5.45. The surface area of water absorbs heat. As water is broken down, more surface area is created and more heat can be absorbed.

As water is more finely broken down, surface area increases and more heat is absorbed. Therefore a spray or fog stream is superior in most fire-fighting situations. An additional advantage for wildfire control is that the spray or fog stream uses less water and covers a larger area.

2. A straight stream may be needed to reach or to penetrate into the fuel. It should be used only when it is needed: to knock the fire out of snags or trees, to cool a hot fire so fire fighters can get at it, or to hit dangerous spots ahead on the line.

3. Hit the base of the flame. It's the heat in the fuel that you are after. If grass or brush is burning, hit the base of the flame; if a tree or snag is burning, hit the bottom first and then work up the trunk. As you spray water on the fire edge, remember to work in close with the finest spray that will do the job—not too much, but just enough. Try to have each drop find a piece of burning fuel to land on.

4. There is an "Off again, on again Flannigan" technique that pays big dividends in water saved. Squirt a little water at the base of the flame. Shut the water off, squirt again, and move along. Then squirt again. If the flame bounces back, give it another shot, but squirting water intermittently uses the least amount to do the job. Practice this technique until you get the feel of it. Don't waste water on cold ground when you move from one spot to another—turn if off. Don't spray the lawn—the closer the nozzle is to the fire, the more effective the water is.

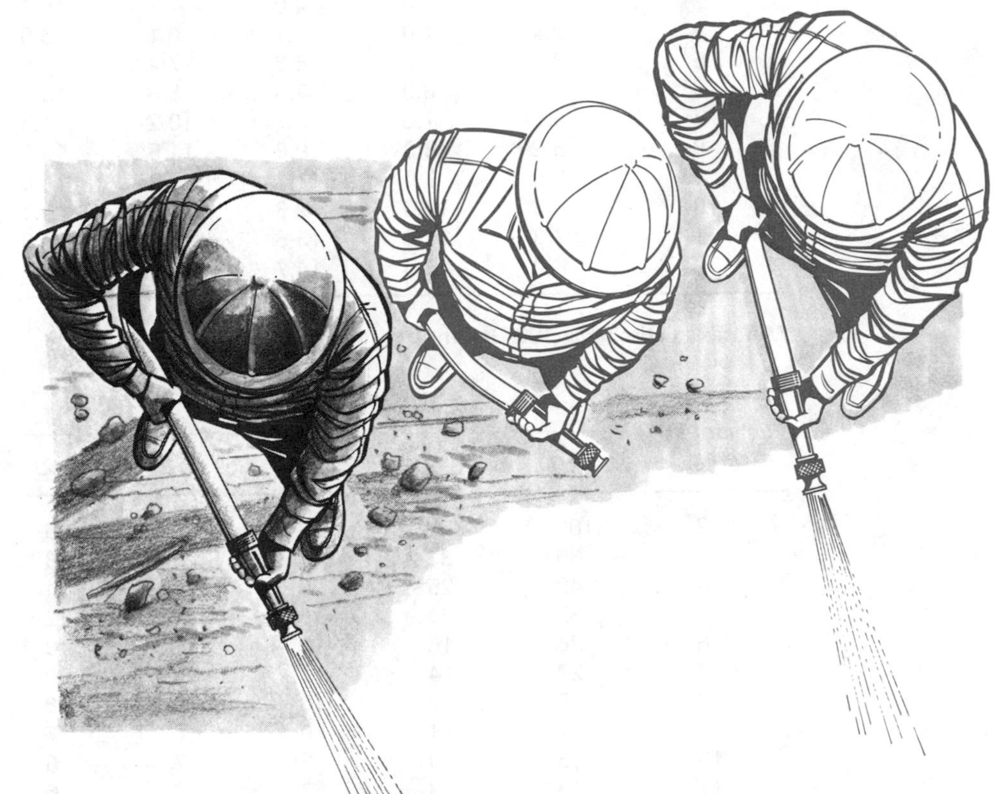

Figure 5.46. Practice turning the nozzle on and off so that water is saved.

5. The nozzleman is the key man in using water. He is the one who must use it to its total effectiveness. He must have control of the pressure, the nozzle, and the shutoff.

The most effective wildfire nozzles are capable of changing easily from fog to straight stream and shutting off quickly.

The number of gallons per minute delivered at the nozzle is dependent on the size of the nozzle opening and the pressure. The higher the pressure at the nozzle, the higher the gallons per minute (gpm). The larger the nozzle opening, the more gallons delivered (see Table 5.6). Pressure is measured in pounds per square inch (psi). This may be as low as 60 to 80 psi for the Elkhart nozzle. Ordinarily, 80 to 100 psi is needed for a fog nozzle, and 50 psi is needed for a straight stream (at the nozzle). A pressure of 30 to 50 psi is adequate for garden hose.

The nozzleman must communicate with the pump operator or the driver by using hand signals to get the pressure he wants. See Figure 5.54 for hand signals for water.

TABLE 5.6

Approximate Discharge Rate in Gallons per Minute (gpm)

Nozzle size in gpm at 100 psi	*Nozzle pressure (psi)*				
	50	*100*	*150*	*200*	*250*
2	1.0	2.0	2.8	3.3	3.9
3[1]	1.7	3.0	4.0	4.7	5.4
4	2.4	4.0	5.1	6.1	6.9
5[2]	3.1	5.0	6.3	7.4	8.5
6[3]	3.8	6.0	7.5	8.8	10.0
7	4.5	7.0	8.7	10.2	11.5
8	5.2	8.0	9.9	11.5	13.0
9	5.9	9.0	11.1	12.9	14.6
10	6.6	10.0	12.3	14.3	16.1
11[4]	7.3	11.0	13.5	15.7	17.6
12	8.0	12.0	14.7	17.0	18.2

Minutes to Discharge 100 Gallons of Water

Nozzle size in gpm at 100 psi	*Nozzle pressure (psi)*				
	50	*100*	*150*	*200*	*250*
2	100	50	36	30	25
3[1]	60	33	25	21	18
4	42	25	20	16	14
5[2]	32	20	16	14	12
6[3]	26	16	13	11	10
7	22	14	11	10	9
8	19	12	10	9	8
9	17	11	9	8	7
10	15	10	8	7	6
11[4]	13	9	7	6	6
12	12	8	6	6	5

[1] "Forester" no. 3 spray tip.
[2] 1/8-inch straight stream.
[3] Garden hose medium spray.
[4] "Fognozl" 4-hole spray tip.

The lowest gpm that gets the job done should be used. Likewise, the lowest pressure that allows the nozzle to perform as intended should be used. Ordinarily, 6 to 10 gpm produces the optimum effect for one nozzle in the usual grass type. Often, 3 gpm is effective on short grass prairies. More gpm may be needed in tall grass or wheat fields. Heavy thatch and ground debris may require a straight stream or narrow fog stream patterns to reach through the fuels.

The lower discharge rates mean longer use time, more fire caught, and more fuel cooled. Practice with nozzle openings and pressures to find the best combinations for your equipment, fuels, and conditions.

Some nozzles at high pressure deliver air as well as water. This combination has the effect of fanning the flame into more action instead of knocking it down. When this happens, lower the pressure. Don't forget to turn the nozzle on and off.

6. More fires have been lost because of too much water than because of not enough. It's not how much water is applied, but how, where, and in what amount the water is applied that make the difference. Equipment used to apply water may vary or change, but the principles governing conservative water use are always the same. There is a saying in California: "If enough water is used to completely extinguish the flame, then too much water has been used."

7. Know how much water is in your tank at all times. Plan the attack to use the water on hand in those locations where it will be most effective. The time it takes to travel for a refill may allow the fire to erase all the work done with the first tank. This is one reason why it is so important to carefully plan the use of the water on board, to employ every possible method to use the water to its best advantage, and to use all the means at hand to assist in containing the fire. If a mother tanker is available for resupply or a line is laid in from a water source, then the amount of water on hand is not so critical, but never waste water because it is plentiful. There always comes a time when it is needed. Running out of water at the head of a fire is not only discouraging but can be downright lethal. Define the availability of water, and plan its use carefully (see Figure 5.47).

8. Some other ways to save water are to (a) eliminate the fingery edge during direct work by firing from point to point and thus having a shorter line to knock out with water, (b) spray water only during lulls in the wind, (c) meet the fire in lighter fuels where less water is required, and (d) leave more work for the mop-up men.

9. The length of line that can be initially contained with a pumper delivering a given amount of water depends on many factors, such as fuels, weather, mobility of the unit on the terrain, and the proficiency of the crew.

A medium pumper (300 gallons) with a capable crew can knock down the fire on $^1/_8$ to $^1/_2$ mile of line per hour in average southern surface fuels. In heavy brush in the West, the distance would ordinarily be less. The estimating standard used by the forest service in Washington and Oregon is: low resistance to control, 0.45 miles of line held per unit hour; medium resistance, 0.28 miles; high resistance 0.20 miles; and extreme resistance, 0.10 miles.

In light grass fuels in the Great Plains, the distance may be 1 to 3 miles per hour for a 1,000-gallon pumper. There are so many variables that even very general estimates are risky. However, through practice, the individual fire control service should have a good idea of their probable performance under various conditions common to their territory. It is ½ mile around a 10-acre square. What is the average size fire in your area?

10. Water alone will not do the job of wildfire control. Follow up with hand tools and patrol. Cold-trail and mop up to make sure the entire edge is dead out.

11. Have the correct fittings for the job: combination nozzles with shutoff, pressure relief valves, applicators for mop-up, etc.

12. Use water additives such as wet water, Firetrol, and Phoschek

where they will improve the capability of water for the fuel encountered and where their cost is justified.

13. Pumper crews are usually well trained in fire control techniques and should have hand tools on the truck. Where the pumper cannot be used on the fire or has done all that it can, these crews should be used as a special hand crew. They are often effective as a hot-spotting crew in advance of a larger line-building crew.

14. Heavy accumulations next to the fire line may burn too hot and will need to be cooled down but should be allowed to burn out. A spray stream is best used and applied according to the heat intensity. Use only enough water to knock the fire down but not out. A straight stream is a water waster in this instance because most of the water will go through the fire to the ground. Heavy fuels such as logs and piles of heavy limbs may require a straight stream to force water into and through the pile. Sometimes these heavy fuels can be mopped up immediately if the fine fuels have been consumed, and thereby long-lasting hot spots and more work later on can be eliminated.

15. Snags burning close to the fire line that cannot immediately be felled can be cooled off and fire can be kept away from the base until these snags are felled. If the entire snag is afire, work from the bottom to the top with a straight stream from several sides. A strong water stream can often throw sparks and embers, especially from the top of a snag. In this case, put water as gently as possible on the top, and then work slowly downward to cool it. Follow this action by working from the bottom upward in an attempt to totally extinguish the fire. Where possible, pry off the bark before applying water.

16. Where plenty of water is available and the surface fuels can be penetrated with a strong straight stream, the line may be made by using the hydraulic or sluicing method. A strong straight jet stream is used, and the nozzle is held about a foot away from the fire edge and perpendicularly about a foot from the ground. The nozzleman will need goggles or a face shield because he will be subject to flying material as well as getting soaked and covered with mud.

The surface material and the top of the soil will be torn up, moistened, cooled, and thrown back into the flames. If the straight stream is used at right angles across the edge and angled from the outside to the edge, it will be formed into a fan-shaped spray toward the fire.

This method wets the soil in a line and knocks down the edge of the fire. It should be followed up with hand tools to make the line secure and continuous. It is sometimes effective in initially containing fires in duff and humus fuels if they are not too deep. If they are over 4 to 6 inches deep, a trench should be dug to mineral soil.

The hydraulic method is comparatively slow and requires an unlimited water supply. At one time, it was a much more used practice than it is now. In fact, in many areas, its use is frowned upon.

Methods of pumper use. Pumpers are used in a number of ways and for several purposes in different parts of the country. Each area seems to have

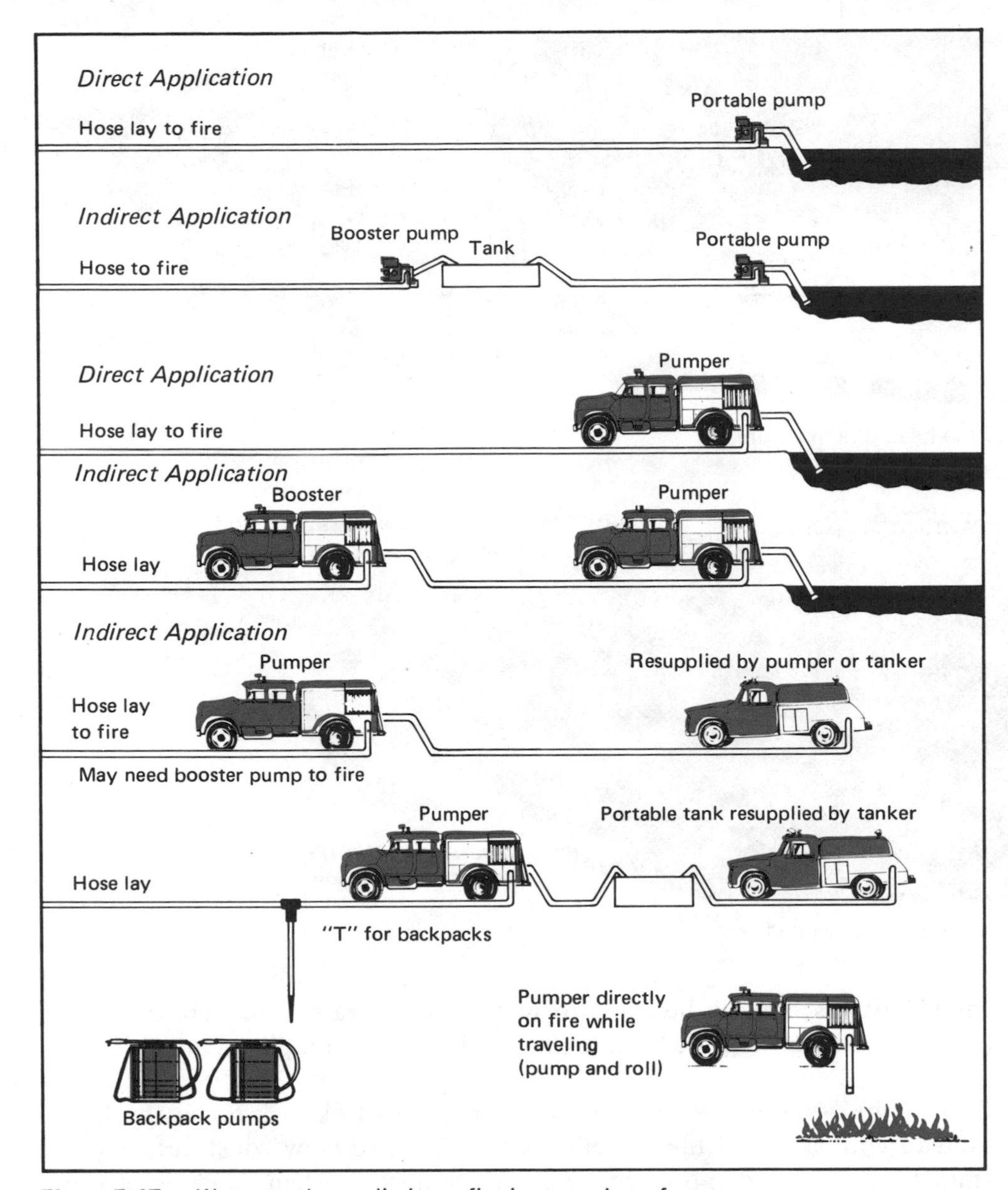

Figure 5.47. Water can be applied to a fire in a number of ways.

developed its particular methods and the practices vary largely because of the type of fuel and the topography as it allows or prevents access to pumpers.

Pump and roll pumpers–direct attack. The pumper is driven directly along the fire edge where the fuel type and topography allow access. Fuel types include grass, grainfields, sagebrush, small brush, and any timber or brush type that is open enough to allow access to the pumper truck. Four-wheel drive trucks considerably extend this use in hilly or rough terrain. Rocky surfaces, gullies, arroyos, large streams, cliffs, and other barriers to wheel traffic limit the use of pumpers. Only slopes of less than 25 percent should be contoured with pumpers and then only if the center of gravity of the unit is low.

Pump and roll operations are principally used in the Great Plains area and in varying amounts elsewhere. If a single pumper is used, it must be followed up by hand tools, backpacks, or both or, at the very least, by a second nozzle off the rear of the pumper. The second nozzle operation has several

Figure 5.48. Water applied directly to the head of a small fire.

Figure 5.49. Water applied directly from inside the burned area.

disadvantages but can be made to work in light grass in moderate or lower conditions if constant watch is maintained over the completed line.

In a great many cases a single pumper is used to knock down the flames and to cool and hold the fire edge while the hand crew constructs a line to mineral soil. Coordination between the pumper and hand crew is of prime importance so that the job load is adjusted between the two. The capacity of the pumper and the time required to refill, the size of the crew, the size of the fire, and its rate of spread will dictate the work distribution. The objective is to move along the perimeter as rapidly as possible but not to allow rekindles or slop-overs behind the crew and to suppress all spot fires immediately.

If the fire is small and fire weather conditions are moderate or lower, then hit the head with a direct attack. This stops the spread, and the flanks are then worked back to the heel to complete the containment. Attack from the inside if there isn't too much smoke and if there is no danger of burning the tires. Attack from the inside is also best if the wind is turbulent and shifting. However, if the fuel is holding the fire from the initial burn, there may be too much heat and flame to work in the burned area, or conditions may be such that it is unsafe to make a frontal attack. The best tactic is then to flank or parallel the edge from outside the burn.

In a flank attack, generally work from the rear along the flanks around the head and tie back to the point of beginning. The point of starting on a flank will depend on the extent of the fire and the amount and kind of manpower

Figure 5.50. Flank attack by direct application of water.

Figure 5.51. Procedure for attacking the fire edge by direct application of water.

and equipment available. If part of the flank appears dead, start where the fire begins. Be sure to have the edge checked back immediately to the rear to make certain it is secure. If the fire should start anew behind you, it may not be long until you are outflanked and in a pocket between two fires. Continue to work on around the head to pinch off the spread, and then tie in the entire perimeter. Check for spots, finish the mop-up, and you're home early. Unfortunately, not too many fires behave this way.

To get into a burning line, reach with a straight stream and aim for the base of a hot spot. Bounce the straight stream off the ground to make more spray and to cool more fuel. As soon as an area of the edge has been knocked down, move into it fast. Then turn toward the head, change to a spray, cover only burning fuel to get started, and use the spray as a protecting shield. Spray down and parallel to the fire edge and cover part of the unburned edge; but use the water mostly on the base of the burning fuel. If you are at right angles across the line from inside, it's possible to spread fire outside the line, or, if you are working across the line from outside, some of the water will be lost on the burned area. Therefore try to work along and parallel to the burning edge. Do not drown the fire, but knock down the flames. Trying to do the whole job with one nozzle is time consuming, is wasteful of water, and doesn't leave a secure line.

Usually, hit the hottest edges first, and then tie in the whole perimeter. If the fuel type changes or there are dead and slow-burning sections, hit the worst places first and then mop up the others. At all times, know how much water is in the tank.

With grass fires the escape route is right next to you all the time—in the burned area. Don't allow the hose to burn; a burned hose is useless.

There are champions of different ways of using hose from the pumper, and they are all quite vociferous in their views. Generally, where it is possible, have the nozzleman ride the truck either on the side, which is preferred, or from a platform on the front (cowcatcher) or the rear. Make sure such platforms have adequate railings high enough so that the nozzleman cannot be thrown off with quick stops or travel over rough terrain. Safety belts should be worn.

Where it is necessary for the nozzleman to walk and carry the hose, remove an absolute minimum from the truck. Don't allow the hose to get tangled in stumps and brush or run over. Close watch must be kept on the hose by both the driver and the nozzleman.

Some suggestions (these are by no means rules) on when to ride or walk are:

- If the fire is in light fuels and easily put out and the terrain allows easy access, have the nozzleman ride and have a third man follow up, or have the driver use the nozzle and follow up with the second man.
- In heavy fuels and difficult driving, have the nozzleman walk and have a third man follow up.
- With many breakouts or limited water supply, the driver is the nozzleman and the other man follows up if only two men are available; otherwise, assign enough men to do the line work.

The assignment of men to pumpers should be flexible and the techniques should be changed to meet different situations.

Two pumpers—direct attack. If two pumpers are assigned on a direct attack, each one may take a flank, provided follow-up is supplied for each one. Each pumper works along his flank with a pincer movement until both meet in front of the head to effect containment.

Ordinarily, the best tactic is to work the pumpers in tandem along one flank and across the head to the opposite flank and then back along it to encircle the fire. The first pumper knocks down only the hot spots and the worst flame. The second finishes the job of containment. With the possible exception of light fuels where it is sometimes possible for the second pumper to secure the fire line, a holding and spot fire crew should always follow the pumpers. With the proper adjustment of work load between the two pumpers and the follow-up crew, this attack is usually the fastest on easily traversed terrain.

In the heavier fuels, enough manpower must be used behind the pumpers to assure that the line is held and spot fires are put out. As the length of line increases, more men will be needed to patrol the completed sections, and the crew will thereby be reduced behind the pumpers. In addition, keep in mind that a fatigue factor begins to work as the line is lengthened.

Pumpers should not get so far ahead of the line crew that rekindles can get out of hand and nullify the work already done. If the pumper must slow down, backtrack to knock down rekindles, or begin using more water to lengthen rekindle time, several men with backpacks following between the pumpers and the holding crew can usually solve this problem.

Another advantage in using two pumpers together is that, when one runs out of water and must go for resupply, the other pumper can stay on the line and continue the forward progress. Therefore it is important to arrange their use so they do not run out of water at the same time.

When two pumpers are used together, one can return to suppress a hot spot that flares up or catch a slop-over. Another method of using two pumpers on direct attack is to send both pumpers from the burn side to the center and in back of the head fire. They work together to split the head fire and then work in opposite directions toward the flanks to knock down the head fire. On reaching the flanks, each one proceeds back along his flank to the heel to tie in the line. Both tankers are followed by holding crews. This method requires precise follow-up with backpacks and shovels. It should not be attempted in other than light fuels with easy access and by well-disciplined crews.

Often a third unit such as a jeep, pickup, four by four, or rancher's spray rig is very useful to chase spots, resupply backpacks, and assist the holding crew, especially in extreme or flash rate of spread fuels.

Pump and roll pumpers–indirect attack. On larger fires where the terrain and fuels are such that the pumpers can work along the flanks, an indirect attack is sometimes used by laying down a water line or lines if the edge cannot be worked directly because of the intensity of the heat and flame. Only the tandem attack should be considered in this operation. The lead pumper lays down a water line and is immediately followed up by another pumper that reinforces the water line where necessary and supports the burning-out and holding crew. The pumpers and follow-up crews work up one flank, across the head, and then down the other flank. They tie in at the rear at the beginning point. If enough equipment and manpower are on hand, a pincer attack can be made from both flanks. In some areas the preferred method is to string the fire first and to immediately follow with the pumper to extinguish the outside edge of the set fire. Care must be taken to let the set fire burn back toward the main fire, or its purpose will be defeated. The water of the first pumper is used to make a dead edge and

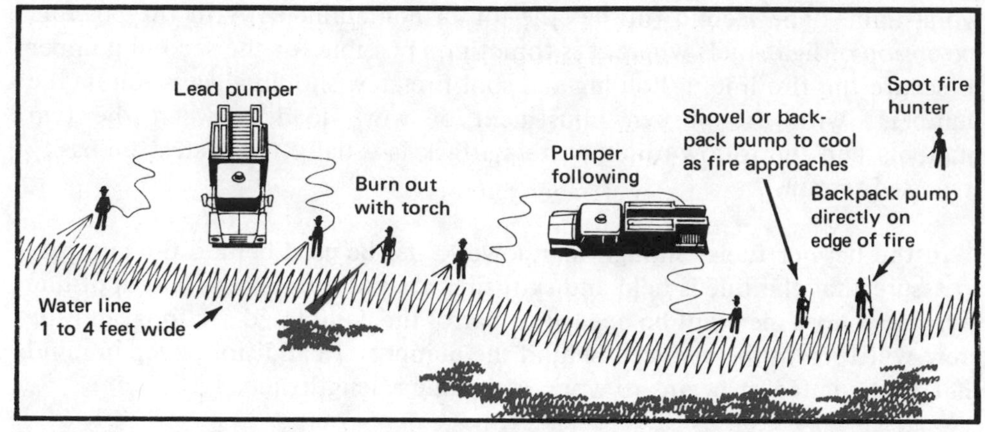

Figure 5.52. Indirect attack by laying down a water line with a pump and roll pumper.

must be followed up with hand tools, supported by building line or cold trailing from the other pumper.

This adaptation of the indirect method is usable in the lighter fuels on terrain that can be easily traversed. In other fuels it would be used as an emergency method only. The preferred method in heavier fuels would use the pumper to support the burnout and holding crew following hand-built, dozer-built, or plowed fire lines.

Strip burning can be used with two tankers that keep down the intensity of the set fire. The lead tanker is on the inside in this case and is followed by a torchman who strings fire at the desired distance from the first water line laid down by the lead tanker. The second pumper blacks out the outside edge of the set fire and assists the holding crew that follows. The amount of hand line will depend on the type of fuel.

Pump and roll pumpers–support of fire against fire. Pumpers also provide excellent insurance during backfiring, strip burning, or burnout from a prepared line, a natural barrier, or a water line. Their action in these cases would be to (1) wet down either side of the fire line as is necessary to reduce spot fires and the effects of radiation; (2) quench the outside edge of the set fire to reduce radiation; (3) lay down secondary water lines toward the head fire for strip firing; (4) knock down the head fire when it reaches the line, which may or may not have been backfired; and (5) support the holding action and control breakovers and spot fires.

Timing becomes important, and the use of water must be planned accurately so that it will be used most effectively especially when the supply is limited to the pumpers and tankers that are available at that time. Unrestricted and rapid movement of crews and vehicles from and to the area must be assured.

Pump and roll pumpers–support to dozers and plows. It is best to operate the machine and pumper together as a team so that their section of the line is made secure as it is built. The pumper may support either a dozer or a plow. Their action may be (1) firing out and controlling the set fire, (2) patrolling after firing out to catch breakovers and spot fires, (3) holding the line where the head fire meets it or the backfire, (4) wetting down either side of the line where necessary, (5) strip burning, (6) assisting and backing up burnout and holding crews, (7) scouting the area in advance of plows or tractors where feasible, and (8) mopping up.

Other uses for pumpers. Some of the other jobs for pumpers are (1) scouting or smoke chasing, (2) transportation and patrol, (3) control of structural fires (if fire department pumpers are not available or if the fire department pumpers need support), (4) supply of water for backpacks and tractors, (5) mop-up, and (6) supply to hose lays from a stationary position. Hose lays may be resupplied by a stationary pump, another pumper, a tanker, a portable tank, or by auxiliary equipment.

Auxiliary pumpers. This category includes vehicles other than the standard wildfire pumpers. These may be pickups, trucks, or trailers equipped with a tank and pump. They may be four-by-two pickups, tankers with volume pumps, mobile concrete mixers, transports, spray rigs, street flushers, oil well trucks, city fire trucks, etc.

This type of equipment should be used in whatever way it can be effective. Nevertheless, auxiliary pumpers should only be used on initial attack in those locations where they can be adequately protected. They would most commonly be used to (1) support and resupply regular wildfire pumpers, (2) control very small fires that are easily reached, (3) control structural fires (if better equipment is available, the auxiliaries would support the regular pumper in the discharge of the high volume of water necessary on structures), (4) patrol established lines or hunt spot fires on terrain where they can be safely and effectively used, and (5) perform other odd jobs not directly on the fire line.

Hose lays. Often the pumper can reach only one point on the edge of a fire or somewhere close to the edge of the fire, so the only way to supply water to the fire area is to lay a line or lines of hose. The hose lay may be made from a pumper or from a portable pump located at a supply of water. As the lay is extended, wyes and hose line tees are coupled into it at intervals so that full coverage of the line can be maintained by branch lines. The branch lines can then be extended as necessary for mop-up.

Simple hose lay. In this process, hose is added to the end of the line as needed. Gated wyes may be spaced at intervals in the line. If gated wyes are not used, the line must be either shut down or kinked, or a hose clamp must be used each time a length of hose is added to the end of the line. Bleeder valves, hose line tees, and hose line tee valves may be added for filling backpacks or adding wyes for branch lines. In any hose lay, much will depend on how much speed is required in applying water to the fire. If the forward spread of the fire is slow, one or two nozzlemen can stop it by spraying their way up one flank and down the other or by installing a wye at the heel and cooling their way up both flanks at one time.

In step one of a simple hose lay the pumper is spotted near the rear of the fire. The first nozzleman uses the booster line to make the heel safe and then proceeds along the right flank. The operator or tanker chief and one crewman connect two lengths of 50-foot CJRL 1½-inch hose to the pumper. The second nozzleman carries three rolls of 1½-inch hose, a shutoff nozzle, and a gated siamese valve to the end of the first two lengths of hose. He attaches one female side to the first two lengths of hose, and he attaches the three lengths of hose that he carried to the male end. He attaches the shutoff nozzle to the end of the three lengths, calls or signals for the line to be charged, and begins work at the desired starting point. If a gated wye is used at the end of the first two lengths of hose, two double female couplings and one double male coupling will be required to convert the wye to a siamese valve.

In the second step of the operation, the operator or tanker chief charges the line and lays two lengths of CJRL 1½-inch hose for the second pumper from the siamese valve. The other crewman carries additional hose forward to the second nozzleman and helps to couple it as directed by the nozzleman. When the first nozzleman ties in with the second nozzleman, he reverses and proceeds along the left flank as far as his booster line will reach. He may then use a backpack pump, assist and spell off the second nozzleman, or use a branch line to advance the action on either the left or the right flank. The second nozzleman continues forward along the right

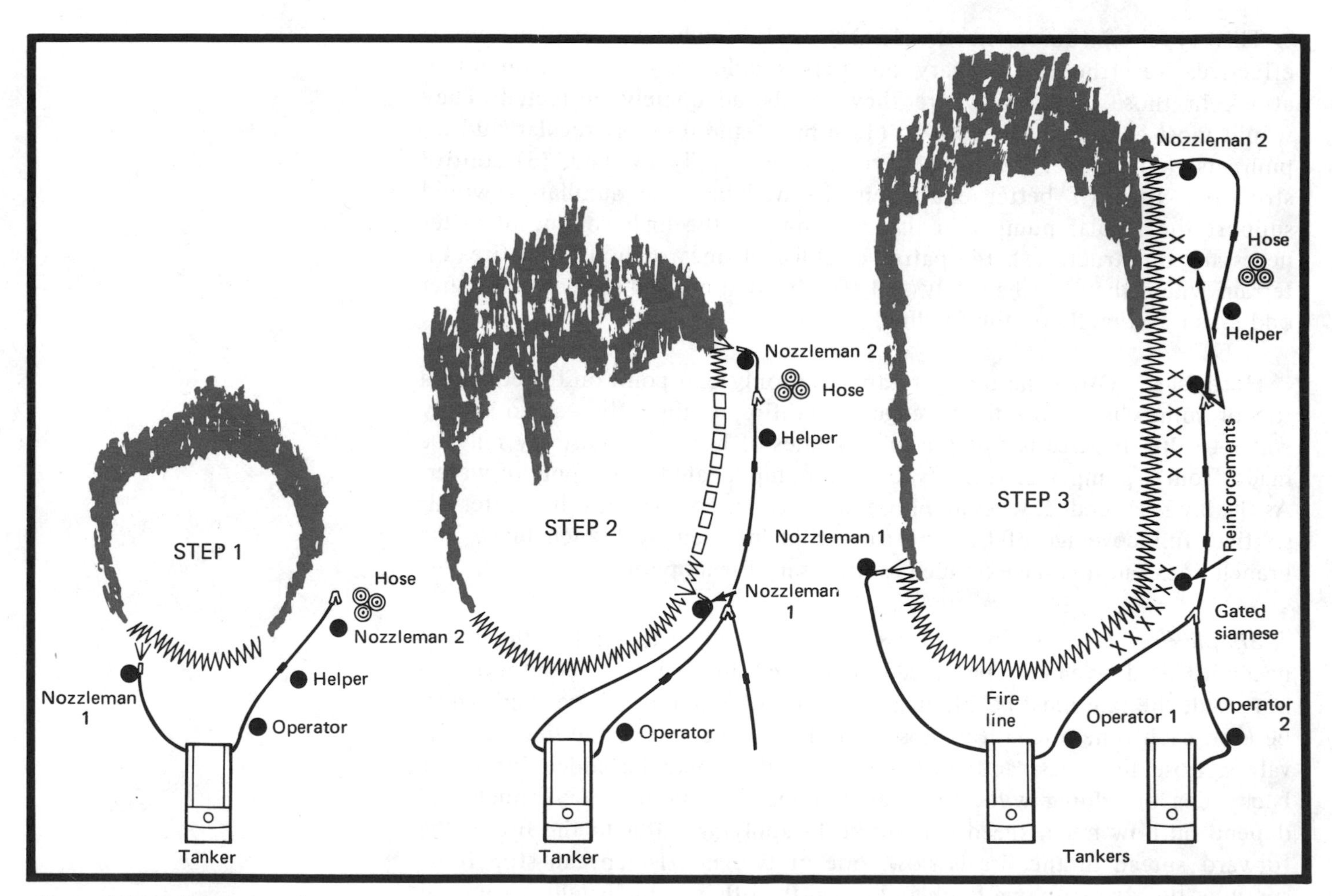

Figure 5.53. Simple hose lay.
Step 1. Nozzleman 1, using a live reel hose, makes the heel of the fire safe. The pumper operator and helper connect two lengths of 1½-inch CJRL hose to pumper. Nozzleman 2 carries three rolls of 1½-inch CJRL hose, shutoff nozzle, and gated siamese valve to end of the two coupled lengths and attaches the siamese valve in reverse.
Step 2. Nozzleman 1 proceeds along fire flank. Nozzleman 2 attaches 1½-inch CJRL hose to siamese, extending main line to desired starting point and installs shutoff nozzle. The pumper operator charges line and lays hose for the auxiliary pumper. The helper carries additional CJRL hose forward and helps to couple it.
Step 3. Nozzleman 1 ties in with Nozzleman 2, and then reverses around other flank. Nozzleman 2 proceeds forward, adding on additional hose lengths as needed. To do this, he cuts off the line behind nozzle with hose clamp or signals the operator. The auxiliary pumper cuts into the main line. The helper and reinforcements carry hose and construct a hand tool line.

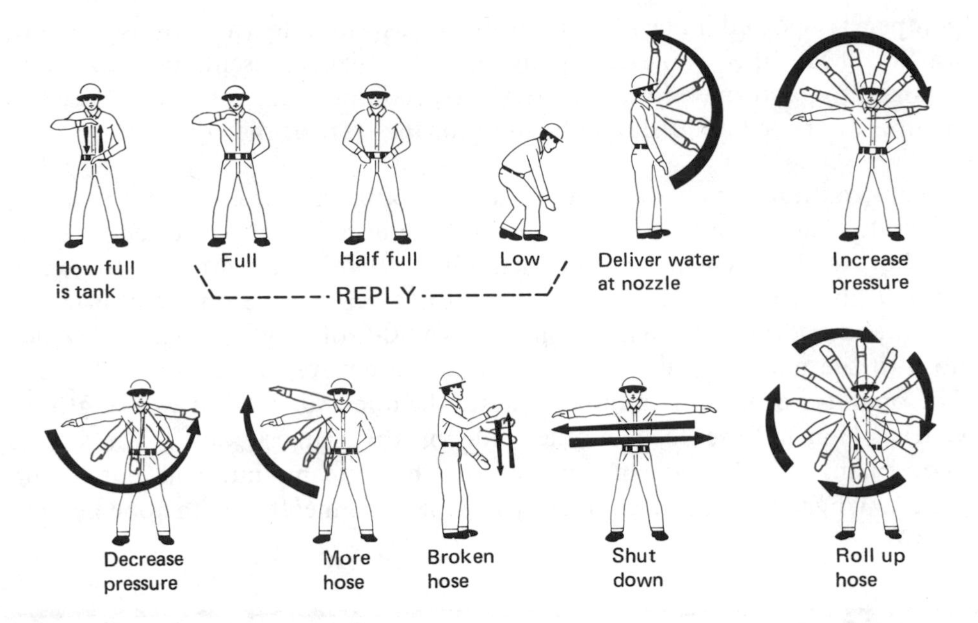

Figure 5.54. Hand signals for water use.

flank, adding lengths of hose as needed. To do this, he cuts off the line behind a gated wye, closes the line with a hose clamp, or signals the operator to shut down the line.

The third step is the arrival of the second pumper. It is immediately attached to the other female side of the siamese valve. The second pumper usually waits to pump into the main line until the first pumper is almost out of water unless the additional water or pressure is needed in the main line. The fourth crew member of the first pumper and the crew of the second pumper carry hose, extending and branching the line as required; they may use hand tools to strengthen the line and secure it; or they may search for spot fires.

Hose is added to the line until a booster pump and portable tank are needed to obtain sufficient pressure for satisfactory nozzle performance. The line may then be continued from the booster pump.

It may be necessary to establish another hose lay on the left flank with paired pumpers, depending on the size and rate of spread of the fire and the availability of equipment. In many instances, sections of either flank will be worked with hand crews or dozers in the initial attack, and, if so, the hose lay is used to support the line construction. There are many variations depending on the fuels, weather factors, accessibility of the terrain, resources involved, timing required, and manpower and equipment available.

Progressive hose lay. If the fire is running faster than one nozzleman can travel, more nozzlemen will obviously be needed to catch it. The progressive hose lay is an adaption of the simple hose lay, but it is designed for speed and safety, especially on hot running fires. One hose team unit after another is put into action as fast as the crews can lay and connect hose. This method allows a direct attack with water at many points along the fire edge at one time. The delivery of water at each nozzle is continuous and is not interrupted as the lay is made. One hose team unit would be constituted as shown in Figure 5.56 and would be composed of two or three men. The first pumper is spotted in a safe location but as close to the fire as possible, usually near the rear depending on the size, present burning edge, topography, and operations necessary to accomplish the job with a hose lay. A suitable anchor point should be used for the start of the lay.

The first nozzleman starts at the anchor point and makes a water trench or scratch line along one flank to protect the hose lay. He continues as far as the booster line will reach and then shuts down and reinforces the water scratch line with hand tools. The second nozzleman, the operator or tanker chief, and another fire fighter connect two 50-foot lengths of CJRL 1½-inch hose to the pumper and attach a gated siamese valve to the end of the hose. The operator of the first pumper charges the line and attaches two lengths of CJRL 1½-inch hose to the other side of the siamese valve. The second nozzleman and a crewman connect one hose team unit and put it into action. When the second pumper arrives, it is connected to the open end of

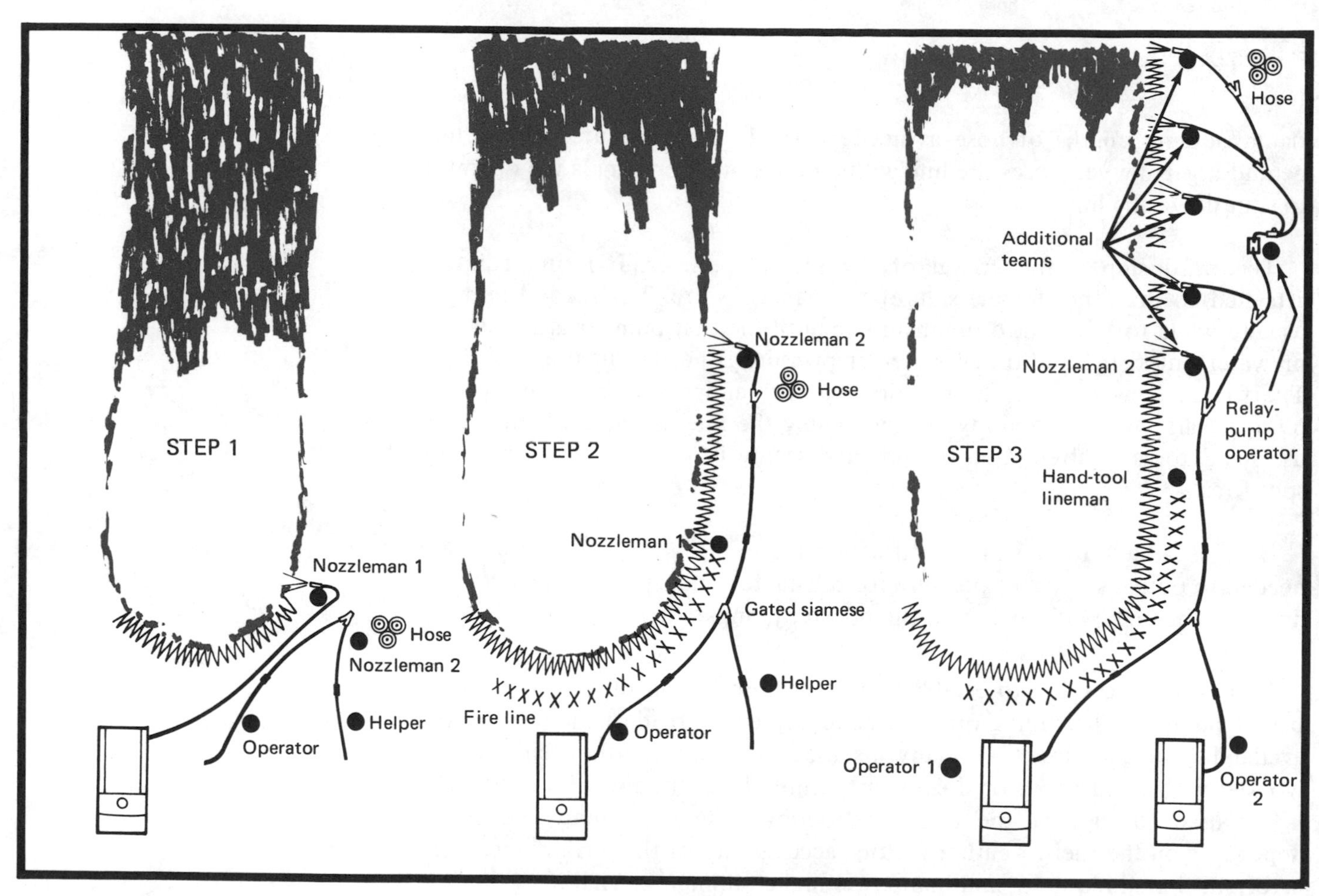

Figure 5.55. Progressive hose lay.

Step 1. Nozzleman 1, using a live reel hose, starts at the heel of the fire and makes a water scratch line along the fire flank to protect the main hose lay. Nozzleman 2, the operator, and the helper connect two lengths of 1½-inch CJRL hose to tanker and attach a gated siamese valve in reverse.

Step 2. Nozzleman 1 continues water scratch line as far as live reel hose will reach; he then shuts down and builds a hand-tool line. The operator charges 1½-inch line and connects two lengths of 1½-inch CJRL hose to reversed siamese for auxiliary pumper. Nozzleman 2 and the helper lay and connect one hose team unit and put it into action.

Step 3. The operator connects the auxiliary pumper to the leg of the main line. Additional two- or four-man teams lay and put into action successive hose team units. As teams finish their water scratch lines, they make them safe by building a hand-tool line and proceed to mop-up or assist in carrying the hose forward. If elevation or distance reduces water pressure below an effective working pressure, two men install a relay tank and portable pump. One man stands by as a relay pump operator.

the main line. It may or may not pump into the line depending on the needs. If a portable tank is to be used, the second pumper would dump into it, and it would not be necessary to install the siamese. Additional two- or three-man units lay and put into action successive hose team units. Each team or unit is assigned a definite length of line. As they finish the water scratch line, they make the line safe with hand tools and proceed to mop up or assist in carrying forward hose and other equipment. Booster line sections are added as required by the situation.

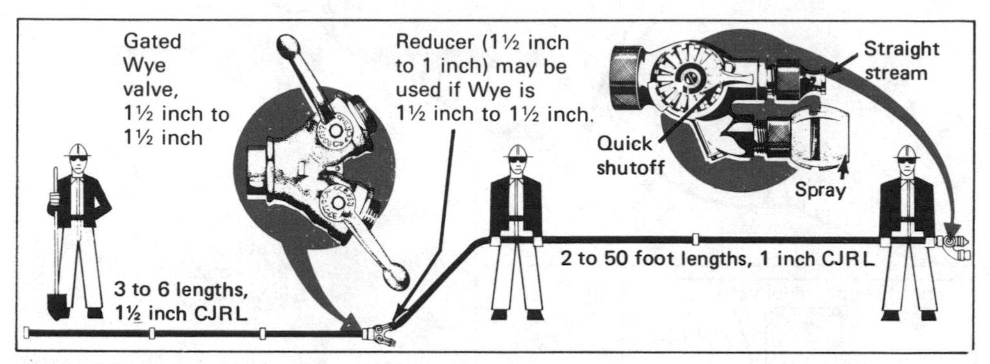

Figure 5.56. One hose team unit on a progressive hose lay.

Booster pumps. If elevation or distance reduces water pressure below effective nozzle pressures, a relay tank and portable pump should be installed. Use of a short 2-foot-long pigtail between the pump and the pressure relief valve-foot valve combination will eliminate breakage of the fittings due to vibration. There are safe working pressures for both pumps and hose. When the limit for either has been approached, install a booster pump and relay tank. A tank can be made by lining a suitable hole dug in the ground with a large piece of canvas or plastic. Harodike bags or staging pods may be used.

The second pump may be installed with a section of draft hose between it and the first pump at the source. Adapters may be needed to connect to one end of the draft hose. However, when it is installed next to the first pumper, the hose must be able to withstand the pressure because it will be somewhat less than twice that of the original pump. Note that the second pump in this configuration must have a lower capacity for pressure than the first pump. The second pump may also be installed in the line without a relay tank where the pressure must be boosted. This installation requires that the pump be carried up the slope to the booster location. When a booster pump is installed directly in the line, careful coordination is required, and the pumps should be constantly attended. Communication between the pumps is highly desirable. Several booster pumps may be employed in one line. The only limitations are the suitability of equipment and the desire to lay the lines. More than three pumps connected directly is usually impractical, but, with tanks, there is no limit except time and distance.

Portable pumps. Like pumpers, portable pumps are used to operate simple and progressive hose lays. They are also used to resupply pumpers and tankers, usually at the water supply source.

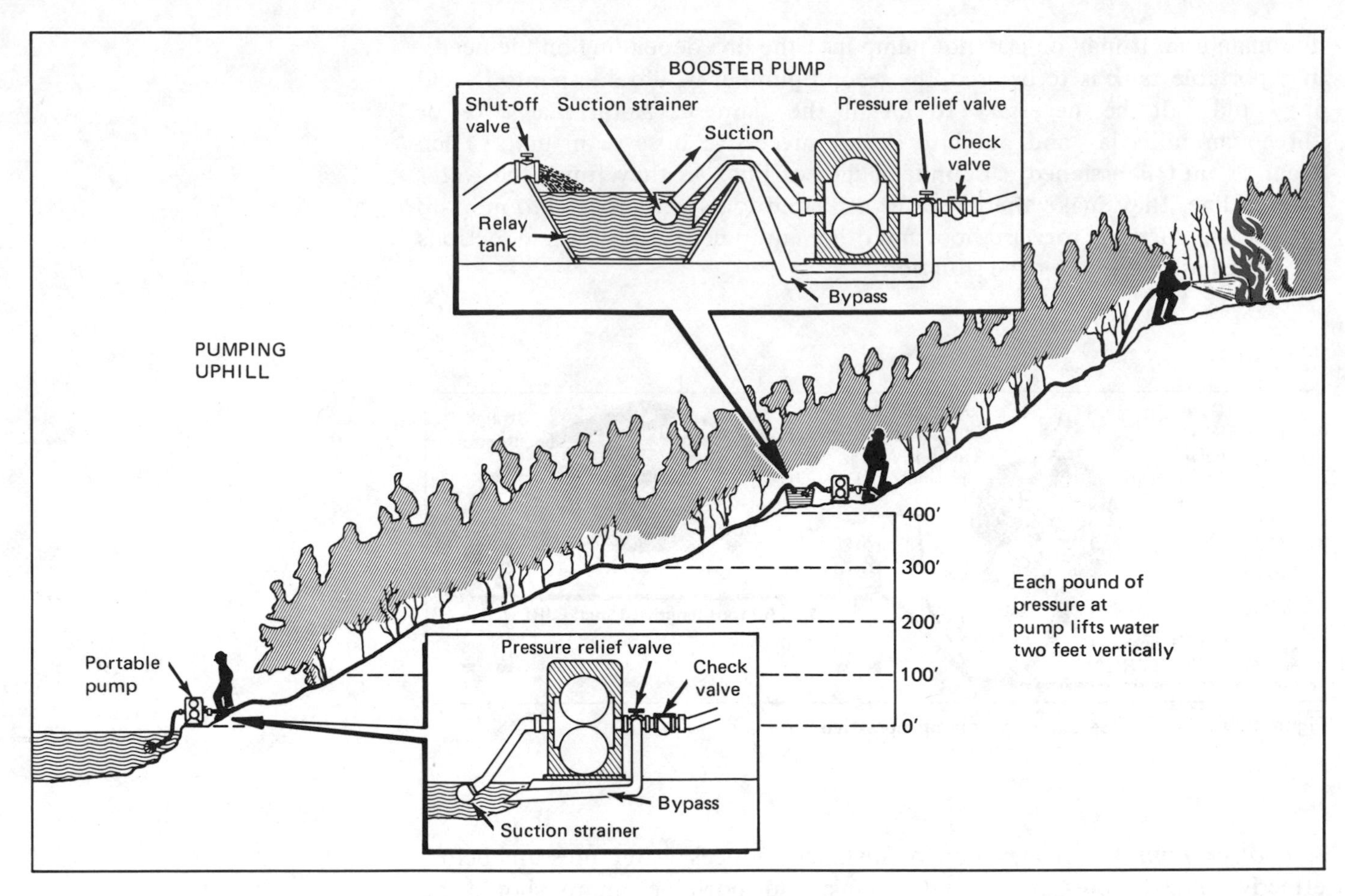

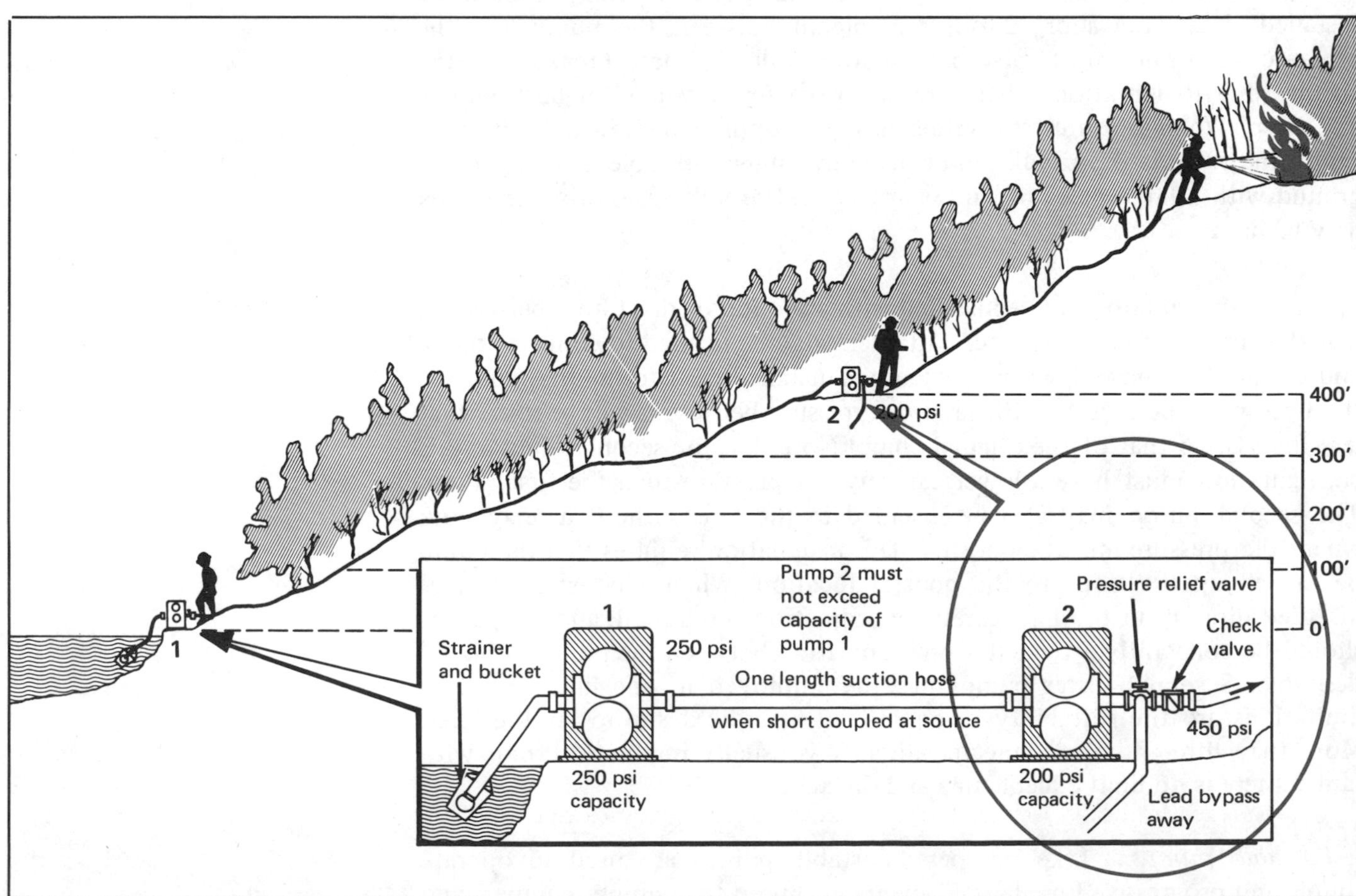

Figure 5.57. Booster pumps for pumping water uphill may be coupled at the source or connected in the line uphill.

In setting up portable pumps, the following should be observed:

- Set the pump on a solid level base such as rock or gravel if it is available or on some plank in a muddy area.
- Operate as close to the water as possible. Every foot of suction lift decreases the volume and pressure at the nozzle.
- Make sure the couplings on the suction hose are tight. If it is continuous, air discharge at the nozzle is caused by leakage on the suction side of the pump.
- Foot valves are essential on the suction hose of centrifugal pumps.
- The strainer must be covered with at least 6 inches of water. To keep it out of sand, gravel, and as much mud as possible, tie the intake end to a log or float to keep it as high in the water as possible but still 6 inches under the surface. Tie the foot valve vertically. Tie a bucket over the end as standard practice. The strainer can also be rested on rocks, submerged boards, or canvas, provided these are securely kept down with weights such as rock.
- If the pump is water cooled (such as a gear pump), lead the cooling water away from the pump to avoid a bog hole. The same is true for the discharge of the pressure relief valve.
- Avoid setting the pump where rolling or falling material can endanger the operator. His hearing will be reduced considerably by the noise of the pump. Avoid setting in depressions where carbon monoxide fumes can accumulate.
- Suction hose will not stand high internal pressure. Always use a check valve on the discharge side to avoid damage to the pump and intake hose. A pressure relief valve is also a necessity for satisfactory operation.
- Use suction hose straight to the water source. Avoid perpendicular bends in suction hose.

Hose-handling techniques. In handling the hose, observe the following:

- Do not drag the hose over rough surfaces or sharp rocks because the outside surface may become frayed, and pinhole leaks may develop. Avoid hose lays over sharp objects as the vibration causes wear.
- Use the best hose next to the pump where the pressure is greatest, and use the poorer hose nearest the nozzle, especially on slopes.
- Lay in the most generally direct route from the pump to the point of use. Lay with a few short shallow curves on steep slopes and around trees and heavy brush so that the hose can be tied at intervals to keep it from slipping. When the hose is full of water, its weight is enough to pull the line downhill unless it is anchored.
- Assign a man to patrol long lines to keep kinks out, tighten couplings, replace gaskets, and mend hose. Pinhole leaks can be mended by tying a knot in a piece of twine and placing the knot over the pinhole. Tie the twine tight, and, as it becomes wet, it will tighten and make a fairly good emergency patch.
- Practice holding the hose in position with three fingers of each hand while using the thumbs and index fingers to turn the female collar

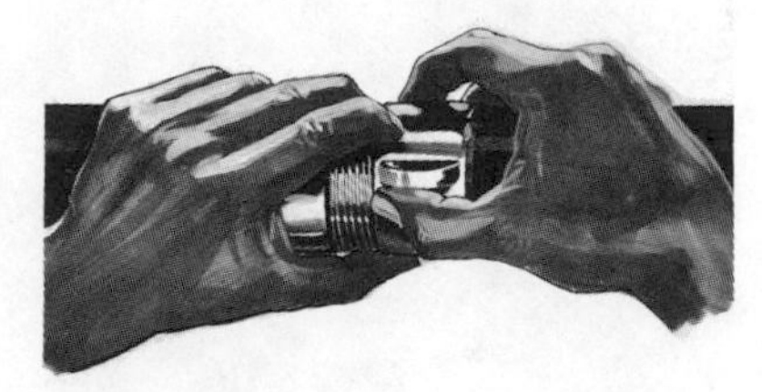

Figure 5.58. Practice holding the hose in position with three fingers of each hand while using the thumb and index fingers to turn the female collar.

on couplings. A little graphite helps. Never use oil or grease. Handle hose threads carefully and keep them repaired.

- Carry rolled hose on packboards or in packsacks. More hose can be carried this way, and it doesn't hurt as much.
- Folded hose can be carried forward by four or five men spaced evenly along its length and carrying it on their shoulders as a team. Rolled hose is a much better way to distribute hose in wildfire control. After use on the fire, roll it as one would a ball of twine. This method is quicker and easier than the doughnut roll and is much better for the hose than dragging it.
- Even unlined linen hose can burn, and CJRL burns more easily. Avoid lays through hot spots.
- Use of hose clamps is better than kinking hose to attach another length or accessories.

Downhill lays. When the fire is below your pumping equipment or water source, the weight of the water itself may build up enough pressure in the hose to break it. Nozzle pressure is about 1 pound more than pump pressure for every 2 feet than the nozzle is below the pump. This is static head. This amount of pressure is added to the pump pressure, so on the longer lays downhill the total pressure may build up to more than the hose can stand.

Gravity systems. The above principle can be put to beneficial use in developing gravity hose lay systems. It is the most trouble free, because there is nothing mechanical to break down. However, there must be a suitable water source at least 75 feet above the highest part of the fire. Even a small creek will provide a surprising amount of water for fire use.

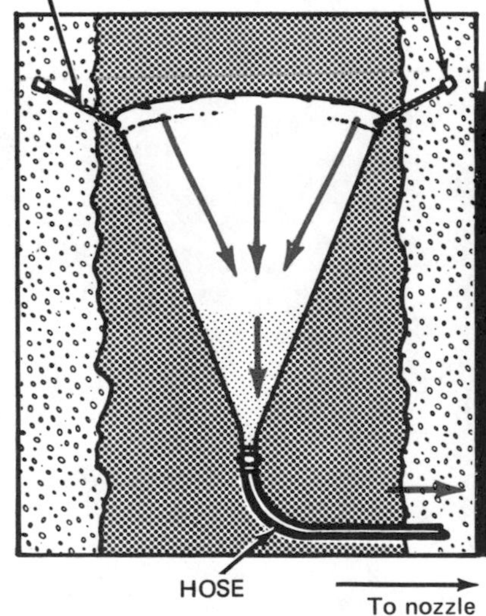

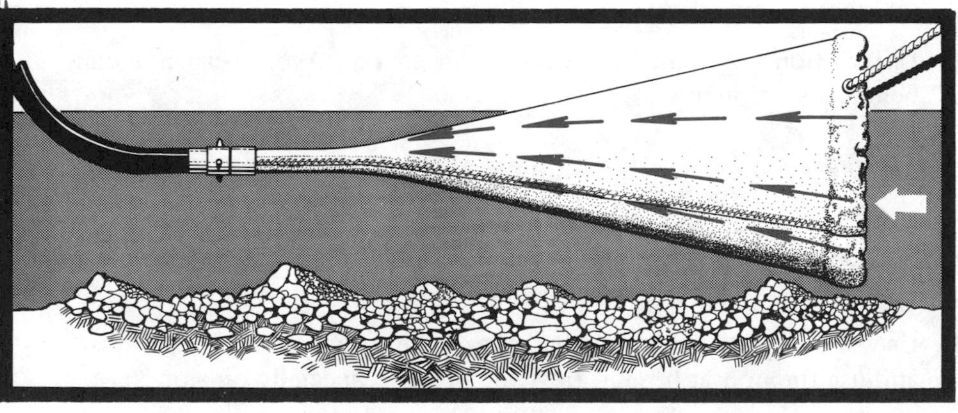

Figure 5.59. A downhill hose lay may start with a gravity sock tied firmly in place at the water source. One pound of pressure is added for each 2-foot decline in elevation. Two different views are shown.

In a gravity hose lay, the gravity funnel or sock must be tied firmly in place with rope and reinforced with rocks if possible. It is best to choose a location with a drop in the stream bed, or it may be necessary to build a small dam and sump. The funnel should be completely submerged. Attach the first length of hose and lead it away and downhill in as direct a course as possible. If the water does not extend the hose, start just below the sock by pinching the hose to fill it with water and continue pinching it at intervals to lead the water through it. Once enough pressure is developed, the water will continue to flow. As the line lengthens and the drop is increased, too much pressure may develop. It may be reduced by (1) using larger nozzles, (2) using a wye valve as a bleeder, or (3) inserting another gravity sock at the end of the first line. So long as the funnel is kept free of debris and the water is flowing into it, the system will work with little attention to it. Branch lines can be installed and accessories can be used in the same way as in other hose lays.

Friction and elevation loss. Friction loss in fire hose is the pressure loss caused by the turbulence created by the friction of the water against the interior surfaces of the hose. Additional pressure at the pump is needed to overcome this loss, while the desired pressure is maintained at the nozzle. Friction loss is increased by increased water flow, reduction of hose diameter, and the roughness of the interior hose surface. Pressure loss at the nozzle is also increased by bends and twists in the hose line. Hose should be laid as straight as possible.

Elevation loss is the pressure drop caused by raising water to higher elevations. This loss is roughly 1 pound for every 2 feet the water is raised. Therefore, to obtain the same pressure at the nozzle as at the pump, 1 pound of pressure must be added at the pump for every 2 feet the water is raised in elevation. Both friction loss and elevation loss are measured in pounds per square inch (psi). Friction loss and elevation loss must be calculated for any type of hose operation.

The formula for calculating this loss is pump pressure in psi equals nozzle pressure in psi plus friction loss in psi plus elevation loss in psi: $PP = NP + FL + EL$. Friction loss is shown in Table 5.7.

Example: A 500-foot hose lay of CJRL 1½-inch hose with a 100-foot rise in elevation from the pump to the nozzle and 50 gpm at 100 psi is needed at the nozzle. Referring to Table 5.7, the friction loss for 50 gpm at 100 psi is 9.2 psi per 100 feet. Therefore, 5 x 9.2= 46 psi, which is the friction loss,

TABLE 5.7
Friction Loss in Fire Hose†

Nozzle flow (gpm)	*¾-inch booster*	*1-inch booster*	*1½-inch lined linen single line*	*1½-inch lined linen two lines*	*1½-inch unlined linen*	*2½-inch unlined linen‡*
10	13.5	3.5	0.5	—	1.0	.145
15	29.0	7.2	1.0	—	2.2	.25
20	50.0	12.3	1.7	0.5	3.6	.38
25	75.0	18.5	2.6	0.75	5.5	.53
30	105.0	26.0	3.6	1.0	8.0	.71
35	140.0*	35.0	4.8	1.4	10.0	.92
40	180.0*	44.0	6.1	1.7	13.0	1.12
45	222.0*	55.0	7.6	2.2	16.0	1.36
50	270.0*	67.0	9.2	2.6	20.0	1.6
60	—	79.0*	13.0	3.6	28.0	2.22
70	—	93.0*	17.3	4.8	37.0	3.1
80	—	108.0*	22.0	6.1	47.0	4.08
90	—	125.0*	27.3	7.6	59.0	5.0
100	—	—	33.0	9.3	72.0	5.9
120	—	—	47.0	13.0	100.0*	8.3
150	—	—	70.0	20.0	—	12.5
300	—	—	—	72.0	—	32.3

*Usually not practical.
†Friction loss is measured in pounds per square inch per 100 feet of hose.
‡A 2½-inch lined linen hose has roughly one-half the friction loss of a 2½-inch unlined linen hose.

or *FL*. The nozzle pressure, or *NP*, is stated as 100 psi. The elevation loss, or *EL*, is 100 feet divided by 2, or 50 psi (1 pound for every 2 feet rise in elevation). Using the formula, 100 *NP* + 46 *FL* + 50 *EL* = 196 *PP* in psi. Therefore 196 psi is needed at the pump to produce 50 gpm at 100 psi at the nozzle, and 10 psi should be used for the loss at each siamese or wye valve.

When two 1½-inch lines are wyed from one line, the friction loss per 100 feet for the length of the double lines will be reduced to approximately one-fourth the friction loss in a single line of 1½-inch hose during discharge of an equal gpm. The nozzle size determines the gpm delivered.

Friction loss may be reduced by:

- Reducing nozzle pressure
- Reducing size of nozzle while maintaining nozzle pressure
- Increasing size of hose
- Changing from unlined to lined hose
- Wye the hose line.

Friction loss calculators in pocket size are available from equipment companies. They are useful items to carry with the pump or pumper.

Backpack pumps. Backpack pumps are one of the most useful and effective wildfire tools. They must be used correctly to be effective. They are the most efficient and economical means of delivering water on the fire when they are skillfully operated.

Either a rigid tank of metal or fiber glass or a collapsible bag is used to carry the water. A short length of small hose attaches the tank or bag to a trombone-type pump to dispense the water. The collapsible bag, usually made of neoprene, has the advantage of being easier on the back and requires less storage space.

Practice proper lifting methods when you are loading the pump on your back, or, better yet, have another crewman help in placing it on your back. This help can prevent a slipped disc or a wrenched back, as the unit weighs over 50 pounds.

When the pump is used, one hand should be placed close to the forward end and held steady, aiming the pump where the water is needed and as close to the base of the flame as possible. Pump with the other hand holding the end of the pump attached to the hose. By holding the forward hand steady, accurate direction is given to the stream. Place the thumb over the nozzle end to obtain a fan-shaped spray. After water is lugged in a backpack for any distance, it becomes that much more precious. Make every drop count, and use it sparingly so that it will go as far as possible.

Backpacks are very helpful in initial attack, especially to stop the spread in lighter fuels, to cool down hot spots along the line, and to knock fire out of snags. They are especially effective if they can be readily refilled. Backpacks are almost indispensible on spot fires, as adjuncts to hand tools in initial attack, and especially in mop-up and patrol.

If plenty of backpacks are available, they are sometimes scattered along the line by whatever transportation is available so that they are on hand when they are needed to combat flare-ups and to mop up. Often the best use of a tanker or a pumper is to supply backpacks when it cannot reach the fire area.

The foregoing information outlines the basic concepts and some hints for successful use of water on wildfires. There is no substitute for practice and drill, which produce teamwork and satisfactory performance. Be sure each fire fighter knows what to do and how and when to do it. The only way to gain satisfactory performance is to practice with your equipment in your area. Remember, equipment is only as effective as the ability of the fire fighter who uses it.

LINE CONSTRUCTION WITH EARTH-MOVING EQUIPMENT

Bulldozers, fire line plows, and other types of heavy earth-moving equipment are extremely effective tools for building fire lines, particularly in heavy fuels and brush. They must be followed with hand tools to finish the line to burn out where necessary, to hold the fire within the line, and to combat slop-overs and spot fires. Once the fire line is built, it is necessary to begin the mop-up from the fire edge toward the center of the fire, and this operation requires hand tools. Often, pumpers or hose lines can be used to assist with the holding action and with mop-up. Much will depend on the kind and volume of fuels. Bulldozers can be used to a limited extent on mop-up operations as discussed in more detail later. Bulldozers and fire line plows will be discussed separately. (See also the chapter on safety and the section in this chapter on fighting fire with fire.)

Heavy equipment should be recognized as another tool and not sufficient in itself, since the line it builds must be held with manpower. A wide bulldozer line or a wide plow line should not give a false sense of security, which has often occurred. All of the actions necessary to establish secure control of the fire area are just as necessary in spite of the wide line.

This equipment is sometimes used in grass and lighter fuels, but usually these fuels are on terrain that can be traversed with pumpers. Sometimes, graders, tractor plows (farm type), discs, harrows, and even combines have been used effectively on all sizes of fires in the lighter fuels. The only limiting factors are their availability and the suitability of the terrain for their use. Combines have been used astraddle the fire edge on wheat field fires. Their use was successful, but follow-up was necessary to extinguish residual fires and spot fires. However, the use of such costly machines in this type of attack cannot be recommended because of the potential for serious damage to them.

Most state and federal fire-fighting agencies, some forest products industries, other full-time fire services, and some volunteer districts will own and operate bulldozers and fire line plows as part of their regular fire suppression equipment. Good operators and mechanics will be employed. Such equipment requires a transport facility such as a tilting bed truck or a truck and lowboy trailer for delivery of the unit to the fire. Many of these fire services also contract with local owners, counties, or municipalities for use of their heavy equipment during fire emergencies. Contracts, arrangements for use, rental rates, responsibility for repair and damage, and dispatch

Figure 5.60. Bulldozer blades are operated either by cables or hydraulic cylinders. Most blades can be adjusted for angle either horizontally or vertically.

should be arranged for in advance of use and preferably in advance of the usual fire season. At this time a condition appraisal of each machine to be used should be made to determine its suitability for fire line work. The appraisal should be made by or with a representative of the fire service who is qualified to determine the condition. Observation of the following items is helpful: (1) general appearance, age, evidence of maintenance, dirty engine, excessive oil leaks, bent control rods, missing equipment, bent sheet metal, torn seats, rusted exhaust stack, and number of hours operated as well as time of last overhaul if available; (2) evidence of wear on rails or chains, track tension, pins and bushings, shoes and bolts, grousers, front idlers, top idlers, lower rollers, and sprockets; (3) condition of hydraulic and cable systems; (4) blade corner bits, cutting edges, and mold boards (replace any of these that are worn down or loose); and (5) condition of the blade. The blade should remain in a hold position when it is raised above the ground. It should raise and hold the front of the tractor when it is operated against the ground, but this will not be possible with a cable-operated dozer. Only those machines that are in good repair should be considered for fire control work because they must be dependable when they are applied to the fire. Machines considered for hire on the spot at or near the fire should also be appraised for condition before they are used. It is essential that contracts of hire clearly state the terms of use, rental fees, and responsibility for repairs and damages.

Bulldozers. Also called dozers, trail builders, and cats, bulldozers are effective fire-fighting tools if they are correctly used, but they are costly to operate and require good operators, good supervision, and adequate service and repair. They will do the work of many men and are said to equal a crew of forty men in some fuel types. However, in excessively rocky areas and in some dense timber stands, especially with many large trees, their progress will be drastically slowed. When they are needed to pioneer ahead of a fire line plow or a fire line crew, they are indispensible. In this capacity they do the clearing work so that the plow or crew can build the line as they follow.

Dozers should not be operated horizontally across slopes of over 45% and then only when the soil type and obstructions will not force the unit to a sidehill position of over 45%. (They can be operated across a slope of over 45% by making a sidehill cut to reduce the slope on which they would operate, but this operation is usually too costly and time consuming.) Locating a fire line across a slope is very rarely good business, either above or below the fire. The top of the ridge or the bottom of the valley is most always a better location. Dozers may be operated uphill on slopes of up to 60% and may be operated downhill on slopes of up to 70%, provided the way is open below and the dozers will not have to come back up. Generally, on slopes of above 35%, the production of the dozer drops off proportionately. The above percentages should be reduced where heavy fuels and rough stump-studded or rocky slopes are encountered.

Line location is as important for tractors as for hand tools, and the same principles of width, depth, and location apply. Locate the line in accordance with the fire control strategy, vegetation, and terrain. Avoid taking the fire where the cat can go, unless it is the best method to use.

A line locator should be assigned to the lead dozer unit. He should be physically fit and have a good understanding of fire behavior and a good knowledge of the capabilities and limitations of dozers. In many situations it pays to have a helper accompany the line locator. The locator scouts from side to side, while the helper blazes or ties marking ribbon on the trees or brush to designate the line chosen by the locator. The location work should stay well ahead of the dozers but not so far that the line location would need to be changed by the time the cats get there because of changes in conditions that would affect the fire's location. The locator should check periodically with the spotter or operator of the lead cat or with the crew boss to advise them of what is ahead and to know what progress the dozers are making.

Line location is basic to the success or failure of the operation. Locations where dozers cannot work effectively should be avoided and completed with hand work. These locations would include areas of large rocks, rock outcrops, excessively steep terrain, or other limitations to the use of bulldozers.

Trench undercut lines, and treat all hazards in the same manner as hand line construction and mop-up.

The dozer line should be built some distance away from the fire edge in all instances even on direct attack unless the dozer is "scrubbing" the line. In a very few instances the dozer might be used on very small fires to push the burning edge into the fire area all the way around the perimeter. This is not a highly recommended practice because of the probable difficulties of mopping up but is used sometimes. As a general rule, the dozer should work far enough away from the fire edge so that there is no possibility of casting burning material to the outside. In this way a pile of dirt and burning fuel is not left inside the line. Also, the fine fuels needed for burnout are not destroyed. As a general rule, cast all dozed material outside the line and scatter it. This may not always be possible, and, if there is any doubt about the dozed material having any fire in it, cast it to the inside but be sure it is scattered and not in piles.

Keep in mind that the area between the fire line and the fire edge must either be burned out or be allowed to burn out as the fire moves to the line,

Figure 5.61. In building a fire line with a bulldozer, debris is pushed *outside* the line to avoid adding fuel to the fire.

or, if the main fire dies out, it must be thoroughly cold-trailed. In any case, the amount and kind of fuel should be such that hot spots can safely burn out inside the line without throwing fire over the line. Never depend on the extra width of a dozer line to hold the fire by itself.

Often it is practical and desirable to build the cat line where it will serve as an access road for tankers and the movement of crews. Grade along the line then becomes important so that four-wheel vehicles can travel along the line. This use is a very valuable assist but should not dictate the location of the fire line. The location for fire control purposes must take precedence.

In hazardous areas it may be necessary for dozers to clear out safety areas or strips for their use as well as for crews. These should be built well in advance of probable need.

Fire lines on steep slopes ahead of the fire are difficult to build and difficult to hold. It is usually much better to locate along a ridge or sometimes on a bench.

Organization of dozer crews will vary with the size of the fire, the kind and amount of fuels, the topography, the practice in that locality, and the personnel available.

At least one man should be assigned to each cat as a swamper. He should stay out of the way of the dozer while it is in operation and also away from falling or rolling rocks, logs, and debris. His job is:

- To handle the winch line and choker, help change blade positions, assist with the maintenence, and otherwise assist the operator of the tractor.
- To communicate by hand signals with the operator.

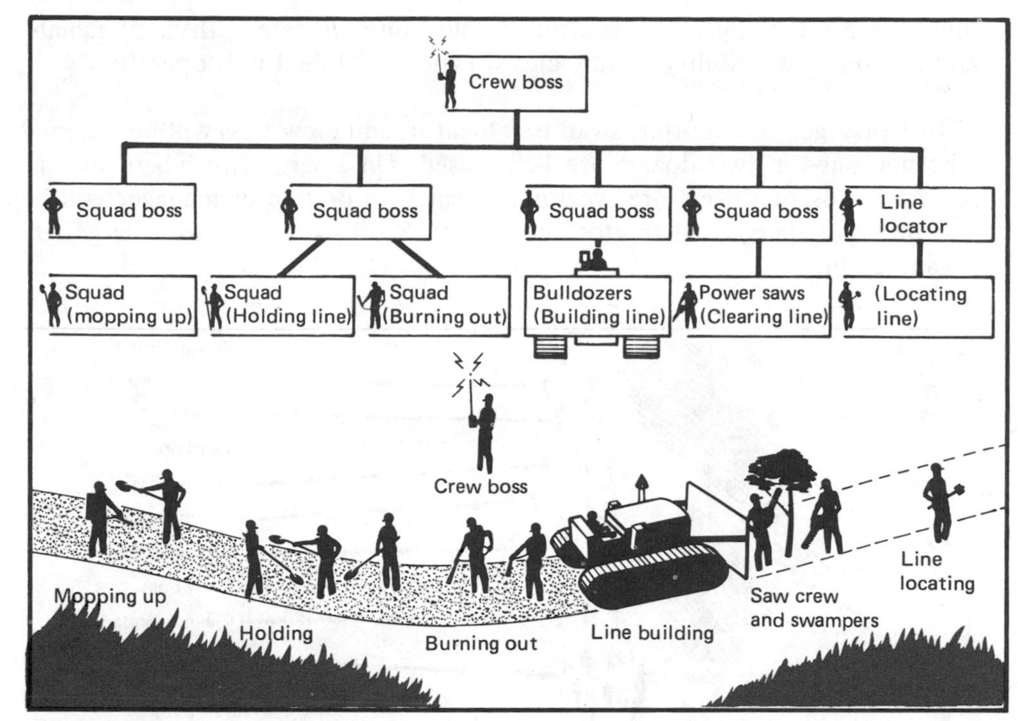

Figure 5.62. Organization of a hand-tool and dozer line crew.

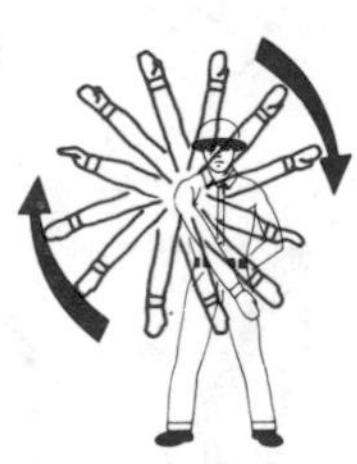

STOP. Back and forth, waist-high swinging motion.

COME AHEAD. Up and down in front of spotter, from waist to arm's length above head.

TURN. Swing flag or light on side to which operator is to turn.

REVERSE OR BACKUP. Full circle in front of spotter.

CAUTION. Wave flag or light in half circle at arm's length above head.

ATTRACT OPERATOR'S ATTENTION. May also use one blast on a police whistle or other suitable substitute.

Figure 5.63. Dozer crew signals from the spotter and swamper to the operator. By gunning his motor twice, the dozer operator can signal that he cannot see the spotter, and by gunning his motor once, he can signal that he wants the helper to come to the dozer.

- To cut away projecting branches and sticks that may jam the machine and cause damage or endanger the operator, to remove rocks and debris from the machine on signal from the operator, and to cut with an ax or chain saw where necessary to move logs and tangled windfalls.
- To work as a spotter if one is not assigned or as a contact between the operator and the spotter.
- To act as an alternate operator.

A spotter is a man working a safe distance in front of the dozer to mark out a specific or final line location for the lead dozer to follow. He is a guide to the operator. He may contact the operator every so often to advise him of what is ahead and the best procedure. He must know what dozers can and cannot do. In some fuel types with good visibility ahead, he may not be needed. His job will then be handled by the swamper or the crew boss. In other situations he may act as both swamper and crew boss and alternate operator for the unit. The spotter should communicate with hand signals, and it is his responsibility to stay clear of the cat while it is in operation.

In heavy going a spotter, swamper, locator, and crew boss will be assigned whether one or two dozers are being used. On a large fire where one to several teams of dozers are working, a cat boss or tractor manager will be assigned to supervise all tractor equipment. See Figure 5.64 for a large fire organization.

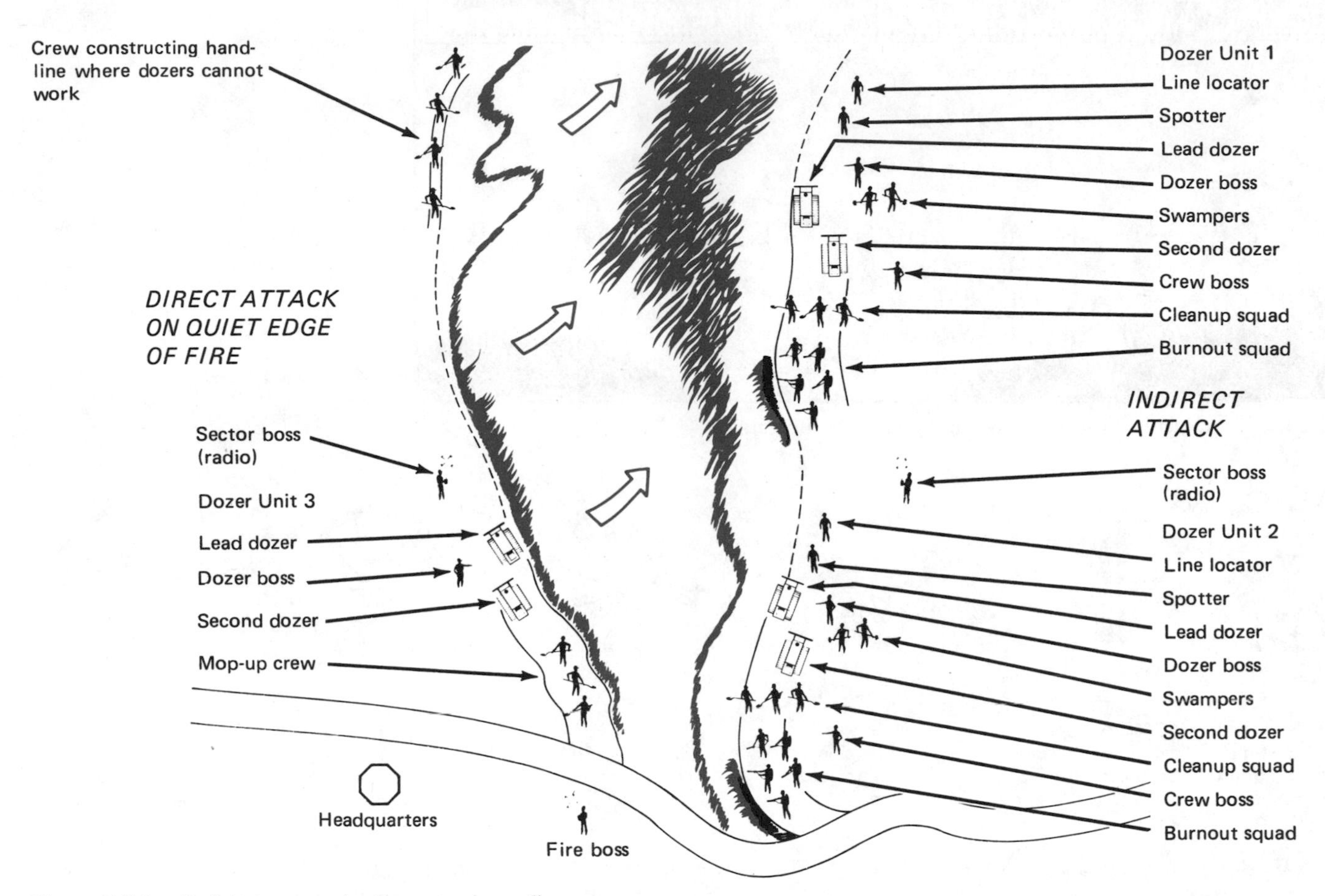

Figure 5.64. Bulldozer organization on a large fire.

Dozer line principles. The following apply to dozer line operations:

1. All unburned fuels moved by the dozer should be cast away from the fire line and scattered to the extent possible. On small fires, all of the burnable material along the edge may be moved into the fire area by dozing at right angles to the fire line. This mixes considerable soil with the fuels that are burning and leaves piles of it inside the line. It also mixes the fine fuels with soil, thereby preventing them from burning freely and igniting other fuels. Consequently, a clean burn is not obtained, but many smoldering fires are left that can later cause trouble. This type of dozing can and usually does create more work than working indirectly and burning out. Mop-up and patrol may last for weeks. Therefore, even with a small fire or on the edges of a large fire that are quiet, the dozer should work far enough from the burning edge so that it will not pick up burning material, and it should shove all material that it does pick up to the outside and scatter it. The same is true for building a line on direct attack—stay a few feet away from the fire edge with the inside edge of the bulldozer blade. However, careful watch must be

maintained to make sure that any material that is on fire and caught up by the blade is cast inside and is scattered. Push debris out as much as possible and operate so this can be done.

2. Where both soil and debris must unavoidably be pushed to the inside, spread and scatter this material well back into the burn side. Windrows and stacks of fuel and dirt are called dozer piles or cat piles and are one of the worst hazards to completely extinguish. They also delay final control of the fire. They hold fire a long time and have caused rekindles that are usually of serious proportions. Elimination of these dozer piles inside the line is one of the secrets of using dozers successfully on fires.

3. In fuel types with down timber such as "jack-strawed" stands of wind-thrown timber, bug-killed stands, or old burns, it may be best to use power saws ahead of the lead tractor to "buck" the material or break it so that the dozer can more easily and quickly move it. Usually one or, at the most, two cuts are all that are necessary. The saw crew should work well out in front of the lead dozer and ordinarily not over two saw crews of two men each should be used. The route of the line to be built should be clearly marked so correlation with the spotter is needed. Ordinarily, the saw crews should cut only windfalls, slash, etc. They should not cut standing trees unless they are too large for the dozer. The saw crew should make one cut only or should make a partial cut on felled trees. This is done on downed logs or trees so that they are cut into 15 to 30 foot lengths that can readily be moved by the dozer. In this way the dozer can push the debris out and away from the line, leaving a fairly clean edge on the fire side.

Chain saw crews should be closely supervised, should be kept well ahead of the dozer, and should avoid doing work the dozer can do. Saw crews are not needed if the lead dozer is large enough to do the job or if it can do the job without creating excessively large piles of debris.

4. Stubby snags can quickly be felled by dozers. The taller snags can sometimes be felled with dozers where the whipping action does not cause too much breakage of the tops and large limbs. Much will depend on the size and condition of the snags involved. Where snag felling is hazardous to dozers, the job should be done by felling crews. Local practices will vary by different sections of the country.

5. As a rule of thumb, it is best to work dozers in pairs so they may assist and reinforce each other. The width of line, the fuels, and the nature of the topography will determine the suitability of the tandem operation. More than two dozers working together is seldom practical. The tandem operation usually has more advantages than disadvantages but may not be possible or desirable. Typically, the lead dozer pioneers the line by doing the rough clearing job. Ordinarily, this dozer is the largest machine or the one in best condition. His job is to rough out a way through and to keep moving because his progress pretty well determines the speed of line construction. He does not attempt to clean the line. The second dozer cleans up the line as much as is practical for cat work and digs a line only to mineral soil. He does what is necessary with the dozer to make a complete line and leaves only those items that can be handled by the cleanup crew that follows. Thus, the second dozer cleans the line to mineral soil, widens the line where necessary, breaks down and scatters cat piles and windrows inside the line,

Figure 5.65. Tandem bulldozer operation on a line built in an indirect attack.

pushes over snags and leaners close enough to the line to cause trouble, and improves the line to the standard desired.

Neither machine should be operated below the other on slopes because of the danger of rolling and falling material. They should not work too close together, nor should they work too far apart. Good judgment and supervision are needed to obtain the right mix in distribution of the work done by each unit so that dozer piles are reduced, the width of line is not excessive, and duplication of work is avoided.

6. The width of line depends on many variables such as the class of fire danger present and expected; the fuels, their volume, kind, position, continuity, compactness, height, and the density of the canopy; the steepness of the slope and the location of the fire on it; and the amount of heat generated against the line through radiation and convection.

In other words, the width of the dozer-built line is directly related to the job it must do in controlling the fire.

Rule of Thumb: The line should be wide enough to hold the fire; the wider the line, the larger the piled-up accumulation of fuel and soil adjacent to it.

The line may be wider in some sections than in others depending on the job it is intended to do. There must be a reason for extra width of line. It may be because the brush is tall and thick, because the canopy needs to be opened up, or to keep the crew from being scorched during burnout.

The usual rule for the width of line is—not less than one half the height of the fuel. With dozers, of course, the minimum width would have to be one

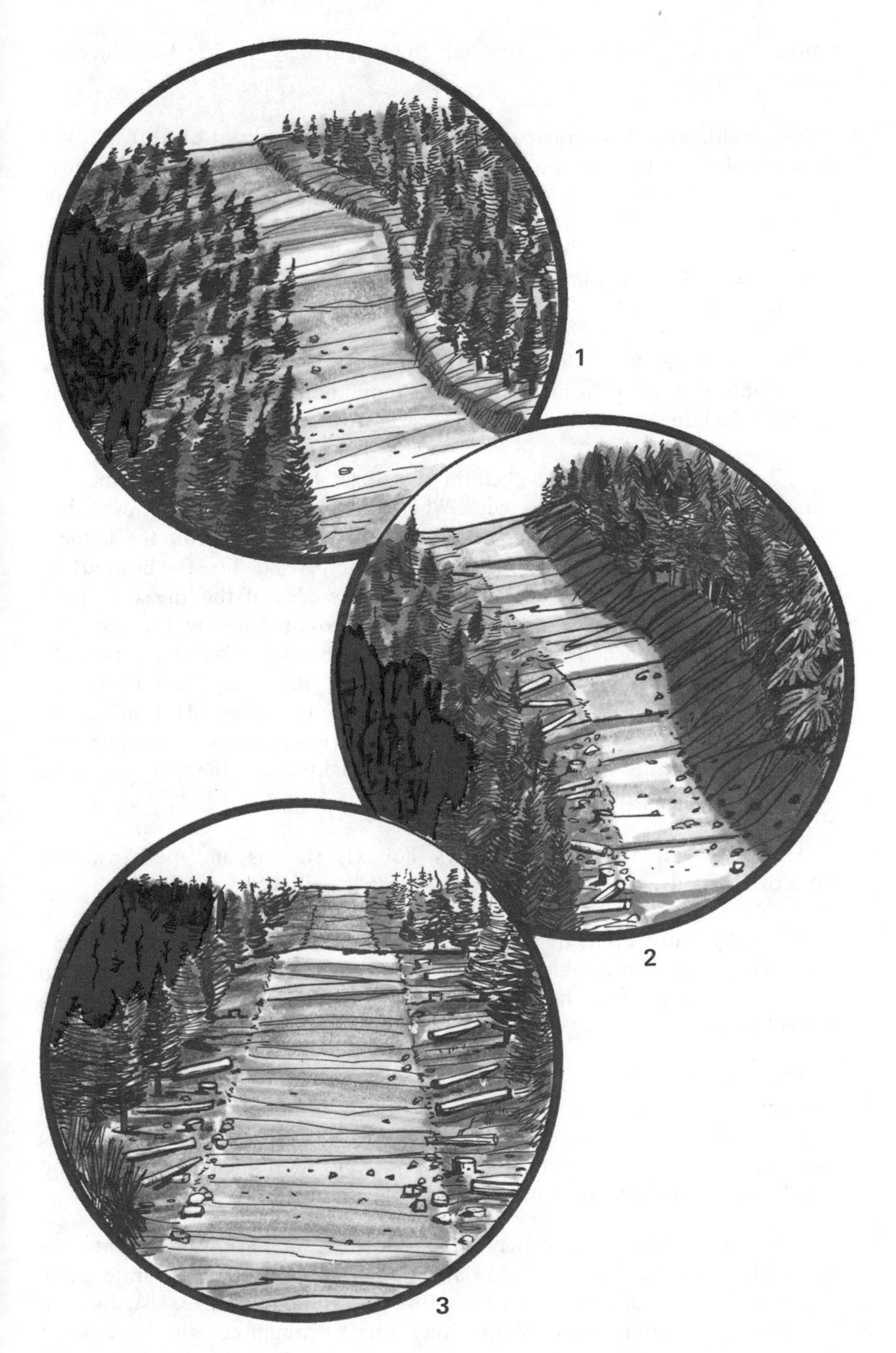

Figure 5.66. The width of a fire line built by a dozer is dependent on the amount of heat that will be created from the burnout fire.

1. The fire line should be wide enough to contain the burnout fire. It is usually impractical to make it wide enough to stop the running fire.
2. The fire line may have to be widened where the slope is steep. Fuel on slopes, especially the larger trees, is pushed downslope (toward an uphill-running fire). The rest of the debris is pushed uphill.
3. If trees are tall and the canopy is thick, the opening may need to be wider. The canopy is opened by felling trees. Keep the dozer line narrow to avoid piling up mountains of debris mixed with dirt.

blade wide. With tall timber, one-half the height may need to be reduced in many instances.

The width of line is principally dependent on the amount of heat that will be created when it is fired out. It is impractical to make it wide enough to withstand the frontal run of the main fire; therefore, enough area must be burned out from the line toward the head to contain the head fire far enough from the line to prevent mass transport of embers over the line. In all cases, the width of the line must be judged on the effects of the burnout fire and not on the effects of the main fire running into the line.

Where a wider line is needed in tall timber, try to make the clearing by felling the trees away from the line. Do a minimum of piling with dozers by keeping the line worked with the dozer as narrow as possible.

7. Use a cleanup crew behind the dozer both to speed up the line construction and to make it secure. Where only one dozer is being used, this crew will ordinarily have more work to do than it would after the tandem operation. The job of the cleanup crew is to prepare the line for burnout by reducing the kind and amount of fuel along the edge of the line so that the chances of radiated heat and mass transport across the line will be reduced. This job is accomplished by: (a) Felling snags on both sides of the line; (b) Felling leaners; (c) Breaking up, tearing apart, flattening, and dispersing accumulations of fuel close to the inside of the line; (d) Loping and scattering tops and branches and cleaning up the lower stems of standing trees by cutting off limbs, moss, and vines; and (e) Making sure the line to mineral soil is continuous and free of surface fuels.

This crew will be using axes, saws, Pulaskis, shovels, and, perhaps, adze hoes or McLeods, depending on the fuel type that they are working in.

Normally, only three to six men are needed in this crew plus the squad boss. They may also double as the saw crew when such a crew is needed in advance of the cat. They may also be part of the burnout crew if the cleanup job is light.

The main point here is that hand tools are needed to follow the dozers in any type of fuel to make sure the line is ready to hold and is ready to burn out. Cleaning up the line with hand crews is faster and cheaper than using dozers. The cleanup crew must stay away from the immediate area around the dozers, since they are constantly backing and maneuvering.

8. The burnout crew may be part of a combination crew that does both cleanup and burnout. On larger fires it will be a separate crew supervised by a squad boss or a crew boss according to its size. Usually, one to three torchmen are used. Many times, one torchman can string plenty of fire to keep everyone busy. Enough men are needed on this crew to do the holding work and chase spot fires. They should use shovels and backpack pumps. A rule-of-thumb ratio is seven spot fire hunters to one torchman, but this ratio may vary considerably with the situation.

If hose lays can accompany the burnout, there is much more assurance of holding the line. The pump crews should be organized separately and should not interfere with the burnout or spot fire hunting, but their actions should be coordinated with the burnout crew.

The burnout may not be done right after the cleanup; it may be done several hours to a day later. For instance, the line may be built at night and the humidity may be too high to get a satisfactory burn. Thus it would be necessary to burn out the next day when conditions have reached the satisfactory point.

If the burning follows the dozers, it should not follow so close that the firing will handicap the dozers operation or jeopardize the line construction and cleanup. Dozers can often compact and mash heavy fuels inside the line so that they can be more easily and effectively fired.

9. Proper supervision must be provided for hired equipment. Proper organization and supervision are the keys to successful operation of equipment.

10. Servicing of equipment must be provided for as soon as the equipment moves onto the fire. Two operators should be assigned to each dozer for each shift, and they should be alternated in the operation of the unit.

11. Bulldozers used on fires should be equipped as follows:

- They should have canopies of sufficient strength to withstand rolling over and to protect them from falling overhead material. This equipment is mandatory in forest types.
- They should be equipped with a winch having suitable and sound cable and blocks. The dozers should be armored underneath, and those that are not should be assigned to easier work.
- They should have quick change blades.
- They should have lights, both front and rear, for night work.
- Radio communication should be established, at least to the sector on which the dozers are working. The dozers should have a radio hand set on each team that is in touch with the crew boss.
- They should have a fire extinguisher, ax, shovel, and water bucket.
- They should have lubricants (oils and greases) and necessary grease guns.
- Two operators should be provided for each shift for each dozer. They may alternate as spotter and operator.
- Fuel filters, track wrenches, crowbars, and necessary small tools should be provided.
- They should have a fuel pump, filter, and means of fuel supply. A 45-gallon barrel of diesel fuel oil will keep the average-sized dozer running one full shift.
- Hydraulic fluid, extra hoses or extra cable, and spare fan belts should be available to the assigned crew.
- Enough fire shelters should be available for the assigned crew.

These are expensive machines, and they cannot be operated long without servicing. Arrangements for service and repair must be made when they start work.

Dozer blades are operated either by hydraulic systems or by cables operated from revolving drums. Either type is effective when it is operated by a skilled operator, but the hydraulic type can perform in more situations than the cable-operated dozer and is most desirable for fire line construction.

Dozers are used in the suppression of wildfires by the direct and indirect (parallel) methods and to construct lines for backfiring or strip burning. They are also used to construct strategic secondary lines, assist hand crews with slop-overs, construct safety areas, open up alternate access routes, clear out old firebreaks, and construct roads and helispots. They are very useful in clearing ahead of a fire line plow. Dozers are also useful in mop-up, but this use is limited and must be carefully supervised. See the section on mop-up in this chapter.

Plows. Fire line plows are pulled by four-wheel drive trucks and by crawler tractors. There are a number of different kinds and sizes, but the ones used with tractors are in the majority and generally are the most satisfactory. In the southern pine flatwoods, plows outnumber pumper units because they are so effective. The medium units are the most numerous. These are pulled by the TD-9, HD-6, and D-4 class tractors. Light units include TD-6 and D-2 class tractors. Heavy units include TD-14 and D-6 class tractors, and these are primarily equipped as bulldozers.

Plows are a principal wildfire tool in the flatwoods and coastal plains of the South, Southeast, and Florida. They are generally used in the Midwest, the lake states, and the northeastern part of the United States. They are in limited use in the Southwest. The fire line plow is best used in fuel types that can be traversed without too much interference from standing timber, where the topography is more or less level to rolling, and where the soils are generally sandy or friable with a minimum of rock. Rocky soils are a deterrent to plows. As the slope increases, the efficiency of the plow decreases.

Since the plow is pulled by a tractor, the fuel type must allow access to that size tractor. Otherwise, the plow unit must be preceded by a bulldozer and/or a saw crew. Thus more resources would be required, and the efficiency of the unit would be decreased. Tractors can usually walk down trees up to 5 inches in diameter if they are not too closely spaced.

In suitable average fuels and soils, most tractors are able to construct a maximum of 3 miles per hour (240 chains), which is equal to a man's walking speed. The speed will vary considerably below the 3 miles per hour on account of soils, stumps, boggy sections, trees, and other obstructions to using plows.

Most plowing is done in third gear; fourth gear can be used in light fuels on sandy soils. Fifth gear should be reserved for walking (traveling), and second gear is recommended for heavier than average fuels or where control of the unit is important, such as in heavy timber or in a stumpy area.

Line construction and line location principles are the same for plows as for any method. The main difference being the considerations of time. The depth of plowing should be as shallow as possible and yet should obtain a clean line down to mineral soil. The shallow line is equally as effective as a deeper one as long as it is clean and continuous. The shallow line puts less drag on the tractor so that plowing is faster, the line is less restrictive to wheel traffic, and less erosion is caused on slopes. The depth should be adjusted while the plow is in motion and should be frequently checked if it is hydraulically operated. Rear view mirrors help with this problem. The parallel and indirect methods are almost always used with plows, since the direct method is rarely used in initial attack.

Since burning out is usually practiced in line construction, the plowed line should be as straight as possible. Deep pockets and turns provide a chance for buildup of fire intensity and consequent probability of slop-over. On the other hand, the line should be bent as necessary to keep stumps, snags, fuel accumulations, and other hazards outside the line. Thus the possibility of spot fires and the mop-up job are reduced.

When the plow is pushing over snags, use the back end of the plow frame, and operate in reverse; thus damage to the radiator grill is avoided, the cooling system is kept farther from the heat of a burning snag, the snag must fall farther to strike the operator if it buckles, and, if the tractor stalls, the danger from fire is not as great. Always watch the top and upper limbs of a snag and be ready to take evasive action. Breaking tops and limbs can become the proverbial "widow maker."

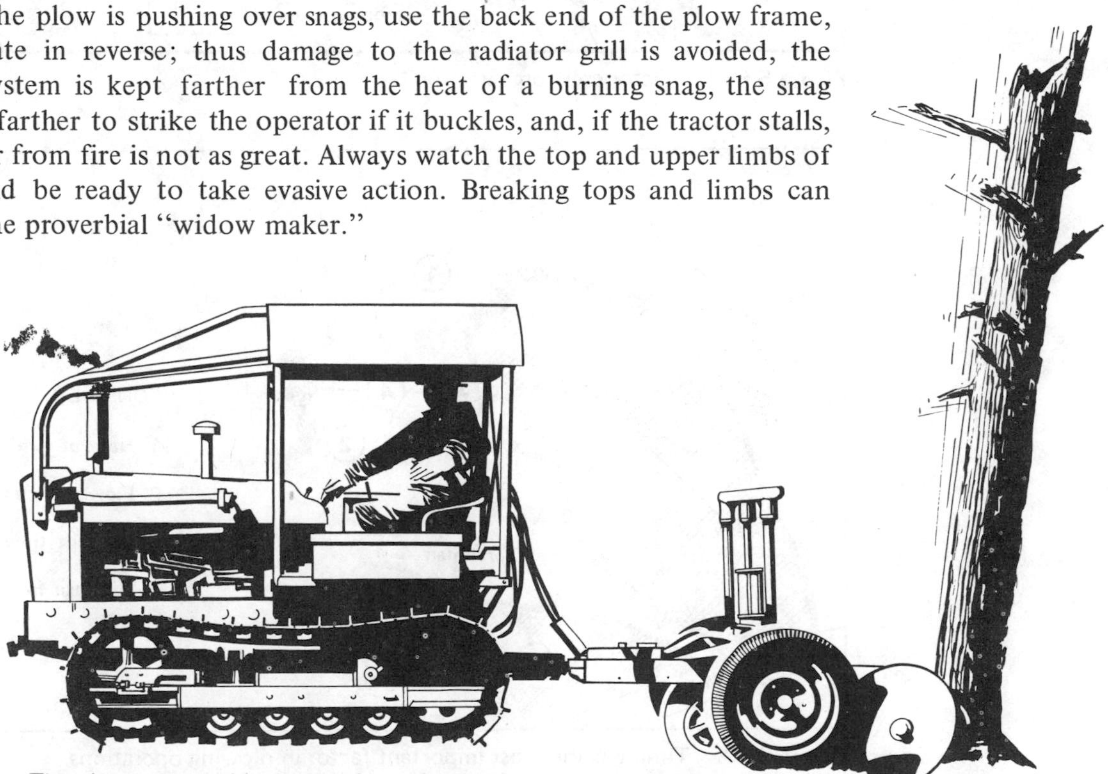

Figure 5.67. The plow, operated in reverse, is used to push over snags.

As with all tractors used on wildfire control, plows should be fitted with satisfactory canopies and roll bars. Many have brush guards fitted in front of the operator to protect him from brush and limbs.

Rule of Thumb: One of the secrets of wildfire control is to make the line secure as it is built. The fire line is useless if the fire can cross it.

Single tractor techniques. The principles of operating a single tractor are much the same as operating with teams of two or more tractors. Almost always, two tractors are the optimum number per team.

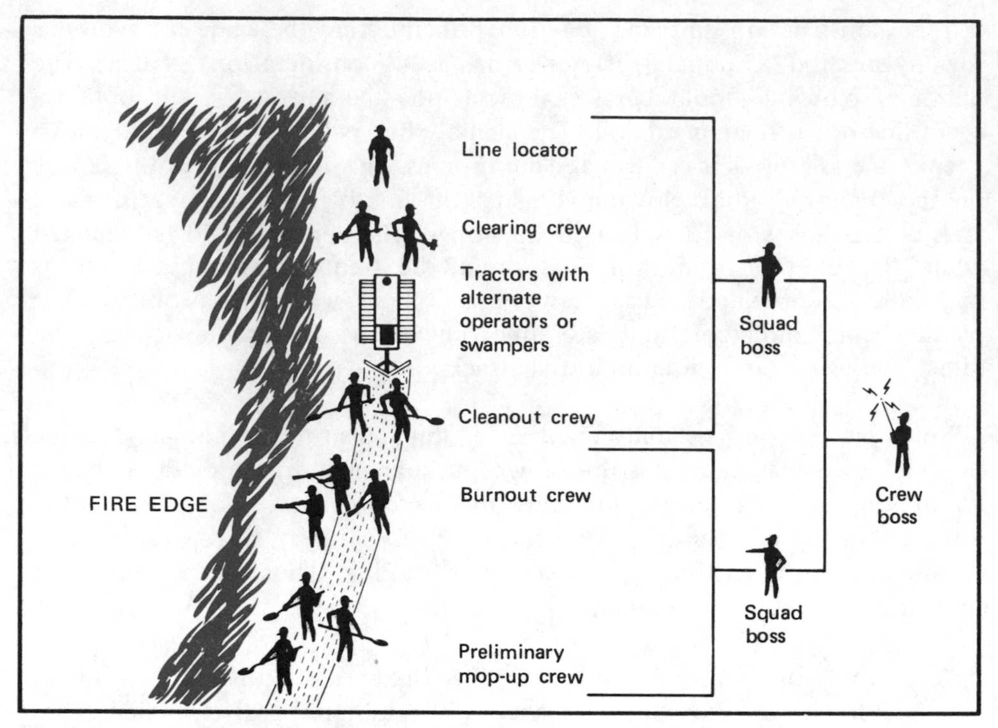

Figure 5.68. Tractor crew organization.

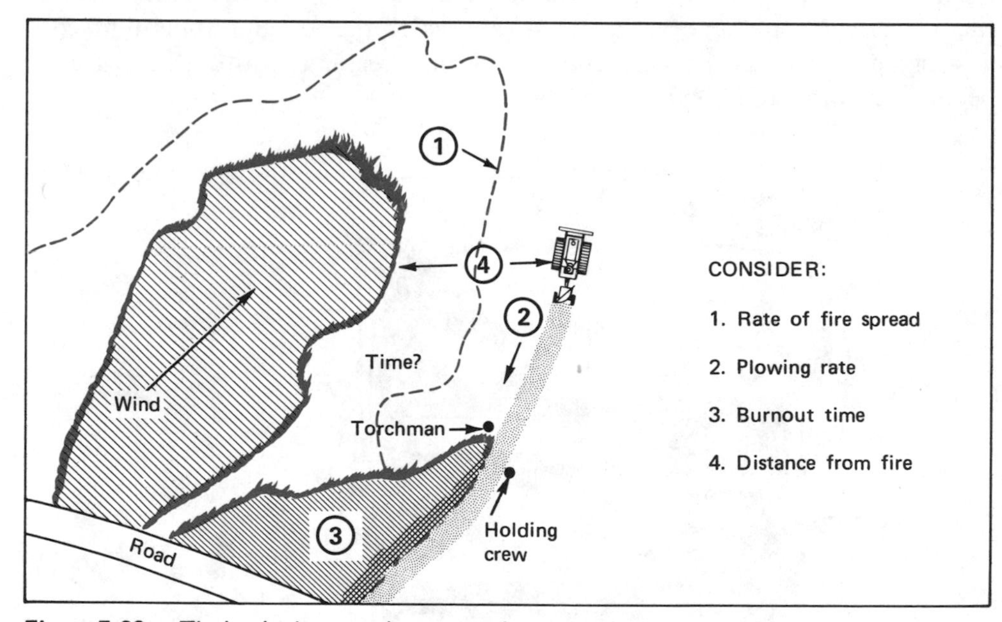

Figure 5.69. Timing is the most important factor in plowing operations.

Most of the time, a tractor crew consists of two men, one operating and the other swamping or firing. Location is handled by either the crew chief or the operator if the fire boss has not yet arrived. A holding crew of one to several men may be required to clean up, hold the line, and chase spot fires, depending on the fuel type and conditions. If a pumper is available, these jobs are accomplished by the pumper crew.

Timing is the most important factor in successful plow operation. Timing must be adjusted to:

1. Rate of fire spread
2. Plowing rate

3. Burnout time and extent of burnout

4. Distance from the fire to (a) heavy or lighter fuels, (b) plantations, buildings, etc., and (c) natural barriers, swamps, etc.

Errors in judging time as related to the above factors result in lost line, a larger burn than is necessary, and damaged or lost improvements.

As with any kind of tools used to control wildfires, the best tactic is to stop the head first if possible. But success in stopping the head depends on the combination of weather conditions, fuels, and topography. Above moderate fire weather conditions, it is not safe to work in front of the head without adequate preparation and sufficient resources. The attack is then concentrated on the flanks to pinch off the head with an oblique movement and then to finish the total perimeter. Generally extreme conditions do not occur as often in fuels where plows are used as they do elsewhere. The oblique line starts on the flank near the head and angles toward and in front of the head fire. Enough area must be left between the line and the head to allow the burning out that accompanies the line building, to cover enough of the area to adequately stop the head fire, and to reduce spotting to a minimum.

If at all possible, the head should be approached on the way in or at least should be seen as soon as possible after the crew arrives at the fire. Natural barriers such as roads may be present so that burning back from them would be the best tactic. Time can be lost in plowing a line when burning out a barrier is the final answer.

Plowing becomes a matter of compromising and judging between the following:

- Plow close enough to the head to save the area from burning and to conserve time but far enough away to ensure safety and an effective burnout.
- Plow fast enough to outrun the head but slow enough to keep the torchman in sight and to see breakovers.
- Get back to the flanks quickly, but do not leave the head until it is secure.

The timing, therefore, is as exacting as it is in any profession.

Where a burnout is not necessary, safety demands that the tractor stay at least 10 to 50 feet from the fire edge, depending on the irregularity of the perimeter and whether the tractor is on the flank or the head. Even this distance does not allow much time for hangups on stumps or other obstructions.

In brush and scrub timber or pure wire grass and where stumps are readily visible, it is best to plow as close to the fire as possible and as fast as possible. Spotting is inevitable, and burnout seldom keeps pace with the plowing. So the direct attack, as nearly as possible, is best with an adequate spot fire patrol. A double line is often more effective than a short time burnout in pure wire grass. In pinc flatwoods the best tactic is to give the fire some room and to burn out from the line.

There is a rule of thumb in Florida: "If it's worth plowing, it's worth burning out."

Multiple-tractor techniques. On a routine fire, each tractor should take a different part of the line, and thus much more fire will be put out much more quickly. This is particularly true if enough pumpers and/or manpower is available to follow up each tractor.

On difficult fires, the best results are usually obtained by working plow units in teams, particularly if it is difficult to hold the line or difficult to construct the line.

Often two parallel plowed lines are made to obtain the width needed if holding is difficult. The rolled-out turf of each line should just touch the brim of the adjacent line. The outside unit follows a short distance behind the inside unit because it establishes the position of the line.

In difficult line construction, the units are operated in tandem. The first unit (usually a bulldozer) takes out obstructions such as big brush and snags and as much surface fuel as time will allow. The second unit makes a clean line to mineral soil. They must work as a team, adjusting the work load between them.

Figure 5.70. In tandem operations, the first unit (bulldozer) clears the line, and the second unit (plow) clears the line to the mineral soil.

If the difficulty is trees and heavy brush, the heaviest tractor takes the lead to walk down obstructions or push them to the outside. Material that has no fire in it is always pushed to the outside. However, if the difficulty is frequent bogging, the favorite method is to send the smaller tractor in the lead to "feel out" the line. It is much easier for the heavier unit to pull the smaller one out of a bog than vice versa. One tractor should be the leader, and both should strive to stay in sight of each other so that they can immediately give aid to each other.

Bulldozers and plows. In the method of using bulldozers and plows described above, only a dozer would be placed in the lead. There are some situations when a dozer is needed, because a plow alone cannot do the job. When a dozer is needed, there is no substitute for it. Conversely, a dozer is very slow by itself in plowing a line except in very light cover where the blade can be dragged to produce a satisfactory line, but this lack of speed often places the machine in the fire.

There are times when tandem dozers are needed to construct a line through heavy timber or across a dry swamp in what is ordinarily plowing country. In these cases, the tandem dozers would be used as described in the previous section on bulldozers.

Dozers can also be used to "scrub" out a fire by working directly on the fire edge. In the South this is sometimes done where the fire has penetrated a dry swamp beyond the point where ordinary tractors can go because of the trees. It may not be practical to burn out the whole swamp from the hill on the other side, so the dozer is applied directly to uprooting trees as necessary to get through the area. The turning and backing usually produce a good line.

Tractor plow units should carry the same equipment as bulldozer units. Service is equally important for these machines. Radio contact with them is a must because of their cost and speed of operation. If one of the tractors on a team is not radio equipped, assign a walkie talkie to at least each team. Organization and supervision are the keys to successful operation of the tractor plow units.

MOP-UP

Mop-up is the process of putting out all fire or putting out all fire in enough of the area around the perimeter so that no spot fires and no breakovers can occur. The amount of area to be mopped up will depend on the fuel, the location of smoldering fire in relation to the perimeter, the slope, and possible changes in weather. Ordinarily, the burned area should be mopped up for a distance of 100 feet from the perimeter toward the center of the fire. In some fuels and on smaller fires, it is necessary to mop up all fire inside the line. In heavy fuels the cost of complete mop-up may be excessive. In these cases, mop up the perimeter and work the necessary distance to the inside. Then patrol the line on a twenty-four-hour basis until

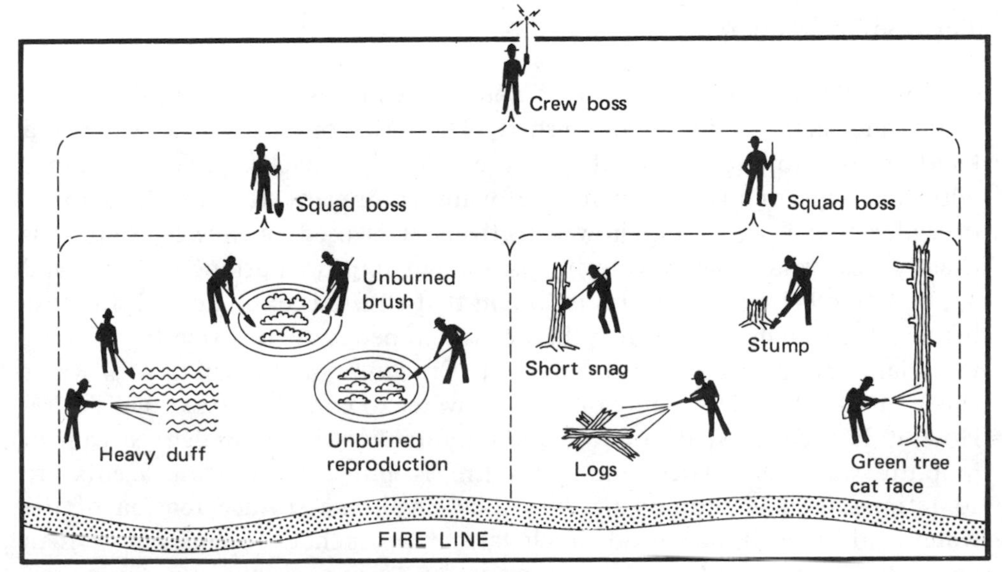

Figure 5.71. Mop-up crew organization.

the entire area has been burned out completely. If all the fuel inside the line cannot be burned up completely of if all the fire cannot be completely extinguished, the fire area must be patrolled until there is no possibility of ignition outside the line.

Mop-up can mean the success or failure of the total fire control operation. It is of little use to confine the fire and then to allow it to break out to cover more acreage because the mop-up was not effective. More fires have been lost because of poor or incomplete mop-up than for any other reason.

Mop-up should begin as soon as the line is complete. In many situations, mop-up may be started during the line construction of the initial attack. Control is not achieved until enough mop-up is accomplished to make sure the fire is permanently confined to a definite area.

Mop-up work is hard, dirty, and tiresome work. It lacks the excitement and challenge of initial control work but is nonetheless important. Here is the real test of the effectiveness of the crew and its leaders. The crew is usually tired from the initial control effort, and the tendency is to slack off on mop-up. Good supervision and leadership are required to obtain an effective job.

Rule of Thumb: The heavier the fuels, the more difficult and longer is the job of mop-up.

Flash fuels require little mop-up but can be misleading. In grass, the cow chips, stubs of bushes or yucca, or small chunks of wood can smolder with little or no smoke until they are fanned by wind. Then they can throw sparks into unburned fuel. These items should be completely extinguished. Water is best for extinguishment, or these items may be dug out and put out. Cow chips need to be broken up and soaked with water.

Much mop-up can be prevented by the line construction crew if they are alert to preventing situations that will cause more work. Examples are piling and stacking of fuels near the line, trying to cover burning materials on top of the ground, failing to prevent fire from getting into snags and stumps, and not getting a complete burnout.

TECHNIQUES OF MOP-UP

If water is available and can be applied, it makes the mop-up job much faster and more easily accomplished. The best combination is water and hand tools. Mop-up with water may employ backpacks, pumpers, pumps with hose lays, or any combination of water equipment. It is not the amount of water that is used, but how effectively it is used. A fine light spray is usually best, and it saves water. Separate and expose burning material, and apply a fine light spray until it is certain that all fire is extinguished. In some instances a straight stream may be needed to penetrate or reach the burning material. Try to make each drop count. Drowning or drenching with water can be ineffective. Never rely on a lot of water to completely extinguish fire. How it is used is what counts. Don't hydraulic with a straight stream in mop-up—water is wasted and little fire is put out. Wetting agents are sometimes used in mop-up because they break the surface tension of the water and allow it to spread much farther and penetrate much more than ordinary water. Hose lays are often made for the express purpose of mop-up.

Figure 5.72. Mopping up with water.

Where water is not available or in limited quantities, hand tools are very effective when they are used knowledgeably. In fact, hand tools should be used with water for the best efficiency in mop-up. The shovel, backpack pump, Pulaski, rake, and saw are the best mop-up tools.

Bulldozers can be very helpful in the initial stage of mop-up, but they can also cause more harm than good if they are used too long and too much. Initially, they can be used to break up concentrations of fuel, spread out and scatter burning material, dig out stumps, and fell snags. The final mop-up must always rely on hand tools and water. Avoid making piles of burning material and soil. Fire in such piles can smolder a long time and break out later on.

Try to eliminate trouble spots before they flare up and endanger the line. Keep fire out of heavy fuels, concentrations of fuel, and unburned islands. If the mop-up cannot be done fast enough to eliminate threats to these locations, they should be burned out, adequate precautions being taken.

Break up concentrations of fuel that are burning. Spreading out will reduce the heat, and much fire will go out without further work. Drag or throw material on fire away from the line, and scatter it well inside the burn. Improve the line and make sure it is continuous.

Turn logs and chunks perpendicular to the slope (up and down) so they cannot roll. On steep slopes, build a trench below logs and stumps to catch the rolling material. Turn logs out of their hot bed of ashes. By turning the log 180 degrees, the fire on the under side is placed on top where most of the heat is convected away from the fuel. Often two or three turnings will completely eliminate all the fire.

Where fire is persistent in logs or where time is important, the burning portions can be chopped off and scattered, rubbed with dirt, or sprayed with water.

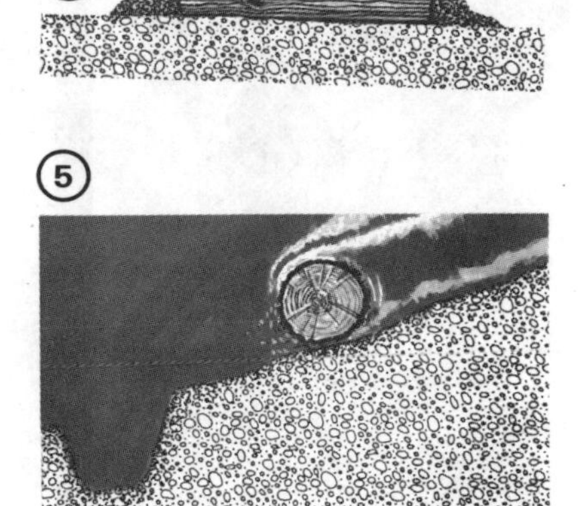

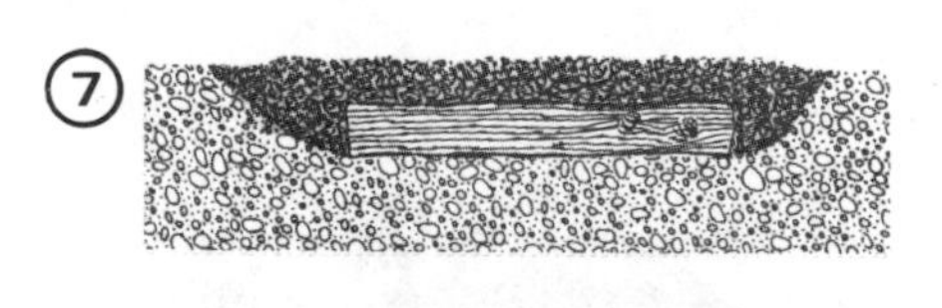

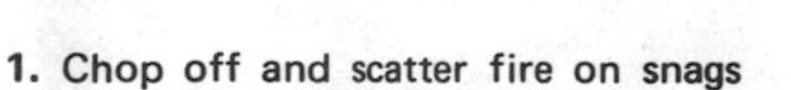

1. Chop off and scatter fire on snags
2. Turn logs out of hot bed of coals
3. Don't try to bury logs on top of ground
4. Do bury in a hole and cover with soil to groundlevel
5. Dig out stump and put out fire
6. Trench to catch roll
7. Separate logs and place up and down slope to prevent rolling
8. Mix cowchips with dirt or water

Figure 5.73. Mopping up with hand tools.

Do not cover burning stumps, logs, or chunks with soil and expect them to go out completely. Usually, the soil drops away as it dries, and the smoldering material underneath breaks into flame, and sparks can be thrown by the wind. Soil may be used to cool the material so that it can be worked, or soil may be used to cover stumps and logs that have not yet caught fire. This soil covering prevents radiated heat from igniting unburned stumps and logs. Sometimes smaller chunks can be buried in a hole dug in mineral soil. Six or more inches of mineral soil should cover the burning material, and the final surface should be level with the ground around it and should be compacted. Many times, fires have broken out to make new runs because of the wrong application of soil in mop-up. It is much better to chop off the fire and completely extinguish all burning material or to allow the fire to burn out.

Snags inside the line and in locations where sparks from them can be cast across the line should be felled away from the line and extinguished. Likewise, snags outside the line and susceptible to catching fire should be felled. Snags act like large roman candles and cause much spotting. Clear all flammable fuels from an area of sufficient size to encompass the snag, and fell the snag into this area. Then put out all fire in it. Snag felling requires special skills and definite safety precautions.

Clear areas of fuel around stumps and other heavy fuels so they cannot catch fire. Break up accumulations of fine deadwood, which, if they are ignited, can cause crowning in overhead fuels.

Dig out stumps already on fire. Cut off roots and completely extinguish stumps that have been burning for some time. Stumps should be thoroughly exposed, as fire can be deeply entrenched in them. Do not try to drown stumps with water.

Remove roots across the line. Fire can travel along roots underground and can break out on the surface many feet away and as much as two weeks later.

Fire in heavy duff may smolder a long time. If the area is small, it can be dug out and scattered on a burned area or on a surface of mineral soil. It can then be allowed to burn out under observation or can be finalized with water. With larger areas, a trench to mineral soil should be made around the outside of the area of burning duff. The area can then be allowed to burn out or can be finalized with water. In this process, it is often necessary to turn all material over, perhaps several times, to make sure all fuel particles are soaked. Deep duff is much like hay or sawdust because these fires cannot be considered to be out until all the fuel is consumed or until it is scattered out far enough that water can be applied to each piece of fuel. These principles apply to leaves, needles, and dry rotten wood accumulations.

When you are in doubt about the existence of fire, feel with the hand. One's reflexes are quick enough to prevent burning the hand. It is one sure way to know that no fire exists. Use the back of the hand because it is more sensitive.

In mop-up, dig out, separate, chop off, isolate, and burn out or extinguish all burning material.

Boneyard. The boneyard is a method developed in Washington and Oregon to make certain the mop-up job is thorough and complete. First, an area is cleared to mineral soil. The size will depend on the amount of material to be worked and the space available. Usually, it is not less than 6 feet by 6 feet but can be much larger.

Figure 5.74. The boneyard is a mop-up method. The boneyard is cleared to mineral soil, and extinguished materials are placed inside this area (they should not touch).

Second, each chunk, branch, or piece of burning material is thoroughly extinguished by chopping off the burning surfaces and then completely extinguishing them with water or rubbing them with soil. Each individual piece is then placed on the bare mineral soil surface so that is does not touch any other piece of fuel. In this way, each piece is isolated, and, if it should reflame, it will have the least chance of igniting other material. Also, any rekindling will be clearly visible.

Often a shallow hole is dug and filled with water. Burning chips and small chunks are soaked in the hole and then removed to the boneyard. This is a sure method and is helpful if water is in short supply.

PATROL AND LOOKOUT

Patrol is often necessary to make sure all fire is out. It should be properly organized, and definite responsibilities should be assigned. Patrol is moving back and forth over every foot of control line and over the area outside of where sparks have fallen to check for and put out any fire. Men should be assigned to a definite patrol route so that the entire perimeter is covered. Well-understood signals and communications should be established between crews still on the fire or with the dispatcher. Patrolmen must be alert and must keep moving. Patrol must be maintained until it is known that the last spark is out. In some areas it may be best to place lookouts at vantage points to detect and report hazardous flare-ups or spot fires to crews working on mop-up and equipped to handle the complete job of mop-up.

Rules of Thumb: Clean them up as you go, or else you are not accomplishing anything. It takes a wind from every point of the compass to test the line before you can safely say the fire is mopped up. The fire that is the most difficult to mop-up is also the one that needs the most attention.

A mental warning buzzer should sound if one of the following conditions exist:

- The initial mop-up job cannot be accomplished with the manpower and equipment on hand by 10 A.M. the next day.
- There are big unburned islands inside the line that cannot be burned out.
- There is low humidity, and some wind is predicted for that night. No dew is possible.
- There is worse wind predicted for the next day, and there is a shift in wind direction.

AIR SUPPORT

The use of aircraft in suppression of wildfires began in California in 1919 and in Canada in 1920. Since that time, many improvements in both equipment and usage have been made so that airplanes and helicopters are standard tools used by state, federal, and other agencies engaged in wildfire control. Although their greatest amount of use is in supporting large fire control campaigns and in detection of fire starts, their most effective use is during the initial stages of a fire with slurry drops to slow the spread until ground crews can reach the fire.

Airplanes are referred to as fixed wing aircraft, and helicopters are referred to as rotary wing aircraft. Both types have their advantages and disadvantages for specific tasks. Generally, the helicopter is more flexible and versatile, but it is much more costly to operate and has less cargo capacity. Few aircraft, relatively speaking, are owned by fire control agencies; most are rented or leased under contract for the fire season.

Both fixed wing aircraft and helicopters are used for detection of fire starts; for observation and scouting on going fires; for slurry or water drops on fires; for transportation of men, supplies, or equipment by direct delivery; and for communciations when they are required. Fixed wing craft are also used for aerial photography, infrared scanning, and parachute delivery of supplies and equipment.

All aircraft naturally have limitations. Therefore, requests by the ground crews for air support should never be made against the rules of safety procedures and limitations of the aircraft (see Chapter 4 on safety).

Detection. Detection is a salient function of aircraft. Programed flights are adjusted to obtain the best coverage in conjunction with fixed lookout towers. Detection is normally done with fixed wing aircraft but is sometimes done with helicopters for intensive patrol of high-hazard zones or areas that are too difficult for fixed wing aircraft. Generally, a plane flying at 1500 feet above the surface has adequate observation for 6 miles on either side.

Use on going fires. Observation and scouting of a going fire are extremely helpful to the planning and operation of ground forces. Aerial observers can view the fire from many directions and positions. The fire edge can be located on a map, and travel routes and other conditions can be shown before delivery to the fire boss. A trained observer may make recommendations to the fire boss on tactics and expected fire behavior as observed from his vantage point. Up-to-date topographic maps are best for such use and the ground crews should use copies of the same map.

Observers can direct crews into the fire edge, advise on hot spots, and discover and report spot fires. Often a plane is needed to relay messages by radio or to drop a message to ground crews without radios. Aircraft usually have their own radio frequency with radios at the base camp, airport, other aircraft, and dispatcher. This is called the "air net" and reduces traffic on both that frequency and the "fire net," the frequency used for ground forces. An aerial observer can also be of excellent assistance in law enforcement work.

Periodic checks of mop-up can often be done more effectively and cheaply by air. Changes in wind direction and intensity can often be detected by aircraft before the effects are felt on the ground. For this reason, aircraft have been termed wind machines.

Transportation. Fixed wing aircraft are used to transport men, supplies, and equipment to airports nearest the fire. Much time is saved in comparison to ground transport.

Smoke jumpers available in the West are employed primarily on the initial attack of the more remote fires and sometimes reinforce ground crews already on the fire. They are highly trained and maintain excellent physical condition, and these attributes make them specialists in the truest sense. They are usually released as soon as other crews arrive so that they can be available if they are needed for other fires.

Cargo planes such as a Twin Beech, DC-3, etc. are used to drop supplies and equipment by parachute. Helicopters are also used for all transportation purposes except cargo dropping, and they are covered in a following section.

Retardant drops. There are a number of retardants used in slurry drops, and most contain a color additive so that each drop can be evaluated. Retardants are used mostly in land-based planes. A mixing plant mixes the retardant with water before it is loaded on the plane. Plain water is employed in float-equipped aircraft and flying boats where suitable lakes are available. This type of craft has the capability of skimming the water surface to pick up the water. Helicopters also use plain water for most operations.

Converted World War II bombers are primarily used for retardant drops. These include TBM's, B24's, B17's, Navy PBY's, and others. Planes especially designed and built for retardant drops are now being produced. Slurry planes carry from 1,000 to 3,000 gallons of slurry, and drops can be made by using part or all of the load. A highly maneuverable plane called the "lead plane" or "bird dog" is employed with slurry planes. It is called a bird dog because it functions in just that fashion. Its function is to fly the route over the drop zone to test air currents and other conditions in advance of the drop. He keeps in radio contact with the slurry plane and supervises his operations. The lead plane may fly in formation with the slurry plane to show him the way in to the drop area, or he may direct him in by radio. Thus the pilot of the slurry plane is given a reference point he can see as he makes the run. The lead plane then checks the results of the drop while organizing the other planes.

Slurry drops are most effective when they are used on initial attack while the fire is small, especially in hard to reach areas. The slurry drops can be very effective in slowing or stopping the forward spread of the fire until the ground crews reach it. As the size of a fire increases, the possibility of making drops on the head is progressively reduced because of the smoke and the turbulence of the convection column.

During the course of a fire, slurry can be used effectively to knock down hot spots and slop-overs along the line. This is often a valuable assist to ground crews in the holding action. Likewise, spot fires can be treated with slurry to assist in their control. The ready capability of retardant drops is very good insurance to have on most any fire.

There are some areas restricted to the use of retardant drops. Steep slopes and deep ravines seriously reduce the effectiveness of retardant drops. When wind speeds exceed 30 miles per hour, retardant drops should not be attempted. During early morning and late afternoon, visibility for pilots is reduced and difficult. Aircraft cannot be used at night except for flights to and from standard airports. Operations in dense smoke or in the vicinity of strong convection columns are not possible. Tall timber, snags, and high obstructions require that the flight be made at a higher elevation, and, consequently, the flight is less effective. Seventy-five feet is considered the minimum height for slurry planes, but most drops are made from 100 to 150 feet over the terrain.

Helicopters. The primary value of "copters" is their ability to provide intimate localized use and their versatility. Their speed can be slowed to hover, they have a short turning radius, and they can land and take off from relatively small landing spots. However, they are more costly to operate and more difficult to obtain than fixed wing craft. Each type of helicopter has a definite limitation on the altitude at which it can operate. This altitude will vary with the temperature.

Helicopters have definite advantages for observation because they can move relatively slowly and can hover. Additionally, they can operate closer to the surface and can turn in smaller spaces; these qualities make them well qualified for close observation work. Equipped with amplifiers, they can communicate directly with ground forces.

Transportation of fire fighters is an important use for helicopters. Crews can often be transported quickly by helicopter to areas where most of their energy would have been used in foot travel. They can be used in rougher and more inaccessible topography than fixed wing craft when helispots have been prepared for their use. They are ideally suited for the transportation of men and equipment over short distances. Fitted with stretchers, they make excellent ambulances for transport of the injured.

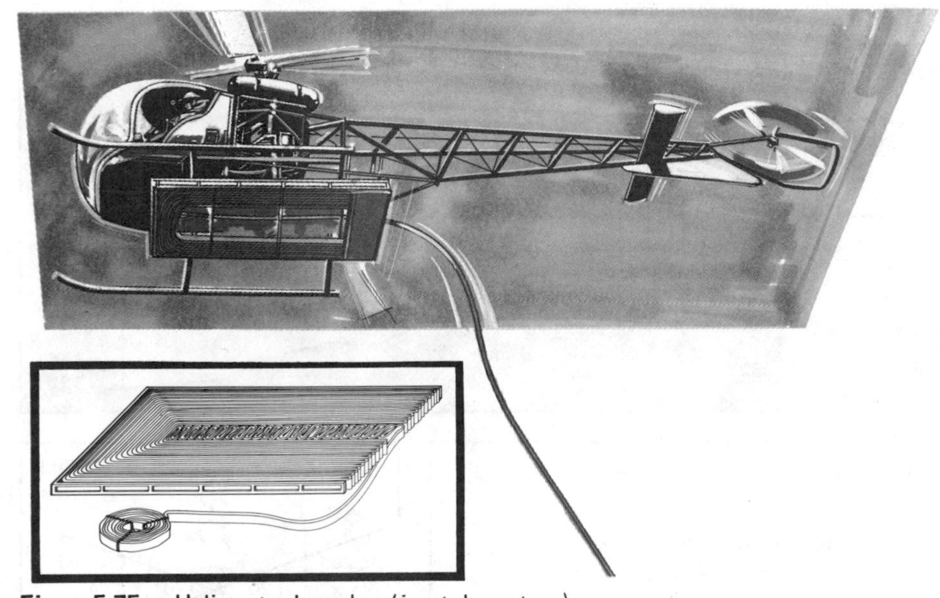

Figure 5.75. Helicopter hose lay (*inset:* hose tray).

Cargoes specifically constituted for helicopter transport are more quickly transferred because the shuttle distance can be considerably reduced. Specialized pump outfits have been developed for helicopter transport, and hose lays can be quickly laid from previously prepared trays. Cargoes can be transported directly to a helispot rather than dropped by parachute.

Bags or tanks have been developed to hang below helicopters to transport and release water on fires. These devices can be filled from lakes or tanks by dipping the container in the water source. Because of the short shuttle distance that may be possible, a copter can often deliver a surprising amount of water in a short time even though the individual trip may produce only a small amount. Helicopters can deliver water loads of 50 to 450 gallons according to the size of the copter and the elevation. Their ability to get into hard to reach areas also adds to their usefulness for retardant drops.

The rotor blast from copters, when they are operating low, can readily spread fire the same as winds caused by the vortex from the wings of an airplane. Operations of aircraft should not continue when the wind speed reaches 30 miles per hour because of the safety factors involved with them.

Helispots. Helispots can often be constructed with a minimum amount of work, or they may already exist and may only need to be marked (see Figure 5.76). Points to consider in developing helispots are:

- Landing areas should be located on exposed knobs if possible so that the pilot has a wide choice of approach and departure options with takeoff and landing into the wind.
- The touchdown space for the copter should be at least 15 feet square, level, of solid surface, and free of rocks, stumps, or any other obstructions. Choose a spot where a dropoff is possible for takeoffs. Clearings, swamps, meadows, and river bars are not desirable unless there is plenty of room to maneuver the craft in takeoffs.

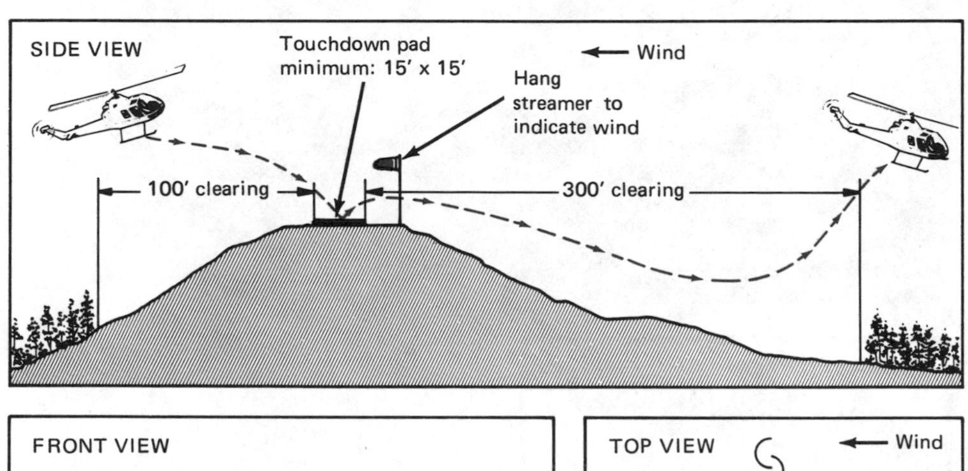

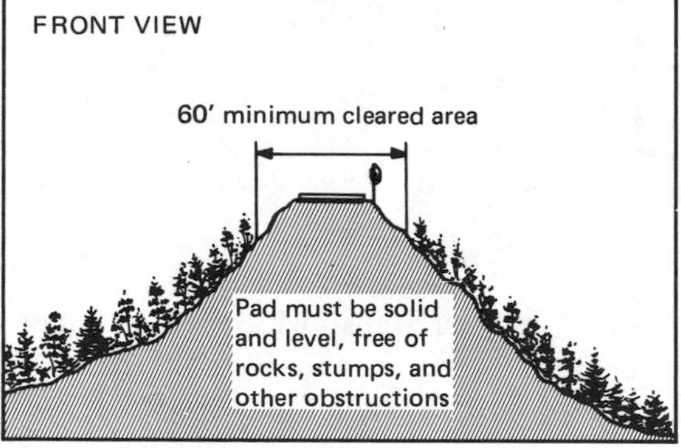

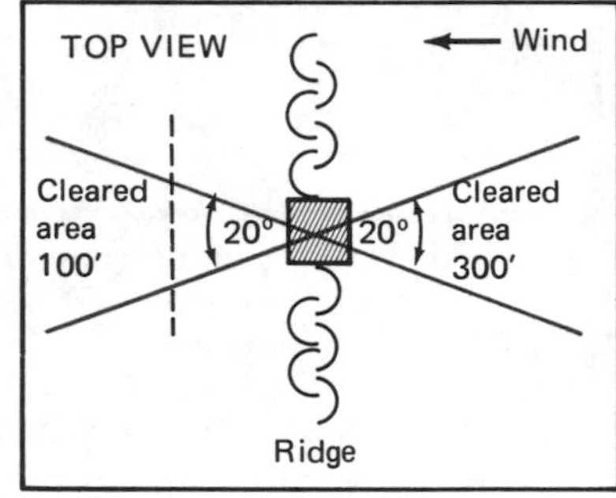

Figure 5.76. Typical helispot.

- Trees and brush should be cleared for a distance of 100 feet below the level of the landing area in the approach zone and 300 feet in the departure zone. Additional clearing is needed for larger craft. Landing areas should not be located on the lee side of ridges or any area where a serious fire hazard may exist.
- Landing areas should be marked to be visible from the air with white strips of cloth in the form of the letter H. The strips should be weighted down with rocks or otherwise attached to the ground.
- Some type of wind indicator is necessary, such as a piece of brightly colored cloth attached to a pole. It should be located well outside the landing area but should be characteristic of the wind at the touchdown area.

- Waiting areas for people should be at least 100 feet from the touchdown pad.

Aerial photos. If aerial photos are available for the fire area, they can be of great value, and they make good maps for planning purposes. Often, Polaroid photos taken from the air can be very helpful. They can be delivered to ground crews by dropping or can be sent after landing.

Infrared mapping. Infrared scanners are employed on the larger fires. The scanner is mounted in a special plane and is a highly technical operation. The photos that it takes show all fire in the area scanned and some of the topography. One of its advantages is that it can take imagery through smoke and haze or at night. Usually, the information is dropped to the fire headquarters for use in planning.

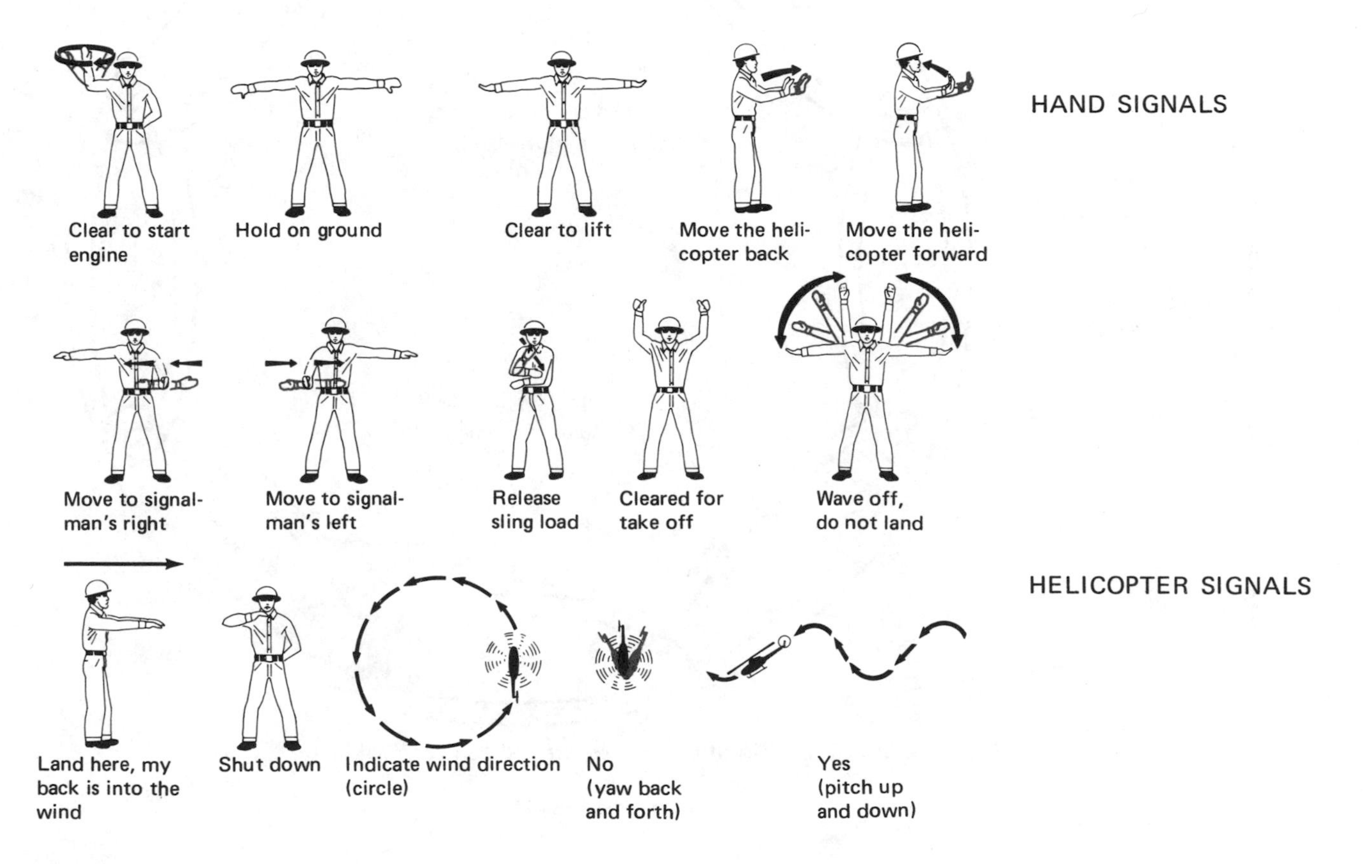

Figure 5.77. Helicopter and hand signals.

Aerial spray planes. In some sections of the country, aerial spray planes are used to drop retardants. Most of them need to be modified to drop enough retardant at one time to be effective. However, they are a tool that could well be considered by many fire districts.

SUMMARY

When the wildfire prevention effort is successful, there are no wildfires out of control, but, when prevention measures fail, suppression is required. So the necessity of controlling wildfires is the result of the failure of the prevention program. This statement may sound oversimplified, but it is a statement of fact that needs to be recognized in wildfire management.

Many wildfires are easy to control and completely extinguish. All wildfires start from a spark or small flame, but, as a fire increases in size, many factors increase the intensity of burning and the rate of spread. In other words, fire begets fire. Therefore, a prime objective of wildfire management is to keep each wildfire that does occur as small as possible. This objective requires planning, preparation, training, and teamwork by all concerned, including the public.

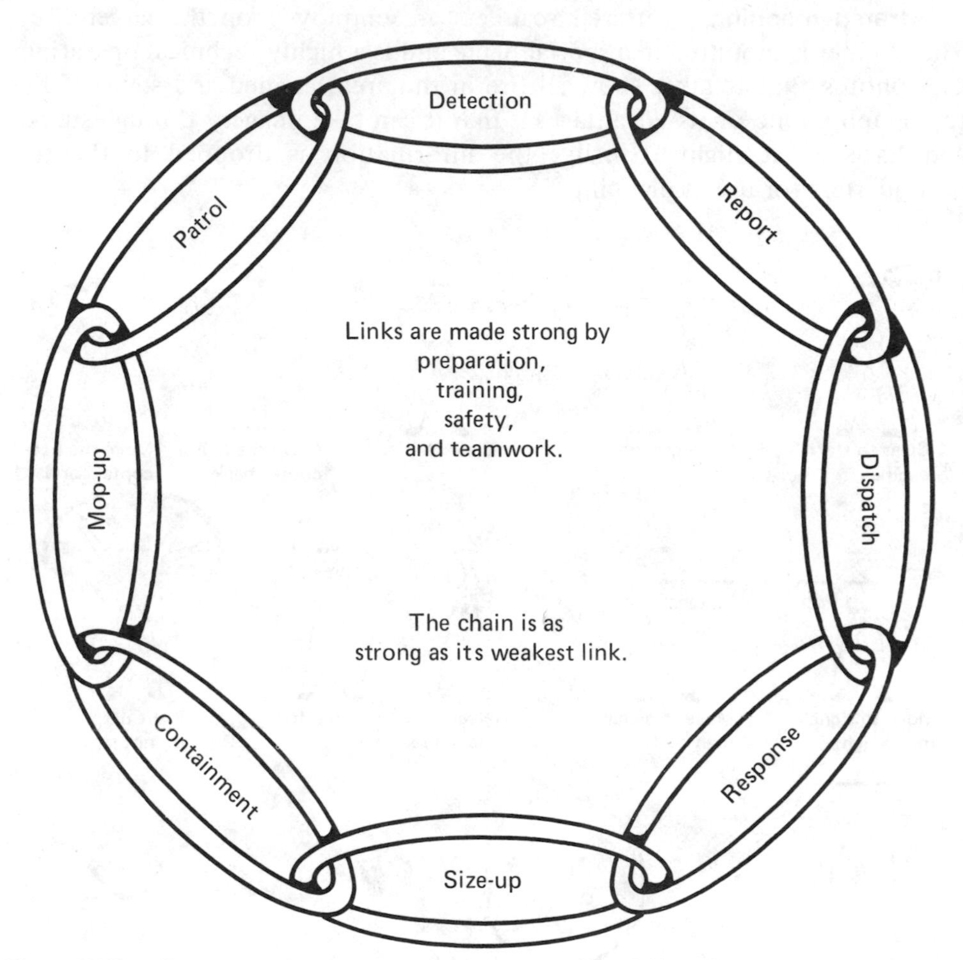

Figure 5.78. The chain of actions required to control a wildfire.

Perhaps 1 to 5 percent of all wildfires in a given area will reach conflagration proportions. The fire manager must be able to appraise each situation that he encounters and be able to recognize when a given fire is beyond the capability of his resources. He must then take the required steps to safely and effectively bring the fire under control. With the large or conflagration type of fire, reinforcements must be obtained quickly to be most effective. In the meantime, the forces on the fire must be used to the best of their capability in controlling those sections of the fire that will be most important to the final containment. As the size of the fire increases, the complexity of the control job increases because of the area involved, the size of the organization needed, and the logistics of the support and service functions. It should be recognized that no two wildfires are exactly alike and that successful control depends on applying basic principles effectively to the situation encountered. The basic principles are defined here as the seven secrets of wildfire control:

1. Be prepared. Presuppression in all its forms sets the stage for successful suppression of wildfires. Presuppression includes equipment preparation and maintenance, organization and management planning, training and practice, skill in handling tools and equipment, knowledge of your territory, development of community responsibility for detection and reporting, development of effective dispatch procedures, and quick competent response by fire fighters.

2. Hit 'em hard—keep 'em small. It's often said that the first five minutes is worth the next two hours in controlling wildfires. There is no substitute for speed of attack and speed of control with safety. All of the items in point 1 above and in point 3 below are involved. However, get there safely. Make sure the pumper arrives in one piece and that the crew is not worn out on arrival.

3. Size-up is important. A correct size-up often means the difference between success and failure. There must be one fire boss and only one fire boss. Appraise the situation from a knowledge of fire behavior and tactics. Use all the aids to control that are available. Use what you have to the best advantage. Better to have too much reinforcement than not enough. Mount a prompt skillful attack on the basis of the size-up of weather, fuels, and topography.

4. Hold the line. Have confidence in your ability. This comes from training, practice, and experience. Make the line secure as it is built. Prevent slop-overs, put out spots, improve the line at hot spots, and start initial mop-up as soon as the line is built. Develop understanding, cooperation, and coordination between fire fighters and between adjacent crews. Make everything click.

5. Don't try the impossible. Remember, there is no mystical or magical tactic that can save the day once the fire has reached the stage where it is beyond the capability of the resources at hand, and this determination must be made in the size-up. Use good judgment in appraising burning conditions and what can be done with the resources at hand. Use tactics effectively, use water wisely, and avoid any breakdown of organization and support.

6. Mop it up. Make sure the fire cannot get away again. As a fire fighter, why start another fire through carelessness or neglect? No fire is safe until the last spark is known to be out.

7. If it's worth doing, it's worth doing safely. Fight fire aggressively, but provide for safety first. Teamwork is a key to success. Accidents destroy morale, nullify hard work already accomplished, and generally cause confusion in the entire operation. Accidents are a minus factor that can be eliminated.

The functioning or nonfunctioning of the little, everyday, mundane things is the difference between success or failure in wildfire control. Tactics are, for the most part, dictated by the situation. There is an old story about a kingdom being lost because of a horseshoe nail. Because of the lack of a horseshoe nail a horseshoe was lost, the horse was lost, the battle was lost, the war was lost, and the kingdom was lost. It's the same with wildfire control. The everyday little ordinary actions are the horseshoe nails that can make or break the fire control operation.

Finally, if you are a fire fighter, be worthy of the title.

TOWARD A BETTER ENVIRONMENT

The years ahead promise to be dedicated to halting and reversing the long decline in the quality of the American environment. The American people are now acutely aware of the alarming threat of pollution to the future of mankind and all other forms of life—pollution of the air we breathe and the water we drink. They are demanding, rightly, that strong and effective action be taken. The drive has begun, at all levels of government and by industry and aroused citizens.

Wildfire prevention and control play an important part in this effort. Fewer and smaller wildfires mean less air and water pollution. Fewer and smaller wildfires mean a more beautiful natural environment.

We all agree on these goals. We can all help to reach them. As a fire fighter, you are a key man in the prevention and control of wildfires. Your ability and your action will have a profound effect on the results that are attained.

GLOSSARY

A

adapter—a hose-coupling device for connecting hose threads of the same nominal hose size but having different pitches and different diameters.

advancing a line—moving the hose line toward a given area from the point where the hose-carrying apparatus has stopped.

aerial attack—use of aircraft in fighting wildfires; aircraft are chiefly used to drop fire-extinguishing solutions in the path of or on wildfires.

aerial reconnaissance—use of either fixed wing or rotary wing aircraft for observing spread of wildfires to facilitate command decisions on tactics needed for suppression.

air-ground detection—a fire detection system combining fixed coverage of key areas by ground detectors with aerial patrol.

air tanker—any aircraft equipped to drop retardants on fires in bulk.

alarm—any audible or visible signal indicating the existence of a supposed fire or emergency requiring response and emergency action on the part of the fire-fighting service; also, the alarm device or devices by which fire and emergency signals are received. The alarm is not used to designate nonemergency courtesy calls, drill calls, or other duties that are not of an emergency nature.

anchor point—an advantageous location, usually a barrier to fire spread from which construction of the fire line should begin. It is used to minimize the chance of fire fighters being flanked by the fire while the line is being constructed.

apparatus—a motor-driven fire truck, or a collective group of such trucks which may be of different types, such as pumper trucks, ladder trucks, etc.

applicator—a special pipe or nozzle attachment for applying foam or water fog to fires, usually designated as "foam applicator" or "fog applicator." The type of applicator can also be identified by the size of the hose employed, e.g., 1½-inch fog applicator.

approved—in fire service terminology, that which is inspected and listed by recognized fire-testing agencies. The term as used in National Fire Protection Association standards means approval by the authority having jurisdiction, such as the fire chief, insurance inspection department, or other agency that enforces standards or regulations.

area ignition—the ignition of a number of individual fires throughout an area either simultaneously or in quick succession; the fires are spaced so that they soon influence and support each other to produce a fast, hot spread of fire throughout the area. *See* simultaneous ignition.

arson—the setting of fires to defraud, or for other illegal or malicious purposes. *See* model arson law.

aspect—the direction a slope faces the sun. Expressed in cardinal direction. *See* exposure.

atmospheric inversion—the atmospheric condition in which the air temperature increases with height, contrary to the usual condition in which the air temperature increases closer to the earth. In a fire situation, it is the effect of a stratum of warm air in the atmosphere holding products of combustion near the ground level, appearing to reverse the normal tendency of heat from a fire to rise.

attaching—act of connecting a nozzle to a hose or hose line to a pump or hydrant outlet.

attack—actual physical fire-fighting operation utilizing available men and equipment; the implementation of tactical plans on the fireground in an aggressive manner.

attack line—line of hose (usually from a pump) used to fight or attack the fire directly in contrast to supply or feeder lines connecting the water supply with pumping apparatus.

average relative humidity—the mathematical average of the maximum and minimum relative humidities measured at a fire danger station from one basic observation time to the next.

average temperature—the mathematical average of the maximum and minimum dry-bulb temperatures measured at a fire danger station from one basic observation time to the next.

B

backfire—intentionally setting fire to fuels inside the control line to contain a rapidly spreading fire when attack is indirect. Backfiring provides a wide defense perimeter and may be further employed to change the force of the convection column. Backfiring is a tactic that makes possible a strategy of locating control lines at places where the fire can be fought on the fire fighter's terms. Except for rare circumstances meeting specified criteria, backfiring is executed on a command decision made through line channels of authority. *See* burning out.

backing—a fire burning against the wind.

backpack—a 5-gallon pump-tank extinguisher carried by straps on the back and usually having a pump built into the nozzle. It is used extensively for grass, brush, and forest fires, and the tank may be either metal or a neoprene sack. It is also called a backpack pump or an Indian pump.

baffle—compartmentation provided in fire department water tanks to prevent shifting of water load when apparatus is in motion.

barrier—any obstruction to the spread of fire; typically, an area or strip devoid of flammable fuel.

base area—a zone which is representative of the major fire problem areas of a protection unit. From this area, the base fuel model and slope class used to calculate the unit's fire danger ratings are chosen. *Compare* fire danger rating area.

base fuel model—the fuel model that represents the cover on the base area. The base fuel model is used in the calculation of the fire danger ratings that are the basis for the unit's suppression readiness.

base station—a fixed central radio dispatching station controlling movements of one or more mobile units on the same radio frequency.

basic observation time—the time established to take the fire danger observations that rate the particular day.

berm—outside or downhill side of a ditch or trench.

blind area—an area that cannot be seen from an observation point under favorable atmospheric conditions.

blowup—sudden increase in fire intensity or rate of spread sufficient to preclude direct control or to upset existing control plans. It is often accompanied by a violent convection and other characteristics of a firestorm.

board of review—board or committee selected to review results of fire control action on a given unit, or the specific action taken on a given fire, in order to identify reasons for both good and poor action and to recommend or prescribe ways and means of doing a more effective and efficient job. Also termed "critique" when the review is made by the fire crew.

booster pump—a pump of less than 500 gpm (U.S.), carried as an integral part of fire apparatus and used to supply streams through hose in ¾-, 1-, and 1½-inch sizes. On pumping apparatus the standard pump (500-1500 gpm) usually supplies the "booster," or small line, hose. In wildfire service, "booster pumps" are not true booster pumps (stationary building fire pumps used to boost the pressure head in water supplied to provide pressures more effective for fire service) but are added to a hose lay to reestablish pressure and volume.

booster reel—a reel for the booster hose mounted on fire apparatus. This reel usually carries a 1-inch or ¾-inch hose and frequently contains an electric rewind mechanism.

break a coupling—to detach two pieces of hose by backing the swivel thread off the nipple thread.

break a line—to insert a gate or some other device into a hose line. The line is broken where it comes from the hose body so that a nozzle or one end of the line can be attached to a pump or hydrant outlet. A line may be broken into doughnut rolls after it is no longer needed.

breakover—a fire edge that crosses a control line or natural barrier intended to confine the fire; also called slop over.

British thermal unit (BTU)—measure of the amount of heat required to raise the temperature of 1 pound of water by 1 degree Fahrenheit.

brush—shrubs and stands of short scrubby tree species that do not reach merchantable size; not a synonym for slash or reproduction.

buildup—cumulative effects of drying during a preceding period on the current fire danger and acceleration of a fire with time.

burning conditions—the state of the combined factors of environment that affect fire in a given fuel association.

burning index (BI)—a number related to the amount of effort needed to contain a fire in a particular fuel type within a rating area.

burning out—setting fire inside a control line to consume fuel between the edge of the fire and the control line when attack is indirect or parallel; also called firing out or clean burning. *See* backfiring and firing out.

bushwacker—a fire fighter assigned to a suburban fire company doing frequent duty fighting brush and grass fires. The term is derived from the practice of beating out brush fires with brooms before the days of portable pumps.

C

calculation of probabilities—evaluation of all exisitng factors pertinent to probable future behavior of a going fire and of the potential ability of available forces to carry out control operations on a given time schedule.

campaign fire—a wildfire of such complexity as to require a large organization and several days or weeks to control.

campfire—started for cooking or for providing light and warmth and that spreads sufficiently from its source to require action by a fire control agency.

captain—the officer in charge of a fire department company or station or any other position of comparable responsibility in the department. This is the highest rank in an individual company unit.

carried wet—booster hose carried full of water during mild weather to speed discharge of water on fire without filling or priming from tank.

cat boss—a person responsible for supervising one or more tractor operators and helpers to get efficient and productive use of the machines in constructing a fire line or in mopping up.

catface—a wound on the surface of a tree where healing has not reestablished the normal cross section. *See* fire scar and fire wound.

cat line—a fire line constructed by a tractor with a bulldozer or scraper.

causes of fires—for statistical reporting purposes the U.S. Forest Service has grouped wildfires into nine general causes: campfires, debris-burning fires, incendiary fires, lightning fires, equipment use fires, railroad fires, smoker fires, fires caused by children, and miscellaneous fires.

center firing—a technique of broadcast burning in which fires are set in the center of the area to create a strong draft. Additional fires are then set progressively nearer the outer control lines as indraft builds up to draw them in toward the center. *See* simultaneous ignition and area ignition.

centrifugal pump—a pump which uses impellers to throw water by centrifugal force. For many uses this is a much more flexible type of pump than a positive displacement pump. Most of the 500- to 1500-gpm fire department pumpers built in the past 20 years are of the centrifugal type.

chain—a standard unit of linear measurement (66 feet) used in surveying.

chain of command—the order of rank and authority in the fire service, for example, commissioner, chief of department, deputy chief, district chief, captain, lieutenant, and fire fighter. The senior man in each grade exercises command upon arrival until relieved by a superior officer. An example of chain of command in wildfires is: fire boss, line boss, division boss, sector boss, crew boss, tanker boss, tractor boss, firing boss, fire fighter.

charged line—a line of hose filled with water ready for use and under pressure.

check valve—a valve designed to permit flow in only one direction. For example, a check valve will open to permit water to flow on demand from a water tank on a piece of apparatus but will close to prevent inadvertent overfilling of the tank from the fire pump. With fire-fighting equipment a "clapper valve" is used as a check valve.

chief—a fire department officer of greater than company officer rank.

class of fire—size of wildfire: class A, a fire of ¼ acre or less; class B, a fire of more than ¼ acre but less than 10 acres; class C, a fire of 10 acres or more but less than 100 acres; class D, a fire of 100 acres or more but less than 300 acres; class E, a fire of 300 or more acres but less than 1,000 acres; class F, a fire of 1,000 acres or more but less than 5,000 acres; class G, a fire of 5,000 acres or more.

clean burning—*See* burning out.

closed area—area in which specified activities such as burning or entry are temporarily prohibited because of acute fire hazard.

cloudy—the adjective classification of the sky when three-fifths or more of the sky is obscured by clouds.

coarse fuels—*See* heavy fuels.

cold trailing—a method of controlling a partly dead fire edge by carefully inspecting and feeling with the hand to detect any fire, digging out every live spot, and trenching any live edge.

combination nozzle—a nozzle designed to provide either a solid stream or a fixed spray pattern suitable for wildfires or flammable liquid fires; an all-purpose nozzle.

combustible—a material or structure that will ignite and burn at temperatures ordinarily encountered in fires; technically, a material that, when it is heated, will give off vapors that, in the presence of oxygen, may be oxidized and consumed by fire.

condition of vegetation—state of growth or degree of flammability of vegetation that forms part of a fuel complex. "Herbaceous condition" is used in referring to herbaceous vegetation alone. In grass areas, minimum qualitative distinctions for stages of annual growth are usually green, curing, and dry or cured.

conduction—heat transfer by direct contact or through an intervening heat conducting medium; transmission through a conductor.

conflagration—a raging, destructive fire, often used to connote a fire with a fast-moving front. *Compare* firestorm.

control—a term used by wildfire protection agencies indicating the overall program of control and suppression of fire losses as well as the control of individual fires. The term is seldom used in connection with structural fire fighting. A wildfire is "under control" when it no longer threatens additional destruction and has reached the phase in which mop-up can begin. *See* suppression.

control a fire—to complete a control line around a fire, any spot fires therefrom, and any interior islands to be saved; to burn out any unburned area adjacent to the fire side of the control lines, and cool down all hot spots that are immediate threats to the control line until the lines can be expected to hold under foreseeable conditions.

control center—a communications or dispatch center used for the control of emergency operations including civil defense and mutual aid. *See* fire headquarters.

control force—personnel and equipment used to control a fire.

control line—an inclusive term for all constructed or natural fire barriers and treated fire edge used to control a fire.

control time—*See* elapsed time.

controlled burning—burning conducted or permitted for the purpose of abating hazards and removing undesirable growth; also, a backfire set to remove fuel from the path of an uncontrolled fire.

controlling nozzle—a shut-off nozzle that permits the nozzleman to open or shut the nozzle at will or to adjust the pattern of the stream.

convection—heat transfer by a circulating medium, either gas or liquid, as air heated by a fire circulates and transfers heat to distant objects.

convective column—rising warm air above a continuing heat source or fire.

corral a fire—to surround a fire and any spotfires therefrom with a control line that becomes the ultimate fire edge.

counterfire—fire set between main fire and backfire to hasten spread of backfire. Also called draft fire. The act of setting counterfires is sometimes called front firing or strip firing, and in European forestry, is synonymous with backfire.

cover type—the designation of a vegetation complex described by dominant species, age, and form.

creeping—fire burning with a low flame and spreading slowly. *See* smoldering, running, spotting, and backing.

crew boss—a supervisory person in charge of usually 5 to 30 fire fighters and responsible for their performance, safety, and welfare for the duration of their assignment; sometimes called a foreman.

crown fire—a fire that advances through tops of trees or shrubs more or less independently of the surface fire. Sometimes crown fires are classified as either running fires or dependent fires in order to distinguish their degree of connection with the surface fire. *See* surface fire and ground fire.

crown-out—fire burning principally as a surface fire that intermittently ignites the crowns of trees or shrubs as it advances.

D

daily activity level—a subjective estimate of the degree of activity of a potential man-caused fire source relative to that which is normally experienced. Five activity levels are defined as: none, low, normal, high, and extreme.

daily burning cycle—a standard 24-hour period beginning at 1000 hours. During this cycle, most wildfires undergo a predictable slowing down and speeding up of burning intensity, depending primarily upon the influence of weather and fuel factors. Also called the burning period or fire day.

damage—the total loss caused by fire including indirect losses such as business interruption, loss of future production, loss of grazing, wood products, wildlife habitat, recreation, and watershed values in forest, brush, or grass fires.

dead fuels—naturally occurring fuels in which the moisture content is governed almost entirely by atmospheric moisture (relative humidity and precipitation).

debris-burning fire—spread from a fire originally set for the purpose of clearing land, or for rubbish, garbage, range, stubble, or meadow burning. This does not include incendiary fires, lumbering fires, or hazard reduction on rights-of-way or common carrier railroads.

deep-seated fire—a fire that has burrowed into baled stocks, grain storage, and other combustibles as contrasted with a surface fire. A fire that has gained headway and built up heat in a structure so as to require greater cooling for extinguishment. Deep charring of structural members; a stubborn fire; a fire burning in deep duff, muck, or peat.

detection—the act or system of discovering and locating a fire.

dew point—the temperature to which a parcel of air must be cooled to reach saturation.

difficulty of control—*See* resistance to control.

direct method—a method of suppression that treats the fire or all its burning edges as a whole by wetting, cooling, smothering, or chemically quenching the fire or by mechanically separating the fire from unburned fuel.

disaster—a catastrophe caused by nature, hostile action, or an abnormal emergency usually developing from neglect or failure to provide a reasonable degree of community fire protection or failure to use modern fire-fighting methods commensurate with modern hazards. Disaster, neglect, and failure are often synonymous, while preparedness is an antithesis of disaster.

discovery—determination that a fire exists. In contrast to detection, location of a fire is not required in discovery.

dispatch—ordering a fire company to respond to a certain location or signal, usually one to which it was not previously assigned to respond on a given alarm.

dispatcher—a person who receives reports of discovery and status of fires, confirms their locations, takes action promptly to provide the men and equipment likely to be needed for control in first attack, and sends them to the proper place. For additional needs, he acts on orders from the fire boss.

diurnal—the periodic changes between daytime and nighttime conditions of temperature, relative humidity, wind, and stability.

division—a unit of a complex fire perimeter between designated relief, drainage, or cultural features organized into two to four sectors for control. Usually planned so that it can be personally and completely inspected by the division boss twice per shift.

division boss—a supervisory staff member responsible for all suppression work on a fire division under general instructions from the fire boss or line boss acting for him.

double doughnut—two lengths of hose rolled side by side or a single length (usually 1½ inch hose) rolled into two small coils for convenient handling.

double female—a coupling device having two female swivel couplings to permit attaching of male hose nipples when lines are laid with couplings in opposite or reverse directions.

double jacket hose—fire hose having two cotton or other fiber jackets outside the rubber lining or tubing.

double male coupling—a device having two male thread nipples for connecting hose and for connecting two female swivel couplings.

doughnut roll—a 50- or 100-foot length of 1½ inch hose or a 50-foot length of 2½ inch hose rolled up for easy handling. There are various ways of forming the doughnut. A convenient one has both couplings close together with the male thread protected by the swivel.

draft—drawing water from static sources into a pump which is above the level of the water supply. This is done by removing the air from the pump and allowing atmospheric pressure (14.7 psi at sea level) to push water through a noncollapsible suction hose into the pump. It is sometimes also necessary to draft from a booster tank.

drift smoke—smoke that has drifted from its point of origin and has lost any original billow form.

drill—practicing fire-fighting evolutions such as laying hose, raising ladders, and operating pumps to develop teamwork and proficiency.

drought—a period of moisture deficiency, extensive in area and time.

dry-bulb temperature—the temperature of the air.

dry lightning storm—a lightning storm with negligible precipitation reaching the ground.

duff—partly decomposed organic material of the forest floor beneath the litter of freshly fallen twigs, needles, and leaves. *See* litter.

E

edge firing—a technique of broadcast burning in which fires are set along the edges of an area and allowed to spread to the center.

ejector—*see* suction booster.

elapsed time—total time used to complete any given step or steps in fire suppression, such as discovery, reporting, travel, control, mop-up, and patrol.

elevation loss (EL)—loss of pressure caused by raising water through hose or pipe to a higher elevation. The loss is roughly equal to 1 pound for every 2 feet of increase in elevation above the pump. It is measured in psi.

energy-release component (ERC)—a number related to the rate of heat release (BTUs per second) per unit area (square foot) within the flaming front at head of a moving fire. The expression differs from intensity but is indicative of how "hot" a fire is burning.

equilibrium moisture content (EMC)—the moisture content that a fuel particle would attain if it is exposed for an indefinite period in an environment of specified constant temperature and humidity. When a fuel particle has reached its EMC, there is no net exchange of moisture between it and its environment.

equipment fire—resulting from the use of equipment.

evolution—an agreed operational sequence requiring teamwork and covering various basic fire-fighting tasks such as the placement of hose and ladders.

exhaust primer—a device using the exhaust cylinder for priming a centrifugal pump.

exposure—property that may be endangered by a fire in another structure or by a wildfire. In general, property within 40 feet of a fire may be considered to involve exposure hazard, although, in very large fires, danger may exist at much greater distances.

extra period fire—not controlled by 10:00 A.M. of the day following the discovery.

F

false alarm—a reported smoke or fire requiring no suppression, e.g., brush burning under control, mill smoke, false smoke. *See* false smoke.

false smoke—any phenomenon likely to be mistaken for smoke, such as gray cliffs, sheep driveway, road dust, or fog.

feeling for fire—examining burned material after fire is apparently out and feeling with bare hands to find any live coals. *See* cold trailing.

female coupling—a swivel coupling made to receive a male hose nipple of approximately the same thread pitch and diameter. *See* male coupling.

fine fuel moisture (FFM)—an adjustment to the one-hour time-lag fuel moisture which compensates for the presence of living plant material in fine fuels.

fine fuels—the complex of living and dead herbaceous plants and dead woody plant materials less than ¼ inch in diameter. *See* flash fuels.

fingers of a fire—the long narrow tongues of a fire projecting from the main body.

fire—rapid oxidation of combustible materials resulting in light and heat. The type of fires with which fire fighters are concerned may be termed "unfriendly" as opposed to hearth fires or furnace fires which are "friendly."

fire behavior—the manner in which fuel ignites, flame develops, and fire spreads and exhibits other phenomena.

fire boss—the person responsible for all suppression and service activities on a wildfire. Primary responsibilities are to develop control plans and organize and direct the fire suppression organization in such manner that the fire is completely and efficiently controlled. He may carry out all responsibilities alone or assign prescribed line and staff duties to subordinates.

firebrand—any source of heat, natural or man-made, which is capable of igniting wildfire fuels.

firebreak—a natural or constructed barrier utilized to stop or check fires that may occur or to provide a control line from which to work; sometimes called a fire lane. It also exists in urban areas where structures have been removed to reduce the likelihood of conflagrations.

fire camp—a camp used to accommodate men and equipment while suppressing a fire; also called base camp, side camp, or fly camp, depending on location and function.

fire coat—a waterproof protective coat having ring snap fasteners for quick donning, wrist protectors, and, frequently, a detachable winter lining. It is also called bunker coat or turnout coat, but never referred to as a "raincoat" which lacks protective features essential to a good fire coat.

fire control—the art of controlling an unwanted or "unfriendly" fire as contrasted with fire prevention, or complete extinguishment. In wildfires fire control refers to all activities to protect an area from fire including prevention, presuppression, and suppression; similar to the general term "fire protection."

fire-control equipment—all tools, machinery, special devices, and vehicles used in fire control, but excluding structures.

fire-control improvements—the structures primarily used for fire control, such as lookout towers, housing, telephone lines, radio stations, and roads.

fire-control planning—the systematic technological and administrative management process of designing organization, facilities, and procedures to protect land area from fire.

fire cooperator—a local person or agency who has agreed in advance to perform specified fire control services and who has received advanced training or instructions in giving such service. Also called cooperator, planned cooperator, fire warden, fire agent, per diem guard, etc.

fire damage—the loss caused by fire, expressed in money or other units. This damage figure includes all indirect losses, such as reduction in future values produced by the area, as well as direct losses of cover, improvements, wildlife killed or consumed by fire, etc.

fire danger—result of both the constant and variable fire danger factors, which affect the inception of, spread of, difficulty in controlling, and the damage caused by fires.

fire danger rating—a fire control management system that integrates the effects of selected fire danger factors into one or more qualitative or numerical indices of current protection needs.

fire danger rating area—a geographical area where the fire danger throughout is adequately represented by that measured at a single fire danger station. It is relatively homogeneous in climate, fuels, and topography.

fire devils—relatively small cyclonic effects observed principally during forest and brush fires and also in free-burning structural fires. These develop as heated gases rise in the form of a vortex, resulting from areas of low pressure into which cooler gases and products of combustion rush. In very large fires, burning debris may be carried upward for considerable distances by such thermal columns. *See* fire whirlwind.

fire district—a rural or suburban fire organization, usually tax supported, that maintains fire companies and apparatus. It is also called a fire protection district. In some areas the two terms are largely synonymous.

fire duty—actual physical engagement in fire-fighting service as distinguished from staff work at headquarters or maintenance divisions; work at an individual fire done by an individual fire fighter or by a company.

fire edge—the boundary of a fire at a given moment.

fireguard—fireman, lookout man, fire patrolman, and others employed for prevention, detection, and suppression of fires. Usually short-term employees.

fire headquarters—a control center for operations against a particular wildfire, established in the field close by the fire. It is also called fire camp.

fire line—the part of a control line that is scraped or dug to mineral soil or that is made by a continuous strip of the surface wet down with water. It is sometimes called a fire trail. It is not to be confused with a control line.

fireman—a fire department member whether permanent (full-paid), call (part-paid), or volunteer (unpaid); a fire fighter. Other names used in referring to fireman are smoke chasers, fire patrolmen, and lookouts.

fire pack—a one-man unit of fire tools, equipment, and supplies prepared in advance to be carried by wildfire fighters.

fire prevention—that part of fire protection activities exercised in advance of the outbreak of fire to prevent such outbreaks and to minimize loss if fire does occur.

fire progress map—a map maintained on a large fire to show at given times the location of the fire, deployment of suppression forces, and progress of suppression.

fireproof—something that is not burnable, or to treat an area, hazard, road, etc., to reduce the danger of fires starting or spreading, e.g., to fireproof a roadside or campground.

fire protection—the science of reducing loss of life and property by fire, including both fire prevention and fire extinguishment by public or private means; also, the degree to which such protection is applied.

fire protection district—a rural or suburban fire district, usually tax supported, that purchases fire protection from a nearby fire department or may own rural fire apparatus. *See* fire district.

fire pump—a stationary water pump listed or approved for private fire service, usually gasoline, diesel, or electric engine. This pump is available with various characteristic curves and volume ratings. It should not be confused with a "pumper" which is an automobile pumping engine used for fire department service. *See* portable pump.

fire safety officer—a staff person responsible for identifying the accident and health hazards to fire suppression forces and for advising the fire boss on means of keeping the hazards at a minimum.

fire scar—a healing or healed wound on a woody plant caused or accentuated by fire; also, the scar made on a landscape by fire. *See* fire wound, catface.

fire season—the time of the year when fires are likely to occur, spread, and do sufficient damage to warrant organized fire control.

fire service—the organized fire prevention and fire-fighting service; its members, individually and collectively; allied organizations assisting fire-fighting and fire prevention agencies.

fire setting—the starting of a fire usually with malicious or illegal intent.

firestorm—violent convection caused by a large continuous area of intense fire, often characterized by destructively violent surface indrafts near and beyond the perimeter, and sometimes by tornadolike whirls.

fire suppression organization—the management structure designed to enable line and staff duties of the fire boss to be carried out with increases in size and complexity of the suppression job; all supervisory and facilitating personnel assigned to fire suppression duty under the direction of a fire boss.

fire tool cache—a supply of fire tools and equipment assembled in planned quantities or standard units at a strategic point for exclusive use in fire suppression.

fire triangle—three factors (oxygen, heat, and fuel) necessary for combustion and flame production. When any one of these factors is removed, flame production ceases.

fire truck—any piece of motorized fire-fighting apparatus.

fire warden—an officer in charge of fire control in a given area, usually a volunteer civilian, though sometimes designates a regular state employee.

fire weather—weather conditions favorable to the start and rapid spread of fire. In wildfires this generally includes high temperatures combined with strong winds and low relative humidity. With structural fires it may also include strong winds combined with low temperatures which will encourage forced heating and low relative humidity in buildings and contents.

fire weather forecast—a weather prediction specially prepared for use in wildfire control.

fire weather station—a meteorological station specially equipped to measure weather elements that have an important effect on wildfire control.

fire whirlwind—a revolving mass of air caused by a fire that may have sufficient intensity to snap off large trees. *See* fire devils.

fire wound—fresh or healing injuries of the cambium of a woody plant caused by fire. *See* fire scar.

firing out—the act of setting fire to fuels between the control line and the main fire in either a backfiring or burning out operation; also called firing. *See* burning out.

first attack—the first suppression work on a fire.

flails—various swatters for smothering and beating out grass fires. These are most effective on short grass fires, though the use of water is preferred.

flame—the light from burning gases and incandescent particles during a fire.

flaming front—that zone of a moving fire within which the combustion is primarily flaming. Behind the flaming front, combustion is primarily glowing.

flammability—the relative ease with which fuels ignite and burn, regardless of the quantity of the fuels.

flank fire—a fire set along a control line parallel to the wind and allowed to spread at right angles to it.

flanking—attacking a fire by working along the flanks, either simultaneously or successively from a less active, or anchor, point and endeavoring to connect the two lines at the head.

flanks—in structural fires the exposures on two sides of a structure as distinguished from the front (usually facing the street) and the rear (usually an alleyway or backyard). In wildfires the parts of a fire's perimeter which are roughly parallel to the main direction of spread.

flare-up—any sudden acceleration of fire spread or intensification of the fire. Unlike blowup, a flare-up is of relatively short duration and does not radically change existing control plans.

flash fuels—fuels such as grass, leaves, draped pine needles, fern, tree moss, and some kinds of slash which ignite readily and are consumed rapidly when they are dry; also called fine fuels. *See* heavy fuels.

flashover—rapid combustion and/or explosion of unburned gases trapped at some distance from the main fire front. This usually occurs only in poorly ventilated location. It is more commonly associated with structural fire behavior.

foam—a chemical fire-extinguishing mixture that forms bubbles on application, greatly increasing the mixture volume. It adheres to the fuel and reduces combustion by cooling and moistening and by excluding oxygen. Compounds introduced into a stream of water by special nozzles or proportioning devices to develop a stream of tenacious foam capable of smothering fires, especially those involving flammable liquids;

the foam produced by such equipment; also, a portable or wheeled foam extinguisher.

fog—a jet of fine water spray discharged by spray nozzles, used to extinguish fire; generally considered most efficient at nozzle pressures of approximately 100 psi. High pressure fog is water spray delivered through gun-type nozzles attached to small hose supplied by pumps discharging at pressures above 250 psi, the normal maximum pressure for standard pumpers. Usually the pressures are between 400 to 600 psi at the nozzle.

follow-up—the act of supporting the first man or men who go to a fire by sending additional men or equipment to facilitate suppression. Sometimes called reinforcement.

forest fire—any wildland fire not prescribed for the area by an authorized plan (as used by national and state fire control agencies). *See* wildfire.

forest protection—prevention and control of any cause of potential forest damage.

forest utilization fire—resulting directly from timber harvesting, harvesting other forest products, and forest and range management, except use of equipment, smoking, and recreation as related to the above activities.

free-burning—the condition of a fire or part of a fire that has not been checked by natural barriers or by control measures.

friction loss (FL)—loss of pressure created by the movement of water in fire hose, pipe, or fittings. It is due to the turbulence caused by moving water against the interior surface of the pipe or hose. Friction loss normally is measured in psi per 100 feet of hose or pipe. *See* pressure loss and elevation loss.

fuel—combustible material adding to the magnitude or intensity of a fire or combining with oxygen to contribute to the burning process. Without fuel, there is no possibility of a fire.

fuelbreak—a wide strip or block of land on which the native vegetation has been permanently modified so that fires burning into it can be more readily extinguished. It may or may not have fire lines constructed in it prior to fire occurrence.

fuel class—a group of fuels possessing common characteristics. In the National Fire Danger Rating system, dead fuels are grouped according to their time lag (1-, 10-, and 100-hour) and living fuels are grouped by whether they are herbaceous or woody.

fuel model—a simulated fuel complex for which all the fuel descriptors required for the solution of the mathematical fire spread model have been specified.

fuel moisture analog—a device which simulates the response of the moisture content of specific classes of dead fuels when exposed in the same environment. Examples are basswood slats which respond like one-hour time lag fuels and ½-inch ponderosa pine dowels which react like ten-hour time lag fuels. An analog may also be constructed of inorganic materials or, in a broad sense, may consist of computational procedure such as used in the National Fire Danger Rating system to determine the one-, ten-, and one-hundred-hour time lag fuel moisture.

fuel moisture content (also fuel moisture)—the quantity of moisture in fuel, expressed as a percentage of the weight when thoroughly dried at 212°F.

fuel moisture indicator stick—a specially prepared stick or set of sticks of known dry weight continuously exposed to the weather and periodically weighed to determine changes in moisture content as an indication of moisture changes in wildfire fuels.

fuel type—an identifiable association of fuel elements of distinctive species, form, size, arrangement, or other characteristics that will cause a predictable rate of fire spread or control difficulty under specified weather conditions. Land areas are divided into units according to their fuel characteristics with respect to rate of spread and difficulty of establishing and holding a control line.

G

gage—a device giving visual indication of pressure, motor speed, etc.

gallon—231 cubic inches of water weighing 8.336 pounds in a U.S. gallon. An Imperial gallon is 1.201 U.S. gallons.

gate valve—a controlling valve for hose or pump outlet or at a large caliber nozzle.

going fire—fire between the time it starts and the time it is declared out.

gpm—gallons per minute, the measure of water flow in fire fighting. It is used to measure the output of fire department pumpers, hose streams, nozzles, hydrants, water mains, etc.

grass fire—fire involving dried grass which may present a severe exposure hazard to valuable property.

gravity tank—a water storage tank for fire protection and sometimes community water service that supplies water by gravity pressure. A water level of 100 feet provides a static pressure head of 43.3 psi minus friction losses in piping when water is flowing.

green fuels—living vegetation of high moisture content that ordinarily will not burn unless it is first dried out by excessive heat or lack of rain.

gridiron—to search for a small fire or evidence by systematically traveling over an area on parallel courses or grid lines.

gross vehicle weight—the actual vehicle weight which is the sum of the weights of chassis, body, cab, equipment, water, fuel, crew, and all other load.

ground fire—a fire at ground level as contrasted to a crown fire or structural fire at upper levels. In a wildfire situation, a ground fire is one that consumes the organic material such as peat or muck below the surface litter or duff. *See* surface fire and crown fire.

gutter trench—a ditch dug on a slope below a fire, designed to catch rolling burning material. *See* undercut line.

H

handie talkie—a two-way radio hand set used for fireground communications.

hard suction—noncollapsible suction hose for drafting water from static sources lower than the pump. In wildfire pumpers, the normal hose sizes used are 2 and 2½ inches.

hazard—a condition of fire potential defined by arrangement, size, type of fuel, and other factors which form a special threat of ignition or difficulty of extinguishment. A *fire hazard* specifically refers to fire seriousness potential and a *life hazard* refers to danger of loss of life from fire.

hazard reduction—any treatment of a hazard that reduces the threat of ignition and spread of fire.

head fire—a fire spreading or set to spread with the wind.

head of a fire—the most rapidly spreading portion of a fire's perimeter, usually to the leeward or up slope.

heat—temperatures above the normal atmospheric, as produced by the burning or oxidation process. While there is some heat at all temperatures above absolute zero where molecular action ceases, the fire fighter is concerned with the abnormal aspects of heat that induce physical discomfort, preventing close approach to the fire and causing exposed combustible properties to reach an ignition point.

heat transfer—movement and dispersion of heat from a fire area to the outside atmosphere by converting water fog particles to steam, thus expanding the volume 1,650 times and creating a slight pressure carrying the heat and heated water vapor outside. The movement of heat from one object to another is by radiation, convection, conduction, and mass transport of embers.

heavy fuels—fuels of large diameter such as snags, logs, and large limbwood, which ignite and are consumed more slowly than flash fuels; also called coarse fuels. *See* flash fuels.

held line—a control line that still contains the fire when mop-up is completed, excluding lost line, natural barriers not backfired, and unused secondary lines.

heliport—a permanent or semipermanent base for helicopters.

helmet—the regulation protective headgear for fire fighters usually carrying insignia of unit and rank, traditionally of leather but also of aluminum and various composition materials. It is also called a fire hat. Hard hats are commonly worn on wildfires.

herb—a plant which does not develop woody, persistent tissue but is relatively soft and succulent. These include grasses, forbs, and ferns.

herbaceous fuels—undecomposed materials, living or dead, derived from herbaceous plants.

herbaceous vegetation condition—the percent, by volume, of the fine fuels which are living.

higbee cut—the first or outside thread of a hose coupling or nipple removed to prevent crossing or mutilation of threads.

high pressure fog—a small capacity spray jet produced at very high pressures and discharged through a small hose having a gun-type nozzle; also, the equipment producing high-pressure fog. This was originally an adaptation of the orchard sprayer for fire fighting, but more recently many fire department pumps have been built with additional pressure stages which provide increased pressure instead of volume.

hitch—one of a variety of methods of securing objects with rope; also, to take a hitch around a hydrant with hose or with a short rope so that the hose will not pull away when the apparatus proceeds toward the fire. Formerly, a hitch was a method of harnessing fire horses as in "a three-horse hitch."

holdover fire—a fire that remains dormant for a considerable time; also called hangover fire or sleeper fire.

hose lay—the arrangement of connected lengths of fire hose and accessories on the ground beginning at the first pumping unit and ending at the point of water delivery. *See* progressive hose lay, simple hose lay.

hose load—methods of loading hose into the hose body of a fire truck such as "accordian load."

hoseman—a fire fighter assigned chiefly to hose duties with a hose, pumper, or pumper-ladder company or any fire fighter performing such duties at a fire; distinguished from "ladder man" or "pump operator." Generally, company officers are excluded from this general term even though they may, from time to time, assist in handling hose.

hose rack—item for storing or drying fire hose.

hose reel—hand-drawn hose reels used in industrial fire protection; permanently mounted reels for fire hose installations; also, booster reels on fire apparatus.

hose size—classified by nominal internal diameters. Nominal diameters are: ¾, 1, 1½, 2½, 3, and 3½ inches. *See* ID and OD.
hotshot crew—an intensively trained fire-fighting crew used primarily as followup for first attack forces.
hot spot—a particularly active part of a fire.
hot spotting—checking the spread of fire at points of more rapid spread or special threat. This is usually the initial step in prompt control with emphasis on first priorities.
humidity—a measure of the water vapor content of the air.

I

ID—the internal diameter of a tube, conductor, or coupling, as distinguished from its OD (outside diameter). Fire hose sizes are classified by a nominal internal diameter. The diameter of fire hose will vary somewhat when carrying water at various pressures.
ignition—the beginning of flame propagation or burning; the starting of a fire.
ignition component (IC)—a number related to the probability that a spreading fire will result if a firebrand encounters fine fuel.
ignition temperature—the temperature at which a fuel ignites and flame is self-propagating.
incendiary fire—willfully set by anyone to burn vegetation or property not owned or controlled by him and without consent of the owner or his agent.
incidence—the range of occurrence of fires.
incipient—a fire of minor consequence or in the initial stages.
increaser—increasing coupling used on hose, pump, or nozzles to permit connection of a larger size of hose.
independent action—suppression action by other than the regular fire control organization or cooperators.
indirect method—a method of suppression in which the control line is located along natural firebreaks, along favorable breaks in topography, or at considerable distance from the fire and the intervening fuel is burned out.
individual assignment method—a system for organizing men to control a fire's perimeter in which each crewman is assigned a specific section of the control line where he is responsible for all suppression jobs from hot spotting to mop-up.
information officer—a staff officer in a fire suppression organization responsible for information releases to the public on a fire situation.
initial attack—the first point of attack and hose lines employed to prevent further extension of fire; safeguarding lives while additional lines are being laid and placed in position; the first attack on a fire.
initiating fire—a wildfire exhibiting reasonably predictable behavior (no crowning or spotting).
insolation—solar radiation received at the earth's surface.
inspection—a fire prevention inspection conducted for the purposes of familiarizing fire department personnel with locations where fires may occur; removal of common or special fire hazards and enforcement of fire prevention ordinances and regulations.
instrument shelter (also thermoscreen)—a naturally or artificially ventilated structure, the purpose of which is to shield temperature-measuring instruments from direct sunshine and precipitation.
intensity—the rate of heat release (BTUs per second) per unit length of fire front (foot).

inversion—a horizontal layer of air through which temperature increases with increasing height.

investigation—in fire fighting, inspection for the presence of smoke, heat, steam or any other indication of fire by an officer; investigation of fire reported extinguished by occupants; checking into suspicious fires and determining the cause and circumstances of a fire of unknown origin. Fire-fighting apparatus should always be dispatched to an investigation of possible fire because hidden fires when they are located are frequently among the most difficult to fight and extinguish.

IR crew—a highly trained interregional fire suppression crew (federal agencies).

iron pipe thread (IPS)—a standard system of threads for connecting various types of rigid piping. These threads are much finer and more difficult to connect in the field than National Standard (NS) fire hose threads. Unfortunately, some fire appliances are equipped with iron pipe thread requiring use of adapters to connect devices of the same nominal size equipped with standard fire threads.

J

jump spot—a selected landing area for smoke jumpers.

K

kink—to bend hose back upon itself several times to cut off flow of water so that another length can be added or a burst section replaced; also, an unwanted bend that restricts the flow of water.

knock down—in fire fighting, to reduce the flame or heat on the more vigorously burning parts of a fire edge.

L

land occupancy fire—started in order to clear land for agricultural purposes, industrial establishment, construction, maintenance and use, or rights-of-way.

lay—a method of placing a hose from fire apparatus such as a "straight lay" or "reverse lay."

lead plane (bird dog)—aircraft used to direct the tactical deployment of air tankers. In many fire control operations of moderate complexity, the lead plane pilot also fills the air attack boss position.

legitimate smoke—smoke from any authorized use of fire, such as locomotives, industrial operations, permitted debris-burning, etc.

life safety—the first responsibility of the fire-fighting service. Although usually the duties of the ladder company, rescue and life safety must always be in the minds of the first unit to arrive.

lift—distance in feet of elevation between a static source of water and the suction chamber of a fire department pumper.

light burning—periodic broadcast burning to prevent accumulation of fuels in quantities that would cause excessive damage or difficulty in suppressing in the event of accidental fires.

lightning activity level—a number, on a scale of 1 to 5, reflecting the frequency and character of a cloud-to-ground lightning either forecasted or observed. The scale is exponential, based on powers of 2; a lightning activity level of 3 indicates twice the lightning as 2; a 4 indicates twice that of 3, etc.

lightning fire—a wildfire caused directly or indirectly by lightning.

lightning risk (LR)—in the National Fire Danger Rating system, a number related to the expected number of cloud-to-ground lightning strikes to which a protection unit is expected to be exposed during the rating day. The lightning risk value used in the calculation of the occurrence index includes an adjustment for lightning activitiy experienced during the previous day to account for possible "holdover" fires.

line—fire hose; a rope (life line); a small piece of rope used to secure hose (hose line); also sometimes used for fire line in a wildfire situation.

line boss—a supervisory officer in a fire suppression organization responsible for executing the fire suppression plan adopted by the fire boss. He is employed in some stages of organization of large forces, and may act as coordinator between two or more divisions or may supervise three to four sector bosses if no divisions have been established.

lined hose—fire hose composed of one or two woven jacketed outside layers with an inside rubber lining to reduce friction loss and eliminate water leakage.

line firing—setting fire to only the border fuel immediately adjacent to the control line. *See* strip firing.

line locator—in wildfire situations, the person responsible for on-the-ground location of fire lines to be constructed.

linen hose—an unlined fire hose, formerly used for first aid standpipes or wildfire service, consisting of a linen or flax fabric without a rubber lining. Under present practice, cotton or synthetic fiber is more commonly used for unlined hose. *See* unlined fire hose.

line scout—a person in a fire suppression organization assigned to scouting duties on the fire line. *See* scout.

litter—the top layer of the forest floor, composed of loose debris of dead sticks, branches, twigs, and recently fallen leaves or needles, little altered in structure by decomposition; also a stretcher for carrying the injured. *See* duff.

live burning—progressive burning of green slash as it is cut. *See* progressive burning.

live line (or reel)—a hose line or reel carried preconnected to the pump, ready for use without the necessity for making connections to pump or attaching nozzle; a charged line containing water under pressure.

living fuels—naturally occurring fuels in which the moisture content is physiologically controlled within the living plant. The National Fire Danger Rating system considers only herbaceous plants and woody plant material which is small enough to be consumed in the flaming front of an initiating fire, such as leaves and needles, and twigs less than ¼-inch in diameter.

lookout—a person designated to detect and report fires from a vantage point; a lookout station, also termed fire tower.

lookout point—a vantage point selected for fire detection.

lookout tower—a structure enabling a person to be above nearby obstructions and to sight and fix the location of fires. It is usually capped by either a lookout house or observatory.

lugging—a condition of engine overload resulting in failure to develop efficient engine speeds, frequently due to faulty operation such as failure to shift down as required for climbing a given grade or failure to develop efficient engine speed when pumping due to an attempt to provide more water and pressure than the engine can handle.

lumbering fire—resulting from any activity connected with the harvesting or processing of wood for use or sale, including those caused by logging railroads which are not common carriers.

male coupling—the threaded hose nipple which fits in the thread of a female swivel coupling of the same pitch and appropriate diameter; a coupling to which nozzles and other appliances are attached. *See* female coupling.

man-caused risk (MCR)—a number related to the expected number of firebrands originating from human activities to which a protection unit will be exposed during the rating day.

man-passing-man method—a system for organizing men in fire suppression whereby each crewman is assigned a specific task on a specific section of the control line, and when that task is completed he passes other workers in moving to a new assignment.

mattock—a hand tool for digging and grubbing, having a narrow hoeing surface at one end of the blade and either a pick or cutting blade at the other end; used in constructing fire lines.

message—a fire department radio message consisting of contact call, response, text, and acknowledgment.

midship pump—a fire pump located under or behind the driver's seat and supported on the truck frame between the front and rear wheels; usually driven by the truck engine.

mixmaster—a person responsible for providing fire retardants to airborne tankers for forest fire fighting.

mobile—a fire department radio unit on mobile apparatus instead of base stations; usually semipermanently attached to the apparatus.

model arson law—a model law recommended by the Fire Marshals' Association of North America and adopted in most states, dealing with the subject of arson.

model fireworks law—model legislation recommended by the Fire Marshals' Association of North America and adopted by many states regulating display of fireworks.

monthly average—the average number of man-caused fires occurring on a protection unit during a specific calendar month.

mop-up—the act of making a fire safe after it is controlled, such as extinguishing or removing burning material along or near the control line, felling snags, trenching logs to prevent rolling, etc.; one of the secrets to successful wildfire control.

mutual aid—in fire-fighting situations, two-way assistance freely given under prearranged plans or contracts by fire departments of two or more areas on the basis that each will aid the other in time of emergency, providing for joint or cooperative response to alarms near their boundaries.

N

National Fire Protection Association (NFPA)—a nonprofit educational and technical association formed in 1896, headquartered in Boston, Mass., and devoted to the protection from fire of life and property through development of fire protection standards and public education. The association now has an international membership of over 18,000 individuals and organizations, including more than 3,000 fire departments.

National Standard (NS) thread—American national standard fire hose coupling screw thread.

natural barrier—any area where lack of flammable material obstructs the spread of a wildfire, e.g., roads, trails, rockslides, plowed fields, changes in cover type, streams, canals, lakes, cliffs, etc.

normal fire season—a season in which weather, fire danger, and number and distribution of fires are about average; period of the year normally comprising the fire season.

nozzleman—a person assigned to operate a fire hose nozzle, usually on a hand line.

nurse tanker—water tank truck used to supply a pumper stationed at a fire.

O

occurrence index (OI)—a number related to the potential fire incidence within a protection unit.

OD—the outside diameter of a tube, conductor, or coupling distinguished from its ID (internal diameter). The outside diameter is an important feature, particularly with reference to coupling design. For example, the hose bowl of a coupling must fit the OD of the hose jacket and the OD of a hose nipple must fit (with specified tolerance) the ID of the swivel coupling of the same basic thread pattern.

officer—in the fire department, a member with supervisory responsibilities, either a company officer such as captain or lieutenant, or a chief officer.

one-hour time lag fuels—fuels consisting of dead herbaceous plants and roundwood less than about ¼ inch in diameter. Also included is the uppermost layer of needles or leaves on the forest floor.

one-hundred-hour time lag fuels—dead fuels consisting of roundwood in the size range of 1 to 3 inches in diameter and very roughly the layer of litter extending from approximately ¾ inch to 4 inches below the surface.

one lick method—a progressive system or method of building a fire line on a wildfire without changing relative positions in the line. Each man does one to several "licks," or strokes, with a given tool and then moves forward a specified distance to make room for the man behind.

overload—gross vehicle weight in excess of the rated gross vehicle weight specified by the chassis manufacturer or in excess of axle ratings or permissible tire and rim loadings.

oxygen—a gas present in the atmosphere in about 21 percent concentration, which, while it is not combustible, is an essential element in combustion and one of the three parts of the fire triangle. It is also the essential gas in respiration, because the oxidation process is basic to life.

P

paracargo—anything intentionally dropped or intended for dropping from any aircraft.

parallel method—an indirect method of suppression in which a fire line is constructed approximately parallel to, and just far enough from the fire edge to enable men and equipment to work effectively, though the line may be shortened by cutting across unburned fingers. The intervening strip of unburned fuel is normally burned out as the control line proceeds but may be allowed to burn out unassisted where this occurs without undue delay or threat to the line.

patrol—to travel a given route to prevent, detect, and suppress fires; also, to go back and forth watchfully over a length of the control line, during or after its construction, to prevent breakovers, control spot fires, or extinguish overlooked hot spots.

peak monthly average—the highest monthly average calculated for a protection unit. *See* monthly average.

perimeter—the outside boundary of a fire area.

pickup—to take up hose and other fire-fighting equipment that have been used at a fire and to return them to the fire apparatus, thus permitting the company or unit to return to quarters.

pin lug couplings—hose couplings where the lugs are pin shaped.

piston pump—a positive displacement pump using 2, 4, and 6 reciprocating pistons to force water from the pump chamber in conjunction with appropriate action of inlet and discharge valves.

pitot tube—a tube having an opening which is inserted into a stream of water from a nozzle and to which a pressure gage is attached indicating the discharge pressure of the stream.

planning chief (plans chief)—a staff officer in a fire suppression organization responsible for the compilation and analysis of data needed for developing suppression plans.

plow line—a fire line constructed by a fire line plow, usually drawn by a tractor.

pockets—the indentations in a fire edge formed by fingers or slow burning areas.

portable generator—a small electric generator, sometimes carried on fire apparatus, providing power for electric tools and floodlights.

portable pump—a small gasoline driven fire pump designed to be carried on a fire apparatus and capable of supplying either a direct stream through 1½-inch hose or relaying a supply to a standard fire department pumper. Available in various pressure-volume ratings. Performance of about 50 gpm at 90 psi and 100 gpm at 60 psi is common.

positive displacement pump—a rotary, gear, or piston type of pump which moves a given quantity of water through the pump chamber with each stroke or cycle. The positive displacement pump is capable of pumping air and thereby is self-priming.

potable water—water fit for human consumption as in public water systems. Fire departments must use extreme care not to allow untreated water to enter mains furnishing potable water. *See* check valve.

potato roll—a quick method of rolling hose for pickup; rolled as a ball of string.

prairie fire—a fast-burning fire involving any dried grass; often presenting a serious exposure hazard to isolated structures.

precipitation—any or all forms of water particles, liquid or solid, that fall from the atmosphere and reach the ground, usually measured to the nearest one-hundredth of an inch.

preconnected—suction or discharge hose carried connected to pump, eliminating delay occasioned when hose and nozzles must be connected and attached at fire.

prefire planning—surveys of target hazards for the purpose of making advance plans of possible fire-fighting operations.

preparedness—in fire fighting, a condition or degree of being completely ready to cope with a potential fire situation; mental readiness to recognize changes in fire danger and act promptly when action is appropriate.

prescribed burning—skillful application of fire to natural fuels under conditions of weather, fuel moisture, soil moisture, etc. that will allow confinement of the fire to a predetermined area and, at the same time, will produce the intensity of heat and rate of spread required to accomplish certain planned benefits to one or more objectives of silviculture, wildlife management, grazing, hazard reduction, etc. Its objective is to employ fire scientifically to realize maximum net benefits at minimum damage and acceptable cost. This is also called controlled burning.

pressure loss—reduction in water pressure between a pump or hydrant and a nozzle due to expenditure of pressure energy required to move water through a hose, including loss from back pressure, due to elevation and losses in fitting. *See* friction loss.

presuppression—activities in advance of fire occurrence to insure effective suppression action; includes recruiting and training, planning the organization, maintaining fire equipment and fire control improvements, procuring equipment and supplies, and working out arrangements with other agencies. *See* prevention and suppression.

prevention—activities directed at reducing the number of fires that start, including public education, law enforcement, personal contact, and reduction of fuel hazards.

prevention guard—one who helps to prevent fires by contacting people in the protection area and inspecting fire prevention measures and fire equipment of industrial operations; also called prevention patrolman.

primary lookout—a lookout point that must be manned to meet planned minimum visible coverage in a given locality. Continuous service is necessary during the normal fire season and lookout man usually is not sent to fires. Sometimes the term designates a person who occupies such a station, though primary lookout man or observer is more specific in this sense.

priming—filling pump with water when pump is taking water not under a pressure head. Necessary for centrifugal pumps. *See* exhaust primer.

priming pump—a small positive displacement pump or device used to prime a centrifugal pump not capable of displacing air in order to bring water into the pump chamber.

progressive burning—disposal of slash by burning as it is piled; also called swamper burning.

progressive hose lay—a hose lay in which double shutoff Y's are inserted in the main line at intervals and lateral lines are run from the Y's to the fire edge, thus permitting continuous application of water during extension of the lay.

progressive method of line construction—a system of organizing men to build fire line in which they advance without changing relative positions in line. It is variously termed move-up, step-up, bump-up, or functional method of line construction.

project fire—usually refers to a fire requiring manpower and equipment beyond the resources of the protection unit on which it originates.

protection—all activities to protect areas, subject to wildfires; includes prevention, presuppression and suppression.

protection boundary—the exterior boundary of an area within which a given agency has assumed a degree of responsibility for wildfire control. It may include land in addition to that for which the agency has jurisdictional or contractural responsibility.

protection unit—a geographical area which is administratively defined and which is the smallest area for which organized fire suppression activities are formally planned.

protective clothing—a general term for the complete ensemble of fire clothing such as firecoat, boots, helmet, turnout pants, gloves, and gas masks or breathing apparatus.

psi (pounds per square inch) — measurement of pressure (pump pressure, nozzle pressure, friction loss in hose, pressure loss or gain due to elevation, etc.).

psychrometer—an instrument used to measure the amount of water vapor in the air in order to determine the relative humidity.

Pulaski tool—an ax with a medium size sharp grub hoe opposite the ax blade.

pumper—a fire department pumping engine with a rated capacity of at least 500 gpm and carrying hose and other fire-fighting equipment (also a pumper company). In a wildfire situation, a pumper is a motorized unit with a capacity of 50 to 1,500 gallons or more which is capable of pumping while traveling, and which may or may not carry other equipment.

pump pressure—water pressure maintained by operation of a fire department pumper in contrast to pressure available from hydrant or gravity supply only. Modern practice requires that except where high pressure fire mains are provided, the correct pump pressure should be provided on every fire department hose stream commensurate with the requirements of the nozzles employed.

punishment—harsh or injurious treatment. In fire-fighting terminology, punishment refers to physical suffering endured by fire fighters, especially from heat, smoke, and gases, and exertion from handling heavy equipment especially during unfavorable weather.

punk—charred or partly decayed material, such as old wood, in which fire smolders unless it is carefully mopped up and extinguished.

Q

quench—to extinguish a fire by soaking the fuel with water.

R

radioactive—various materials giving off alpha, beta, and gamma particles and short-wave radiations, either due to inherent activity, such as in radium, or due to exposure to materials that are radioactive.

railroad fire—resulting from maintenance of the right-of-way or construction or operation of a common-carrier railroad.

rake—to sweep a fire stream about an area with rapid movement to achieve maximum cooling and to knock down flame; a hand tool used in wildfire control.

rate of spread (ROS)—the relative activity of a fire in extending its horizontal dimensions. It is expressed as rate of increase of the total perimeter of the fire, as rate of forward spread of the fire front, or as rate of increase in area, depending on the intended use of the information. Usually it is expressed in chains per hour or acres per hour for a specific period in the duration of the fire.

rating period—the period of time during which a fire danger rating value is considered valid or representative for administrative or other purposes; normally a 24-hour period, extending from midnight to midnight.

rear of a fire—the portion of the edge of a fire opposite the head; the slowest moving part of the fire.

reburn—subsequent burning of an area in which fire has previously burned but has left flammable fuel that ignites when burning conditions are more favorable; an area that has reburned.

recreation fire—resulting from any recreation use, except smoking.

reducing—the act of attaching smaller hose as a leader to larger diameter hose.

reel—a frame on which hose is wound, now chiefly used for "booster" or small hose reel (¾- or 1-inch hose) supplied by water tank on the apparatus; also, a hand-drawn 2-wheel frame for 2½-inch hose used in industrial plants.

reflective tape—reflective material commonly attached to fire fighters' clothing, tools, and appliances to make them visible at night or in buildings under a source of light.

regulation—a controlling or directing principle or rule set forth in writing. In general, regulations are prescribed by regulatory bodies under authority granted by law but are not laws that have been acted upon by a legislative body. For example, a municipal fire department established under terms of law may have authority to promulgate regulations governing fire hazards not inconsistent with law.

rekindle—reignition due to latent heat, sparks, or embers or due to presence of smoke or steam.

relative humidity—the percentage of moisture in the air compared with the maximum amount of moisture that air will hold at a given temperature. Increase in temperature increases the amount of moisture that air will hold. When relative humidity is below 30 percent out of doors, the fire danger is considered to be serious.

relay—the use of two or more pumpers to move water a distance which would require excessive pressures in order to overcome friction loss if only one pump were employed at the source.

relief valve—a pressure-controlling device which bypasses water at a fire pump to prevent excessive pressures when a nozzle is shut down.

rescue—the saving of a life endangered by fire or accident. The term is generally employed to describe assisting persons unable to help themselves, although it is loosely used to describe assisting persons in danger or trouble.

reserve—apparatus not in first-line duty, but available in case first-line apparatus is undergoing repair or available for use by off-duty fire fighters in abnormal emergency conditions.

residence time—the time required for the flaming zone of a fire to pass a stationary point; also, the width of the flaming zone divided by the rate of spread of the fire.

resistance to control—the relative difficulty of constructing and holding a control line as affected by resistance to line construction and by fire behavior; also called difficulty of control. *See* resistance to line construction.

resistance to line construction—the relative difficulty of constructing a control line as determined by the fuel, topography, and soil.

resources—the manpower and equipment available for control of a wildfire; the natural resources of an area, such as timber, grass, watershed values, recreation values, and wildlife habitat.

respirator—a simple filter mask for individual protection against smoke and fumes, usually without full face mask and possibly lacking laboratory approval for fire service use; not recommended for fire fighting. *See* self-contained breathing apparatus.

response—the act of responding to an alarm; also, the entire complement of men and apparatus assigned to an alarm.

resuscitator—an approved mechanical device for assisting the respiration of unconscious persons, also containing the features of the inhalator.

retardant—any substance that, by chemical or physical action, reduces flammability of combustibles. The rate of spread of the flame front is thereby slowed or retarded.

revolutions per minute (RPM)—used in reference to the speed of engines driving fire department pumps as shown on a tachometer.

risk—the chance of a fire starting as determined by the presence and activity of a causative agent; also, the causative agent itself. In the National Fire Danger Rating system, a number related to the potential number of firebrands to which a given area will be exposed during the rating day.

risk source ratio—in the National Fire Danger Rating system, that percent of the man-caused fires which have occurred on a protection unit that can be charged to a specific risk source.

rocker lug coupling—a hose coupling in which the lugs used for tightening or loosening are semicircular in shape and designed to pass over obstructions.

rotary gear pump—a positive displacement pump employing closely fitting rotors or gears to force water through the pump chamber.

roundwood—boles, stems, or limbs of woody material; that portion of the dead wildfire fuels that is roughly cylindrical in shape.

route—in fire-fighting terminology, a predesigned route to be followed by individual fire companies responding from quarters to the location of a fire. Routes are selected with due consideration to distance, speed of response, traffic safety, paths of travel of other emergency vehicles, and effective approach patterns to the scene of a fire or emergency. When companies are necessarily off-route, the dispatcher may broadcast caution to responding units by radio.

route card—an index card used by a fire alarm dispatcher and frequently carried on rural fire apparatus giving specific directions for responding to individual rural properties and frequently including a description of the property, water supplies available, and any special information pertinent to fire-fighting and rescue operations; also called running cards.

run—response to fire or alarm. This term is a carry-over from the days when fire fighters with hand-drawn apparatus ran to the fire with their engines. Even after the horse-drawn apparatus was introduced, the number of riders was initially limited. In some cases, officers were required to run ahead of their apparatus to clear the way. The number of runs per unit remains an important measure of activity because it is an indication of workload and indicates the frequency with which a unit is out of quarters and must be covered by other companies.

running—behavior of fire spreading rapidly with a well-defined head. *See* smoldering, creeping, spotting, and backing.

running fire—a wildfire spreading rapidly with a well-defined head or front.

rural fire protection—fire protection and fire-fighting problems that are outside of areas under municipal fire prevention and building regulations and that are usually remote from public water supplies. It should be noted, however, that in numerous instances municipal limits have been extended to include large areas which are sparsely settled, while, in other instances, water supplies and other municipal services and utilities cover extensive areas outside of municipal corporation boundaries. Also, an increasing number of counties are now providing municipal services including public fire prevention or protection. Rural fire districts (RFD) organized and operated by counties and townships cover an ever-increasing area for fire protection.

S

safety island—an area used for escape in the event the line is outflanked or in case a spot fire causes fuels outside the control line to render the line unsafe. In firing operations, crews progress so as to maintain a safety island close at hand, allowing the fuels inside the control line to be consumed before going ahead.

safety officer—an officer responsible to the plans chief for the safety and welfare of all fire-fighting personnel.

saw boss—a supervisory officer in a fire suppression organization responsible for the activities of saw crews (using hand or power saws) in cutting snags or logs on a fire or part of a fire.

scout—a staff worker in a fire suppression organization assigned duties of gathering and reporting timely information such as existing location and behavior of a fire, progress in control, and the physical conditions that affect the planning and execution of the suppression job.

scratch line—an unfinished preliminary control line hastily established or constructed as an emergency measure to check the spread of a fire.

seasonal risk class—an objective ranking of protection units within an administrative group, based on the incidence of man-caused fires for at least the past five years.

secondary line—any fire line constructed at a distance from the fire perimeter concurrently with or after a line already constructed on or near to the perimeter of the fire.

sector boss—a staff officer responsible for all suppression activities of two or more crews on a specific sector of a fire.

self-contained breathing apparatus—a device enabling an individual to have air or oxygen independent of the atmosphere in which he is working. It consists of a tank of air or oxygen, mixing valve and regulator, face piece, piping and harness, and it is usually rated for 30 minutes of service.

service chief—a staff officer in a fire suppression organization responsible for procuring, maintaining, and distributing men, equipment, supplies, and facilities at the times and places specified by the suppression plan for the fire.

set—an individual incendiary fire; the points of origin of an incendiary fire; material left to ignite an incendiary fire at a later time; individual lightning or railroad fires, especially when several are started within a short time; also, burning material at the points deliberately ignited for backfiring, slash burning, prescribed burning, and other purposes.

shoulder carry—a method of carrying hose on the shoulders.

shrub—a woody perennial plant differing from a perennial herb by its persistent and woody stem, and from a tree by its low stature and habit of branching from the base.

shut-off nozzle—a common type of fire hose nozzle permitting flow of the stream to be controlled by the men at the nozzle rather than only at the source of supply; a controlling nozzle.

siamese—a hose fitting preferably gated for combining the flow from two or more lines of hose into a single stream; one male coupling to two female couplings.

simple hose lay—consecutive coupled lengths of hose without laterals. The lay is extended by inserting additional lengths of hose in the line between pump and nozzle.

simultaneous ignition—broadcast burning or backfiring by which the fuel on an area to be burned is ignited at many points simultaneously and the sets are so spaced that each receives timely stimulation by radiation for the adjoining sets. By such techniques, all burn together quickly and a hot, clean burn is possible under unfavorable burning conditions where single sets would not spread. *See* area ignition.

size-up—the evaluation of the officer in charge enabling him to determine a course of action, including such factors as time, location, nature of occurrence, life hazard, exposures, property involved, nature and extent of fire, weather, and fire-fighting facilities.

skid hose load—a load of hose especially arranged on top of a standard hose load to permit dropping the working line at the fire. A skid load may consist of two 1½-inch lines connected to a 2½-inch hose load by reducing wye.

slash—debris left after logging, pruning, thinning, wind, fire, or brush cutting, including logs, chunks, bark, branches, stumps, and broken understory trees or brush. If not disposed of properly, when it dries, it becomes a serious fire hazard.

slip-on tanker—a complete pumping unit including tank, pump, and plumbing that can easily be loaded or removed from a pickup truck or other suitable vehicle.

slope—the variation of terrain from the horizon; the number of feet rise or fall per 100 feet measured horizontally, expressed as a percentage.

slope class—a code which designates the most common slope encountered in the primary fire problem area on a protection unit.

small line—a line supplied from the water tank on the apparatus, originally a ¾-inch rubber covered line, now more commonly a 1-inch or 1½-inch hose supplied by a water tank. Pumpers and tank trucks for rural use frequently have two reels of ¾-inch or 1-inch small line besides two lines of 1½-inch hose, making it possible to supply four small lines from the water tank without waiting for a hydrant connection. A number of fire departments are replacing rubber-covered small line hose with cotton-rubber lined hose which they believe is easier to handle.

smoke haze—haze caused by smoke alone and not by water vapor, dust, or other particles.

smoke jumper—a well-trained and equipped fire fighter who travels to wildfires by aircraft and parachute.

smoker fire—caused by a smoker's matches or by burning tobacco in any form.

Smokey—the fire prevention bear; the symbol of the Cooperative Forest Fire Prevention program.

smoldering—behavior of a fire burning without flame and barely spreading. *See* creeping, running, spotting, backing.

snag—a standing dead tree or part of a dead tree from which at least the leaves and smaller branches have fallen; often called stub, if less than 20 feet tall.

soft suction—an erroneous but commonly accepted term for a short length of large diameter hose used to connect a fire department pumper with a hydrant. No suction is involved because the hose is useful only when the pumper receives water at pressures that are above atmospheric pressures.

solid stream—water used for fire fighting; discharged by an open round orifice at sufficient pressure to provide a stream having impact and range; also a straight stream.

spaghetti—an extensive snarl of charged hose lines from various fire companies, usually in the street adjacent to the fire area.

spanner—a metal wrench device used in tightening and freeing hose couplings.

span of control—the maximum number of subordinates who can be directly supervised by one person without loss of efficiency. In fire suppression the number varies by activity, but is usually in the general range of 3 to 8.

Sparky—copyright name for a Dalmation fire dog used as a symbol for the home fire prevention campaign sponsored by the National Fire Prevention Association (NFPA).

speed of attack—elapsed time from the beginning of fire to arrival of the first suppression force.

spot burning—a modified form of broadcast slash burning in which only the greater accumulations are fired and the fire is confined to these spots.

spot fire—fire set outside the perimeter of the main fire by flying sparks or embers. A major problem in conflagrations involving structures having wooden shingle roofs and requiring patrolling of areas downwind from the main fire as well as in wildfires.

spotting—placement of fire companies or fire-fighting equipment for effective operation and attack on a fire; fire-producing sparks or embers carried by the wind starting new fires beyond the zone of direct ignition by the main fire.

spray—water applied through specially designed orifices in the form of finely divided particles in order to more readily absorb heat and smother fire, and to protect exposures from radiated heat and to carry water toward an inaccessible fire; a jet from a spray nozzle or water curtain.

spread component (SC)—in the National Fire Danger Rating system, a number related to the forward rate of spread of the head of a fire.

squad (straw) boss—a working leader responsible for the efficient and productive work of 3 to 8 fire fighters.

standard coupling—a fire hose coupling having American National Standard (NS) fire hose threads.

standard drying day—a day which produces the same net drying as experienced during a 24-hour period under laboratory conditions where the dry-bulb temperature is maintained at 90°F and the relative humidity at 20 percent.

standards—established rules of the National Fire Protection Association (NFPA), frequently adopted by insurance agencies as a basis for their regulations and used as a guide for municipal, state, or provincial fire laws, ordinances, and regulations.

standby crew—a group of men especially organized, trained, and placed for quick suppression work on fires.

standing by—a fire-fighting unit on the scene of a fire awaiting orders or investigating a report or suspicion of fire, but not actively engaged in fire fighting; also, a radio report from a chief officer or fire-fighting unit to alarm headquarters indicating that the unit has arrived at the location to which it was dispatched.

state of weather—in the National Fire Danger Rating (NFDR) system a code used for the entry on the 10-Day Fire Danger Weather Record Form which expresses the amount of cloud cover, kind of precipitation, and/or restrictions to visibility being observed at the fire danger station at the basic observation time. It is significant in the NFDR calculations because it indicates whether fuel moisture values should be corrected to compensate for the effects of additional heating on sunny days.

static pressure—water pressure head available at a specific location when no fire flow is being used so that no pressure losses due to friction are being encountered. The static pressure is that pressure observed on the pumper inlet gage before any water is taken from the hydrant. By noticing the percentage of difference between the static pressure and the residual inlet pressure when the first stream is placed in operation, the potential flow available from the hydrant can be closely estimated.

static water supply—a supply of water at rest which does not provide a pressure head for fire fighting, but which may be employed as a suction source for fire pumps, such as water in a reservoir, pond, or cistern.

steam—invisible water vapor at temperatures of 212°F or more; also, condensing white water vapor often interspersed with the smoke of a fire indicating the water has taken effect and the temperature has been reduced, converting invisible steam formed by heating of water from hose streams or sprinklers above 212°F (H_2O in gaseous form) into visible water vapor.

straight stream—a stream projected straight from the nozzle as contrasted with a fog or spray cone. A straight stream may be provided by a solid stream orifice or by adjusting a fog jet into a straight stream pattern.

strainers—wire or metal guards used to keep debris from clogging pipe or other openings made for removing water; they are used in pumps and on suction hose to keep foreign material from clogging or damaging pumps.

strategy—planning and disposition of fire-fighting units to control a fire. *Compare* tactics.

stream—a jet of water thrown from a nozzle and used to extinguish or to control a fire.

strength of attack—the number of men and machines with which a fire is attacked.

strip burning—setting fire to a narrow strip of fuel adjacent to a control line and then burning successively wider adjacent strips inside as the preceding strip burns out; also called strip firing. In addition, the term refers to burning only a relatively narrow strip or strips of slash through a cutting unit and leaving the remainder.

strip firing—setting fire to more than one strip of fuel and providing for the strips to burn together; frequently done in backfiring against a wind where inner strips are fired first to create drafts which pull flames and sparks away from the control line. *See* line firing.

suction—procedure for taking water from static sources located below the level of the pump by exhausting air from the pump chamber and using atmospheric pressure (14.7 psi) to push water through noncollapsible suction hose into the pump.

suction booster—a type of jet syphon device usually supplied from the tank on fire apparatus and used to bring water to a pumper from greater distances and to higher elevations than is possible with suction depending upon atmospheric pressure; also called ejector.

suction hose—a hose reinforced against collapse due to atmospheric pressure and used for drafting water into fire pumps when a partial vacuum is created in the pump, causing the atmospheric pressure to push water through the hose upward into the pump.

suction lift—in fire service, the number of feet of vertical lift from the surface of the water to the center of the pump impeller. For example, in testing, fire department pumpers are required to discharge their rated capacity at 150 pounds net pump pressure at a 10-foot lift. The suction gage would indicate the vertical suction lift in inches of mercury when the pump was primed with no appreciable water flowing.

suction pipe—permanent devices provided with fire pumper threads attached to speed drafting operations in many locations where there are static water sources suitable for fire protection use, such as at piers and wharves, bridges over streams, or highways adjacent to ponds. Suction pipes are also permanently installed to supply private fire pumps which depend upon suction sources; also called dry hydrants.

sunny—classification used when half of the sky or less is obscured by clouds.

supply officer—a person responsible for supplying and distributing the supplies and equipment needed to suppress a fire, and for servicing tools and simple equipment; supervised by the service chief.

suppressant—an agent used to extinguish the flaming and glowing phases of combustion by direct application to the burning fuel.

suppression—all the work of extinguishing or confining a fire beginning with its discovery. *See* control, prevention, presuppression.

suppression crew—two or more men stationed at a strategic location, either regularly or in emergency, for initial action on fires. Duties are essentially the same as those of individual firemen.

surface fire—fire that burns surface litter, other loose debris, and small vegetation. *See* ground fire and crown fire.

swamper—an ax man who cuts and clears away brush, limbs, small trees, and down timber; may also use saws.

T

tactics—methods of employing engine, ladder, and other fire company units in an efficient, coordinated manner in the field in order to achieve satisfactory results with the forces employed and to deny the fire any potential operations. *Compare* strategy.

tailboard—the back step of a hose or pumper truck on which fire fighters stand while riding on the apparatus. The tailboard is also used to carry doughnut rolls of wet hose while returning to quarters and is the station for fire nozzles and various hand extinguishers.

tanker—water tank truck fire apparatus; may be equipped with small line hose reels and either a "booster" or standard pump. Tankers frequently carry a portable pump. Tank capacities normally should not exceed 1,500 gallons (6.25 tons of water); however, water tanks on pumping engines are normally limited to 1,000 gallons or less.

tanker boss—a person in a fire suppression organization responsible for supervising usually three to five tanker units to get efficient and productive use of water in either direct attack or mop-up work on fires.

ten-hour time lag fuels—dead fuels consisting of roundwood in the size range of ¼ to 1 inch and very roughly the layer of litter extending from just below the surface to approximately ¾ inch below the surface.

thermal column—column of smoke and gases given off by fires moving upward because heated gases expand and become lighter and rise while cooler air bringing additional oxygen is drawn in toward the base of a fire. The magnitude and intensity of a fire can often be judged by observing the thermal column; also called convection column.

thermal radiation—the emission of radiant energy waves in forms transmitting or producing heat. There are many wave lengths or frequencies of energy, such as light, sound, radio, X-rays, etc.

thermal turbulence—atmospheric disturbance sometimes rather violent above and downwind of a major fire, due to the effect of the rising column of heated gases being displaced by cooler air; also caused by heating of the surface of the earth by solar radiation.

thermoscreen—*See* instrument shelter.

thread—the specific dimensions of a screw thread employed to coupled fire hose and equipment. American national standards have been adopted for fire hose coupling threads in ¾-, 1-, 1½-, 2½-, 3½-, 4-, 4½-, 5- and 6-inch sizes. Threads having other dimensions than those specified are known as nonstandard threads.

three-way radio—radio equipment with transmitters in the mobile units on different frequencies from the central station, permitting conversation in two directions using two adjacent wave lengths; also, mobile car-to-car communication.

time lag (TL)—the time necessary for a fuel particle to lose approximately 63 percent of the difference between its initial moisture content and its equilibrium moisture content.

tips—nozzle tips for changing the size of orifice of a hose stream.

total risk—the sum of lightning and man-caused risk values; cannot exceed a value of 100.

towerman—a lookout man stationed at a tower.

tractor boss—a person responsible to the sector boss for supervising all of the activities of tractors assigned to him for line construction.

tractor-drawn—a fire-fighting vehicle having a separate four-wheel automotive vehicle with a "fifth-wheel" connection for drawing a two-wheel trailer carrying the fire-fighting equipment.

trailer—a semitrailer vehicle designed for towing by a tractor or power unit.

trailer pump—pumping equipment usually mounted on a two-wheel trailer for towing by automotive vehicles; sometimes used in industrial fire protection.

training officer—the chief or other officer under the chief of department responsible for organizing and conducting a complete training program for the fire department.

travel time map—a map showing the length of time required to reach various parts of an area by a fireman or crew from specified locations.

tree farm—a managed tract of forest or woodland where trees are cultivated as an agricultural product for lumber, poles, pulpwood, Christmas trees, or nurseries, and distinguished from uncultivated"wildlands." In many rural communities, tree farms represent important values requiring fire protection.

trench—formerly used as synonym for fire line, a more preferred term. *See* gutter trench.

turnout coat—a waterproof, lined fire coat with snap fastenings which may be donned quickly when turning out on a fire alarm. *See* fire coat.

two-way radio—radio equipment with transmitters in the mobile units on the same frequency as the central station, permitting conversation in two directions using the same frequency in turn.

U

undercut line—a fire line in the form of a trench to catch rolling embers below a fire on a slope; also called underslung line. *See* gutter trench.

unfriendly fire—an extension of a useful fire intended for heat or for industrial or other useful purposes, which spreads to combustible materials of value not intended as fuel.

unlined fire hose—hose commonly of cotton or synthetic fiber construction without rubber tube or lining. Such hose is provided attached to first aid stand pipes in buildings. Similar hose is often used for wildfire fighting because of its light weight. At a given flow, pressure loss due to friction in unlined hose of a stated diameter is approximately twice that of lined fire hose. *See* lined hose.

unnormalized man-caused risk—the sum of the partial risk factors computed for the risk sources active on a protection unit.

V

visibility distance—maximum distance at which a smoke column of specified size and density can be seen and recognized as a smoke by the unaided eye.

visible area—the ground or vegetation thereon that can be directly seen from a given lookout point under favorable atmospheric conditions.

visible area map—a map showing the specific territory in which either the ground surface or the vegetation growing thereon is directly visible, to practical distances, from a lookout point.

volatiles—readily vaporized organic materials which, when mixed with oxygen, are easily ignited.

volunteer fireman—legally enrolled fire fighter under the fire department organization laws, who devotes time and energy to community fire service without compensation other than Workman's Compensation or other similar death and injury benefits.

walkie-talkie—a two-way radio hand set used, in fire service operations, for fireground communications.

water hammer—impact energy due to sudden shutting of fire nozzles, proportional to the mass multiplied by the square of the velocity.

water supply map—a map showing location of supplies of water readily available for pumps, tanks, trucks, camp use, etc.

weighted monthly occurrence—a number used to determine the seasonal risk class for a protection unit, calculated by multiplying the peak monthly average by two and adding that product to the seasonal monthly average.

wet-bulb temperature—the temperature of a properly ventilated wet-bulb thermometer.

wetting agent—a chemical that reduces the surface tension of water and causes it to spread and penetrate more effectively.

wildfire—an unplanned fire requiring suppression action; fires burning uncontrolled in any surface vegetation fuels, such as crops, grass, brush, or forest.

wild land—uncultivated land, not fallow lands.

wind speed—wind, in miles per hour, measured at 20 feet above the ground or the average height of the vegetative cover, and averaged over at least a 10-minute period.

woods fire—fire burning woodland fuels including small trees and endangering homes, buildings, and camps in wooded areas.

woody vegetation condition—a code reflecting the moisture content of the foliage and small twigs (less than ¼ inch) of living woody plants.

woven jacket fire hose—fire hose of conventional construction, woven on looms from fibers of cotton or synthetic fibers. Most fire department hose is double jacketed, that is, has an outer jacket protecting the inner one against wear and abrasion.

wye—a hose connection with two outlets, preferably gated, permitting two connections of the same coupling diameter to be taken from a single supply line. Thus a 1½-inch wye permits the supply of two 1½-inch lines from a single 1½-inch supply line; also, reducing wyes permit several smaller diameter lines to be supplied from one big line.

SUGGESTED READINGS

I have found the publications that are listed below to be of particular value both from standpoints of research and practice. I recommend them to interested readers. In addition to the readings suggested here, there is a wealth of material, often in notebook form, published for use in wildfire-fighting training courses by the USDA Forest Service and the U.S. Department of Interior Bureau of Land Management. While this material is not available to the general public, it may be obtained, with some effort, by both professional and volunteer fire fighters. State forestry agencies are often helpful in this regard.

Barrows, J.S. "Fire Behavior in the Northern Rocky Mountain Forests," *Station Paper No. 29.* Missoula, Montana: Northern Rocky Mountain Forest and Range Experiment Station, USDA Forest Service, 1951.

Clar, C. Raymond, and Leonard R. Chatten. *Principles of Forest Fire Management.* Sacramento: California Division of Forestry, 1966.

Countryman, Clive M. *This Humidity Business: What It Is About and Its Use in Fire Control.* Riverside, Calif.: Forest Fire Laboratory, USDA Forest Service, 1971.

Deeming, John E., James W. Lancaster, Michael A. Forsberg, R. Wilson Furman, and Mark J. Schroeder. "National Fire Danger Rating System," *Research Paper RM 84.* Fort Collins, Colorado: Rocky Mountain Forest and Range Experiment Station, USDA Forest Service; and Colorado State University, 1972.

Department of Lands, Forests, and Water Resources, British Columbia Forest Service. *Handbook on Forest Fire Suppression.* Victoria, B.C., 1969.

Department of Lands, Forests, and Water Resources, British Columbia Forest Service. *Handbook on Use of Bulldozers for Fighting Forest Fires.* Victoria, B.C., 1963.

Fisher, William C., and Charles E. Hardy. *Fire Weather Observers' Handbook.* Ogden, Utah: Intermountain Forest and Range Experiment Station, USDA Forest Service, 1972.

Florida Forest Service. *Fire Fighters' Guide.* Tallahasee, 1964.

Neuns, Alva G. *Water vs. Fire.* Berkeley: California Forest and Range Experiment Station, USDA Forest Service, 1950.

Pennsylvania Division of Protection, Department of Forests and Waters. *Pennsylvania Forest Fire Wardens' Manual.* Pennsylvania, 1968.

Schroeder, Mark J., and C.J. Buck. "Fire Weather," *Agricultural Handbook 360.* USDA Forest Service, 1970.

United States Department of Agriculture Forest Service. *Fireman's Handbook.* Washington, D.C., 1970.

United States Department of Agriculture Forest Service, Regions 1,2,3,4,5, and 6. *Fire Fighting Overhead Notebooks.* 1960 to 1972.

United States Department of Agriculture Forest Service, Division of Fire Control. *Water-Handling Equipment Guide and Amendments.* Washington, D.C.: 1972.